TABLES AND FORMULÆ

USEFUL IN

SURVEYING, GEODESY,

AND

PRACTICAL ASTRONOMY,

INCLUDING

ELEMENTS FOR THE PROJECTION OF MAPS,

AND

INSTRUCTIONS FOR FIELD MAGNETIC OBSERVATIONS.

Third Edition, Revised and Enlarged.

WASHINGTON:
GOVERNMENT PRINTING OFFICE.
1873.

No. 12.

PROFESSIONAL PAPERS

OF

THE CORPS OF ENGINEERS

OF

THE UNITED STATES ARMY.

PUBLISHED

BY AUTHORITY OF THE SECRETARY OF WAR.

HEADQUARTERS CORPS OF ENGINEERS.

1873.

OFFICE OF THE CHIEF OF ENGINEERS,
Washington, February 24, 1873.

GENERAL: In preparing this third edition of a volume compiled in 1849 for the use of the Corps of Topographical Engineers, when an officer in that corps, I have made such additions and corrections as experience has suggested and the requirements of the service seemed to demand.

Intended more especially for field-use by officers engaged in surveys or explorations, and aspiring to no further merit than that of accuracy, utility, and convenience, it is submitted with the hope that it may continue to be favorably received by those who may have occasion to refer to it.

The revision was undertaken at the suggestion of others, and not without reluctance. I beg indulgence for its defects.

Very respectfully,

THOS. J. LEE.

Brigadier-General A. A. HUMPHREYS,
Chief of Engineers, United States Army.

In the additions and corrections introduced into this volume, besides the authorities named in the text, I have availed myself, in the article on the Gauging of Rivers, of notes by MAJOR ABBOT, *Corps of Engineers, on the practical gauging of rivers, printed in the proceedings of the Essayons Club at Willet's Point.*

The article on Trigonometrical Leveling is taken principally from Appendix No. 7 *of the Coast-Survey report for* 1868, *by* ASSISTANT R. D. CUTTS, *United States Coast Survey.*

I have also made use of Appendices Nos. 9, 10, *and* 11 *of Coast-Survey report of* 1866, *by* ASSISTANT C. A. SCHOTT, *United States Coast Survey, in the revision of the portions relating to the use of the transit-instrument, of the zenith-telescope, and the determination of astronomical azimuths.*

The article on Longitude by Lunar Culminations was prepared in 1858 *for the Topographical Bureau by* PROFESSOR BARTLETT, *United States Military Academy.*

I am indebted to CAPTAIN ERNST *and* LIEUTENANT MERCUR, *Corps of Engineers, and to* ASSISTANTS WOODWARD *and* WRIGHT, *Survey of the Lakes, for suggestions and corrections; to* LIEUTENANT MERCUR *for the article on Probable Errors, &c.; and to* CAPTAIN RAYMOND, *Corps of Engineers, for the valuable contribution on Magnetic Field Observations, which forms the Appendix; and I have to regret that* LIEUTENANT-COLONEL WILLIAMSON *was prevented by absence from the country from preparing a suitable set of hypsometrical tables, which he had consented to do.*

T. J. L.

CONTENTS.

PART I.—MISCELLANEOUS.

PART II.—GEODESY.

PART III.—ASTRONOMY.

APPENDIX.

ERRATA.

Page 10. For "See XXXIV" read "See XXXVII."

Page 85. Hassler's expansion of iron bar, for "0.000240687260" read "0.000250687260."

Page 288. For E read ε.

Page 288. For p, p', &c., read p', p'', &c.

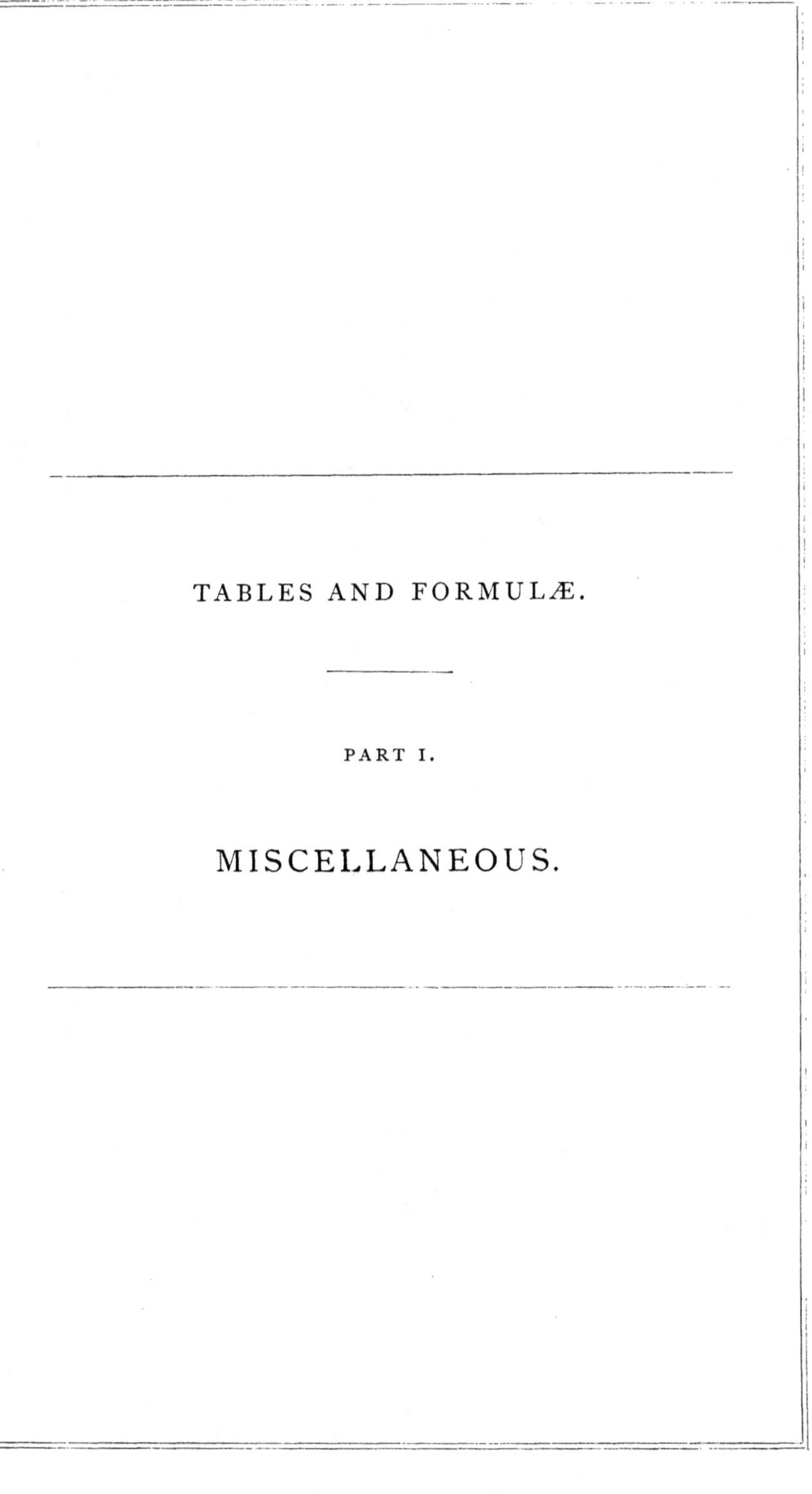

TABLES AND FORMULÆ.

PART I.

MISCELLANEOUS.

TRIGONOMETRY.

I.—*Equivalent Expressions.*

$$\sin^2 x + \cos^2 x = 1$$

$$\begin{aligned}
\sin x &= \cos x \tan x \\
&= \frac{\cos x}{\cot x} \\
&= \sqrt{1 - \cos^2 x} \\
&= \frac{1}{\sqrt{1 + \cot^2 x}} \\
&= 2 \sin \tfrac{1}{2} x \cos \tfrac{1}{2} x \\
&= \frac{\tan x}{\sqrt{1 + \tan^2 x}} \\
&= \frac{1}{\operatorname{cosec} x}
\end{aligned}$$

$$\begin{aligned}
\cos x &= \frac{\sin x}{\tan x} \\
&= \sin x \cot x \\
&= \sqrt{1 - \sin^2 x} \\
&= 1 - 2 \sin^2 \tfrac{1}{2} x \\
&= \cos^2 \tfrac{1}{2} x - \sin^2 \tfrac{1}{2} x \\
&= \frac{1}{\sec x}
\end{aligned}$$

$$\begin{aligned}
\tan x &= \frac{\sin x}{\cos x} \\
&= \frac{1}{\cot x} \\
&= \frac{\sin x}{\sqrt{1 - \sin^2 x}} \\
&= \frac{1 - \cos 2x}{\sin 2x} \\
&= \frac{\sin 2x}{1 + \cos 2x}
\end{aligned}$$

I.—*Equivalent Expressions*—Continued.

$$\cot x = \frac{1}{\tan x}$$

$$\sec x = \frac{1}{\cos x}$$

$$\operatorname{cosec} x = \frac{1}{\sin x}$$

$$\operatorname{versin} x = 1 - \cos x$$

$$= 2 \sin^2 \tfrac{1}{2} x$$

$$\text{co-versin}\ x = 1 - \sin x$$

$$\operatorname{chord} x = 2 \sin \tfrac{1}{2} x$$

$$\sin (A \pm B) = \sin A \cos B \pm \sin B \cos A$$

$$\cos (A \pm B) = \cos A \cos B \mp \sin A \sin B$$

$$\sin 2A = 2 \sin A \cos A$$

$$\cos 2A = 2 \cos^2 A - 1$$

$$= 1 - 2 \sin^2 A$$

$$= \cos^2 A - \sin^2 A$$

$$2 \cos^2 \tfrac{1}{2} A = 1 + \cos A$$

$$2 \sin^2 \tfrac{1}{2} A = 1 - \cos A$$

$$\tan (A \pm B) = \frac{\tan A \pm \tan B}{1 \mp \tan A \tan B}$$

$$\tan \tfrac{1}{2} A = \sqrt{\frac{1 - \cos A}{1 + \cos A}}$$

$$= \frac{1 - \cos A}{\sin A}$$

$$\sin A \pm \sin B = 2 \sin \tfrac{1}{2} (A \pm B) \cos \tfrac{1}{2} (A \mp B)$$

$$\cos A + \cos B = 2 \cos \tfrac{1}{2} (A + B) \cos \tfrac{1}{2} (A - B)$$

$$\cos A - \cos B = 2 \sin \tfrac{1}{2} (A + B) \sin \tfrac{1}{2} (B - A)$$

$$\sin^2 A - \sin^2 B = \sin (A + B) \sin (A - B)$$

$$\cos^2 A - \sin^2 B = \cos (A + B) \cos (A - B)$$

I.—*Equivalent Expressions*—Continued.

$$\tan A \pm \tan B = \frac{\sin (A \pm B)}{\cos A \cos B}$$

$$\cot A \pm \cot B = \frac{\sin (A \pm B)}{\sin A \sin B}$$

$$\frac{\sin A + \sin B}{\sin A - \sin B} = \frac{\tan \frac{1}{2} (A + B)}{\tan \frac{1}{2} (A - B)}$$

$$\frac{1 \pm \sin A}{1 \mp \sin A} = \tan^2 (45° \pm \tfrac{1}{2} A)$$

$$\frac{1 \pm \sin A}{\cos A} = \tan (45° \pm \tfrac{1}{2} A)$$

II.—*Solution of Plane Triangles.*

In the following formulæ, A, B, C, represent the angles, and a, b, c, the sides opposite, respectively.

1. Any plane triangle:

$$a^2 = b^2 + c^2 - 2\, b\, c \cos A$$

$$\frac{\sin A}{a} = \frac{\sin B}{b} = \frac{\sin C}{c}$$

$$\frac{\tan \frac{1}{2} (A + B)}{\tan \frac{1}{2} (A - B)} = \frac{\cot \frac{1}{2} C}{\tan \frac{1}{2} (A - B)} = \frac{a + b}{a - b}$$

$$\sin \tfrac{1}{2} A = \left\{ \frac{(s - b)(s - c)}{b\, c} \right\}^{\frac{1}{2}}$$

$$\cos \tfrac{1}{2} A = \left\{ \frac{s (s - a)}{b\, c} \right\}^{\frac{1}{2}}$$

$$s = \frac{a + b + c}{2}$$

II.—*Solution of Plane Triangles*—Continued.

2. Right-angled triangles:

Making $A = 90°$ in the preceding, they become

$$a^2 = b^2 + c^2$$

$$b = a \sin B = a \cos C,$$

$$c = a \sin C = a \cos B$$

$$\tan B = \frac{b}{c}$$

$$\tan C = \frac{c}{b}$$

III.—*Solution of Spherical Triangles.*

a, b, c, represent the arcs, and A, B, C, the angles opposite.

1. Oblique spherical triangles:

$$\frac{\sin A}{\sin a} = \frac{\sin B}{\sin b} = \frac{\sin C}{\sin c}$$

$$\begin{cases} \cos a = \dfrac{\cos b \sin (c + \varphi)}{\sin \varphi} \\ \cot \varphi = \tan b \cos A \end{cases}$$

$$\begin{cases} \cos A = \dfrac{\cos B \sin (C - \varphi)}{\sin \varphi} \\ \cot \varphi = \tan B \cos a \end{cases}$$

$$\begin{cases} \cot a \tan b = \dfrac{\sin (C + \varphi)}{\sin \varphi} \\ \cot \varphi = \dfrac{\cot A}{\cos b} \end{cases}$$

Napier's Analogies.

$$\tan \tfrac{1}{2} (a + b) = \tan \tfrac{1}{2} c \frac{\cos \frac{1}{2} (A - B)}{\cos \frac{1}{2} (A + B)}$$

$$\tan \tfrac{1}{2} (a - b) = \tan \tfrac{1}{2} c \frac{\sin \frac{1}{2} (A - B)}{\sin \frac{1}{2} (A + B)}$$

$$\tan \tfrac{1}{2} (A + B) = \cot \tfrac{1}{2} C \frac{\cos \frac{1}{2} (a - b)}{\cos \frac{1}{2} (a + b)}$$

$$\tan \tfrac{1}{2} (A - B) = \cot \tfrac{1}{2} C \frac{\sin \frac{1}{2} (a - b)}{\sin \frac{1}{2} (a + b)}$$

III.—*Solution of Spherical Triangles*—Continued.

$$\sin^2 \tfrac{1}{2} a = \frac{\sin S \sin (A - S)}{\sin B \sin C}$$

$$\cos^2 \tfrac{1}{2} a = \frac{\sin (B - S) \sin (C - S)}{\sin B \sin C}$$

$$\tan^2 \tfrac{1}{2} a = \frac{\sin S \sin (A - S)}{\sin (B - S) \sin (C - S)}$$

$$\sin^2 \tfrac{1}{2} A = \frac{\sin (s - b) \sin (s - c)}{\sin b \sin c}$$

$$\cos^2 \tfrac{1}{2} A = \frac{\sin s \sin (s - a)}{\sin b \sin c}$$

$$\tan^2 \tfrac{1}{2} A = \frac{\sin (s - b) \sin (s - c)}{\sin s \sin (s - a)}$$

In which S and s represent the half-sum of the three angles diminished by 90° and the half-sum of the three sides, respectively.

2. Right-angled spherical triangles, a being the hypothenuse:

$\cos a = \cos b \cos c$

$\cos a = \cot B \cot C$

$\cos B = \sin C \cos b$

$\cos C = \sin B \cos c$

$\tan b = \tan a \cos C$

$\tan c = \tan a \cos B$

$\cot B = \cot b \sin c$

$\cot C = \cot c \sin b$

$\tan b = \tan B \sin c$

$\tan c = \tan C \sin b$

$\sin b = \sin a \sin B$

$\sin c = \sin a \sin C$

IV.—*Multiple Arcs.*

$$\sin 2x = 2 \sin x \cos x$$

$$\sin 3x = 2 \sin x \,.\, \cos 2x + \sin x$$

$$\cos 2x = 2 \cos x \,.\, \cos x - 1$$

$$\cos 3x = 2 \cos x \,.\, \cos 2x - \cos x$$

$$\tan 2x = \frac{2 \tan x}{1 - \tan^2 x}$$

$$\tan 3x = \frac{\tan x + \tan 2x}{1 - \tan x \,.\, \tan 2x}$$

V.—*Trigonometrical Series.*

$$\sin A = A - \frac{A^3}{2.3} + \frac{A^5}{2.3.4.5} - \frac{A^7}{2.3.....7} + \text{etc.}$$

$$\cos A = 1 - \frac{A^2}{2} + \frac{A^4}{2.3.4} - \frac{A^6}{2.....6} + \text{etc.}$$

$$\tan A = A + \frac{A^3}{3} + \frac{2\,A^5}{3.5} + \frac{17\,A^7}{3^2.5.7} + \text{etc.}$$

$$\text{arc } A = \sin A + \frac{\sin^3 A}{2.3} + \frac{3 \sin^5 A}{2.4.5} + \frac{3.5 \sin^7 A}{2.4.6.7} + \text{etc.}$$

$$= \tan A - \tfrac{1}{3}\tan^3 A + \tfrac{1}{5}\tan^5 A - \tfrac{1}{7}\tan^7 A. + \text{etc.}$$

$$\log \sin A = \log A + \log\left(1 - \frac{x^2}{6} + \frac{x^4}{120} - \text{etc.}\right)$$

$$= \log A - M\left(\frac{x^2}{6} + \frac{x^4}{180} + \frac{x^6}{2835}\right)$$

$M =$ logarithmic modulus

$= 0.43429\ 45.........$

$\log M = 9.63778\ 43113.....$

Differentials of Trigonometrical Lines.

$$d \sin x = + d\,x \cos x$$

$$d \cos x = - d\,x \sin x$$

$$d \tan x = + \frac{d\,x}{\cos^2 x}$$

$$d \cot x = - \frac{d\,x}{\sin^2 x}$$

$$d \sin^2 x = + 2\,d\,x \sin x \cos x$$

$$d \cos^2 x = - 2\,d\,x \sin x \cos x$$

$$d \tan^2 x = + \frac{2\,d\,x \tan x}{\cos^2 x}$$

$$d \cot^2 x = - \frac{2\,d\,x \cot x}{\sin^2 x}$$

VI.—*Ratio of the Circumference of a Circle to its Diameter.*

$$\pi = 3.14159\ 26535\ 898\ldots\ldots$$
$$\log \pi = 0.49714\ 98726\ 941\ldots\ldots$$

The radius being unity, the number of degrees in an arc equal to radius $= r^\circ = \frac{180^\circ}{\pi} = \frac{1}{\text{arc } 1^\circ} = 57^\circ.29578 = 57^\circ\ 17'\ 44''.8.$

The number of minutes $= r' = \frac{10800'}{\pi} = \frac{1}{\text{arc } 1'}$, or $\frac{1}{\sin 1'} = 3437'.74677.$

The number of seconds $= r'' = \frac{648000''}{\pi} = \frac{1}{\sin 1''}$ $= 206264''.80625.$

$$\log r^\circ = 1.75812\ 26324\ 09172$$
$$\text{comp} \log r^\circ = 8.24187\ 73675\ 90828$$

$$\log r' = 3.53627\ 38827\ 92816$$
$$\text{comp} \log r' = 6.46372\ 61172\ 07184 = \log \sin 1'$$

$$\log r'' = 5.31442\ 51331\ 76459$$
$$\text{comp} \log r'' = 4.68557\ 48668\ 23541 = \log \sin 1''$$

Let a be the length of an arc of a circle whose radius is 1, and a'' the number of seconds in that arc, as

$$r'' = \frac{1}{\sin 1''} \text{ and } R : r'' :: a : a'' \text{ or } a'' = r''\,a\,;\ \ a = a'' \sin 1''$$

In an equation, therefore, any arc a of a circle whose radius is 1 is expressed in seconds by changing a into $a'' \sin 1''$.

Signs of Trigonometrical Lines.

Quadrants.	Sin.	Cos.	Tan.	Cot.	Sec.	Cosec.
1, 5, 9,	+	+	+	+	+	+
2, 6, 10,	+	−	−	−	−	+
3, 7, 11,	−	−	+	+	−	−
4, 8, 12, &c.	−	+	−	−	+	−

VII.—*Weights and Measures of the United States.*

The standards of length and weight of this country and Great Britain are theoretically identical. The United States gallon and bushel represent old English measures.

The standard of linear dimensions, adopted by the Treasury Department in the construction of standards for distribution to the custom-houses and States, is a brass scale of 82 inches in length, made in London by Troughton, which formed part of the instruments collected in 1815 by Mr. Hassler for the Survey of the Coast, and was supposed identical with the Schuckburg scale, one of the old English standards. The standard temperature is 62° Fahrenheit, and the yard-measure is between the 27th and 63d inches of its scale. This length has not been legalized by act of Congress. (See XXXIV.)

Linear Measure.

The *unit of linear measure* is the *yard*. The yard is divided into 3 feet, and the foot subdivided into 12 inches. The multiples of the yard are the *pole* or *perch*, the *furlong*, and the *mile*; but the pole and furlong are now scarcely ever used, itinerary distances being reckoned in miles and yards.

The following are the relations:

Inches.	Feet.	Yards.	Poles.	Furlongs.	Miles.
1	0.083	0.028	0.00505	0.00012626	0.0000157828
12	1.	0.333	0.06060	0.00151515	0.00018939
36	3.	1.	0.1818	0.004545	0.00056818
198	16.5	5.5	1.	0.025	0.003125
7920	660.	220.	40.	1.	0.125
63360	5280.	1760.	320.	8.	1.

$$\log 5280 = 3.7226339$$

$$\log 1760 = 3.2455127$$

VII.—*Weights and Measures of the United States*—Continued.

Square Measure.

In *square measure* the yard is subdivided, as in general measure, into *feet* and *inches;* 144 square inches being equal to a square foot. For land-measure the multiples of the yard are the *pole*, the *rood*, and the *acre.* Very large surfaces, as of whole countries, are expressed in square miles.

The following are the relations of square measure:

Sq. feet.	Sq. yards.	Poles.	Roods.	Acres.	Sq. miles.
1.	0.1111	0.00367309	0.000091827	0.000022957	
9.	1.	0.0330579	0.000826448	0.000206612	
272.25	30.25	1.	0.025	0.00625	
10890.	1210.	40.	1.	0.25	
43560.	4840.	160.	4.	1.	
27878400.	3097600.	102400.	2560.	640.	1.

Measure of Capacity.

The *units of capacity measure* are the *gallon* for *liquid* and the *bushel* for *dry* measure. The gallon is a vessel containing 58372.2 grains (8.3389 pounds avoirdupois) of the standard pound of distilled water, at the temperature of maximum density of water, the vessel being weighed in air in which the barometer is 30 inches at 62° Fahrenheit. The bushel is a measure containing 543391.89 standard grains (77.6274 pounds avoirdupois) of distilled water, at the temperature of maximum density of water, and barometer 30 inches at 62° Fahrenheit.

The gallon is thus the wine-gallon, (of 231 cubic inches,) nearly; and the bushel, the Winchester bushel, nearly.

The temperature of maximum density of water was determined by Mr. Hassler to be 39°.83 Fahrenheit.

DRY MEASURES.

Pint		$=\frac{1}{64}$ bushel.
Quart	= 2 pints	$=\frac{1}{32}$ bushel.
Peck	= 8 quarts	$=\frac{1}{4}$ bushel.
Bushel	= 4 pecks	= 1 bushel.

LIQUIDS.

Gill		$=\frac{1}{32}$ gall.
Pint	= 4 gills	$=\frac{1}{8}$ gall.
Quart	= 2 pints	$=\frac{1}{4}$ gall.
Gallon	= 4 quarts	= 1 gall.
Barrel	= $31\frac{1}{2}$ gallons	= $31\frac{1}{2}$ galls.
Hhd.	= 2 barrels	= 63 galls.

The only *legalized unit* of weight or measure is a *troy-pound*,

VII.—*Weights and Measures of the United States*—Continued.

(act of May 19, 1828,) copied by Captain Kater, in 1827, from the imperial troy-pound of England, for the use of the Mint of the United States, and there deposited. This pound is a standard at 30 inches of the barometer and 62° of the Fahrenheit thermometer.

The standard *avoirdupois-pound*, as determined by Mr. Hassler, is the weight of 27.7015 cubic inches of distilled water. It is greater than the troy-pound in the proportion of 7000 to 5760; that is, the avoirdupois-pound is equivalent in weight to 7000 grains troy.

Weights.

AVOIRDUPOIS.			TROY.		
Dram		= $\frac{1}{256}$ lb.	Grain		= $\frac{1}{5760}$ lb.
Ounce	= 16 drs.	= $\frac{1}{16}$ lb.	Pennyweight	= 20 grs.	= $\frac{1}{240}$ lb.
Pound	= 16 ozs.	= 1 lb.	Ounce	= 24 dwt.	= $\frac{1}{12}$ lb.
Quarter	= 25 lbs.	= 28 lbs.	Pound	= 12 ozs.	= 1 lb.
Hundred-wt.	= 4 qrs.	= 112 lbs.			
Ton	= 20 cwt.	= 2240 lbs.			
Short ton		= 2000 lbs.			

VIII.—*Miscellaneous.*

Length.—Gunter's chain = 66 feet = 4 poles = 100 links of 7.92 inches.

1 fathom = 6 feet; 1 cable-length = 120 fathoms.

1 hand = 4 inches; 1 palm = 3 inches; 1 span = 9 inches.

Solid.—1 cubic foot = 1728 cubic inches.

1 cubic yard = 27 cubic feet = 46656 cubic inches.

1 reduced foot (board-measure) = 1 square foot × 1 inch thick = 144 cubic inches.

1 perch of masonry = 1 perch (16½ feet) long × 1 foot high × 1½ foot thick = 24.75 cubic feet; 25 cubic feet has generally been adopted for convenience.

1 cord fire-wood = 8 feet long × 4 feet high × 4 feet deep = 128 cubic feet.

1 chaldron coal = 36 bushels = 57.25 cubic feet.

Paper.—24 sheets = 1 quire.

20 quires = 1 ream = 480 sheets.

IX.—*Weights and Volumes of various Substances.*

METALS.

Substances.	Cubic foot.	Cubic inch.
	Pounds.	*Pounds.*
Brass { Copper....67 / Zinc......33 }	488.75	.2829
Brass, gun-metal	543.75	.3147
Copper, cast	547.25	.3179
plates	543.625	.3167
Iron, cast	450.437	.2607
gun-metal	466.5	.27
wrought bars	486.75	.2816
Lead, cast	709.5	.4106
rolled	711.75	.4119
Mercury, 60°	848.7487	.491174
Steel, plates	487.75	.2823
soft	489.562	.2833
Tin	455.687	.2637
Zinc, cast	428.812	.2482
rolled	449.437	.2601

WOODS.

Substances.	Cubic foot.	Cubic feet in a ton.
	Pounds.	
Ash	52.812	42.414
Cedar	35.062	63.886
Chestnut	38.125	58.754
Hickory, pig-nut	49.5	45.252
shell-bark	43.125	51.942
Lignum-vitæ	83.312	26.886
Mahogany, Honduras	{ 35. / 66.437	64. / 33.714
Oak, Canadian	54.5	41.101
English	58.25	38.455
live, seasoned	66.75	33.558
white, dry	53.75	41.674
upland	42.937	52.169

IX.—*Weights and Volumes of various Substances*—Continued.

WOODS—Continued.

Substances.	Cubic foot.	Cubic feet in a ton.
	Pounds.	
Pine, yellow	33.812	66.248
Spruce	31.25	71.68
Walnut, black, dry	31.25	71.68
Willow, dry	30.375	73.744

MISCELLANEOUS.

Substances.	Cubic foot.	Cubic feet in a ton.
	Pounds.	
Air	.075291	....
Brick, fire	137.562	16.284
mean	102.	21.961
Coal, anthracite	89.75	24.958
	102.5	21.854
bituminous, mean	80.	28.
cannel	94.875	23.609
Cumberland	84.687	26.451
Coke	62.5	35.84
Cotton, bale, mean	14.5	154.48
pressed	20.	114.
	25.	89.6
Earth, clay	120.625	18.569
common soil	137.125	16.335
gravel	109.312	20.49
dry sand	120.	18.667
loose	93.75	23.893
Granite, Quincy	165.75	13.514
Susquehanna	169.	13.254
Limestone	197.25	11.355
Marble, mean	167.875	13.343
Mortar, dry, mean	97.98	22.862
Water, fresh	62.5	35.84
salt	64.125	34.931
Steam	.036747	....

X.—*The Army-Ration.*

TABLE SHOWING THE WEIGHT AND BULK OF 1000 RATIONS.

One thousand rations of—	Net weight.	Gross weight.	Bulk.	100 rations consist of—
	Pounds.	*Pounds.*	*Barrels.*	
Pork	750.	1218. 75	3. 75	75 lbs. or }
Bacon	750.	903. 19	4. 90	75 lbs. }
Flour	1125.	1234. 06	5. 74	112.5 lbs. or }
Pilot-bread	750.	921. 69	9. 03	75 lbs. or }
Do	1000.	1228. 91	12. 05	100 lbs. in the field. }
Beans	155.	177. 32	0. 71	8 quarts, or }
Rice	100.	114. 50	0. 46	10 lbs. }
Coffee, green	100.	122.	0. 65	10 lbs.
roasted	80.	108.	0. 83	8 lbs.
Sugar	150.	161.	0. 6	15 lbs.
Vinegar	92. 5	107. 50	0. 33	4 quarts.
Candles	15.	17. 50	0. 09	1½ lbs.
Soap	40.	46. 89	0. 19	4 lbs.
Salt	33. 75	38. 63	0. 16	2 quarts.

Forage.

14 lbs. hay or fodder	per horse per day	hay, when pressed, 11 lbs. to cub. ft.
12 quarts oats, or		32 lbs. to bushel, 25.71 to cub. ft.
8 quarts corn		56 lbs. to bushel, 45.02 to cub. ft.

Three beeves or 15 sheep consume the forage of 2 horses.

Weights of Grain per Bushel.

Wheat	60 lbs.	Oats	32 lbs.
Corn and rye	56 lbs.	Barley	48 lbs.

A box 16 × 16.8 × 8. inches contains	1 bushel	dry measure.
12 × 11.2 × 8. "	½ bushel	
8 × 8.4 × 8. "	1 peck	
6 × 6 × 6.4 "	1 gallon	liquid measure.
4 × 4 × 3.6 "	1 quart	

XI.—*Metric System.*

By an act of Congress, approved July 28, 1866, the *metric system* of weights and measures is made optional in the United States; and the act provides that the tables in a schedule annexed shall be recognized "as establishing, in terms of the weights and measures now in use in the United States, the equivalents of the weights and measures expressed therein in terms of the metric system; and said tables may be lawfully used for computing, determining, and expressing, in customary weights and measures, the weights and measures of the metric system."

Schedule annexed to act of July 28, 1866.

MEASURES OF LENGTH.

Metric denominations.	Values in metres.	Equivalents in denominations in use.
Myriametre	10000.	6.2137 miles.
Kilometre	1000.	0.62137 mile, or 3280 feet and 10 inches.
Hectometre	100.	328 feet and 1 inch.
Decametre	10.	393.7 inches.
Metre	1.	39.37 inches.
Decimetre	0.1	3.937 inches.
Centimetre	0.01	0.3937 inch.
Millimetre	0.001	0.0394 inch.

MEASURES OF SURFACE.

Metric denominations.	Values in square metres.	Equivalents in denominations in use.
Hectare	10000	2.471 acres.
Are	100	119.6 square yards.
Centare	1	1550 square inches.

MEASURES OF CAPACITY.

Metric denominations and values.			Equivalents in denominations in use.	
Names.	No. of litres.	Cubic measure.	Dry measure.	Liquid or wine measure.
Kilolitre or stere.	1000.	1 cubic metre	1.308 cubic yards	264.17 gallons.
Hectolitre	100.	0.1 cubic metre	2 bus. and 3.35 pks.	26.417 gallons.
Decalitre	10.	10 cubic decimetres	9.08 quarts	2.6417 gallons.
Litre	1.	1 cubic decimetre	0.908 quart	1.0567 quarts.
Decilitre	0.1	0.1 cubic decimetre	6.1022 cubic inches	0.845 gill.
Centilitre	0.01	10 cubic centimetres	0.6102 cubic inch	0.338 fluid-ounce.
Millilitre	0.001	1 cubic centimetre	0.061 cubic inch	0.27 fluid-drachm.

XI.—*Metric System*—Continued.

WEIGHTS.

Metric denominations and values.			Equivalents in denominations in use.
Names.	Number of grammes.	Weight of what quantity of water at maximum density.	Avoirdupois weight.
Millier or tonneau..	1000000.	1 cubic metre	2204.6 pounds.
Quintal	100000.	1 hectolitre	220.46 pounds.
Myriagramme	10000.	10 litres.....................	22.046 pounds.
Kilogramme, or kilo	1000.	1 litre......................	2.2046 pounds.
Hectogramme	100.	1 decilitre	3.5274 ounces.
Decagramme	10.	10 cubic centimetres	0.3527 ounce.
Gramme...........	1.	1 cubic centimetre	15.432 grains.
Decigramme.......	0.1	0.1 cubic centimetre	1.5432 grains.
Centigramme	0.01	10 cubic millimetres.........	0.1543 grain.
Milligramme.......	0.001	1 cubic millimetre...........	0.0154 grain.

ADDITIONAL METRICAL EQUIVALENTS.

1 surveyor's chain in metres.. = 20.11662 .... log = 1.3035550
1 metre in surveyor's chain .. = 0.04971 .... log = 8.6964450

1 square foot in square metres = 0.09290 .... log = 8.9680221
1 acre in hectares .... = 0.40467 .... log = 9.6071100
1 square mile in hectares..... = 258.994 .... log = 2.4132900

1 square metre in square feet. = 10.76410 .... log = 1.0319779
1 hectare in acres .... = 2.47109 .... log = 0.3928900
1 hectare in square miles..... = 0.00386 .... log = 7.5867100

1 cubic foot in steres .... = 0.02831 .... log = 8.4520332
1 cord in steres.............. = 3.62445 .... log = 0.5592432

1 stere in cubic feet .... = 35.31561 .... log = 1.5479668
1 stere in cords.............. = 0.27590 .... log = 9.4407568

1 grain in grammes .... = 0.064798 .... log = 8.8115680

2

XII.—*Foreign Measures of Length.*

TABLE OF RELATIONS BETWEEN THE LINEAR MEASURES OF SEVERAL COUNTRIES, WITH CORRESPONDING LOGARITHMS.

Metre.	France.	England and Russia.	Prussia and Denmark.	Bavaria.	Saxony.	Baden and Switzerland.	Austria.	Spain and Mexico.
	Paris feet.	*Feet.*	*Feet.*	*Feet.*	*Feet.*	*Feet.*	*Vienna feet.*	*Feet.*
1	3. 078444 0. 4883313	3. 280899 0. 5159929	3. 186199 0. 5032730	3. 426310 0. 5348266	3. 531197 0. 5479220	3. 333333 0. 5228787	3. 163446 0. 5001605	3. 537877 0. 5487427
0. 3248394 9. 5116687	1	1. 065765 0. 0276616	1. 035003 0. 0149417	1. 113000 0. 0464954	1. 147072 0. 0595907	1. 082798 0. 0345475	1. 027612 0. 0118292	1. 149242 0. 0604114
0. 3047945 9. 4840071	0. 938293 9. 9723384	1	0. 971136 9. 9872801	1. 044320 0. 0188337	1. 076290 0. 0319291	1. 015982 0. 0068859	0. 964201 9. 9841676	1. 078325 0. 0327498
0. 3138535 9. 4967270	0. 966181 9. 9850583	1. 029722 0. 0127199	1	1. 075359 0. 0315536	1. 108279 0. 0446490	1. 046178 0. 0196058	0. 992859 9. 9968875	1. 110375 0. 0454697
0. 2918592 9. 4651734	0. 898472 9. 9535047	0. 957561 9. 9811663	0. 929922 9. 9684464	1	1. 030612 0. 0130954	0. 972864 9. 9880521	0. 923281 9. 9653339	1. 032562 0. 0139161
0. 2831901 9. 4520780	0. 871785 9. 9404093	0. 929118 9. 9680709	0. 902300 9. 9553510	0. 970297 9. 9869046	1	0. 943967 9. 9749567	0. 895856 9. 9522385	1. 001892 0. 0008207
0. 3000000 9. 4771213	0. 923533 9. 9654525	0. 984270 9. 9931141	0. 955860 9. 9803942	1. 027893 0. 0119479	1. 059359 0. 0250433	1	0. 949034 9. 9772817	1. 061361 0. 0258630
0. 3161109 9. 4998395	0. 973130 9. 9881708	1. 037128 0. 0158324	1. 007193 0. 0031125	1. 083094 0. 0346661	1. 116250 0. 0477615	1. 053703 0. 0227183	1	1. 118361 0. 0485822
0. 2826553 9. 4512573	0. 870139 9. 9395886	0. 927364 9. 9672502	0. 900597 9. 9545303	0. 968465 9. 9860839	0. 998112 9. 9991793	0. 942184 9. 9741360	0. 894165 9. 9514178	1

XIII.—*Table of Relations between Itinerary Measures of Several Countries, with the Corresponding Logarithms.*

France.	England.	Prussia and Denmark.	Austria.	Russia.	Spain and Mexico.	Germany.	England and France.
Myriametre = 10000 M.	Statute mile = 5280′.	Mile = 24000′.	Mile = 24000′.	Verst = 3500′.	Jud. league = 15000′.	Geo. mile, 15 = 1 deg.	Naut. league, 20 = 1 deg.
1	6. 213824 0. 7933590	1. 327583 0. 1230617	1. 318103 0. 1199492	9. 373997 0. 9719248	2. 358584 0. 3726514	1. 347680 0. 1295869	1. 796907 0. 2545256
0. 1609315 9. 2066410	1	0. 213650 9. 3297028	0. 212124 9. 3265903	1. 508571 0. 1785659	0. 379570 9. 5792924	0. 216884 9. 3362279	0. 289179 9. 4611666
0. 7532485 9. 8769383	4. 680554 0. 6702972	1	0. 992859 9. 9968875	7. 060950 0. 8488631	1. 776600 0. 2495897	1. 015138 0. 0065251	1. 353518 0. 1314639
0. 7586663 9. 8800508	4. 714219 0. 6734097	1. 007193 0. 0031125	1	7. 111736 0. 8519756	1. 789379 0. 2527022	1. 022440 0. 0096376	1. 363253 0. 1345764
0. 1066781 9. 0280752	0. 662879 9. 8214341	0. 141624 9. 1511369	0. 140613 9. 1480244	1	2. 516092 0. 4007266	0. 143768 9. 1576620	0. 191691 9. 2826008
0. 4239831 9. 6273486	2. 634556 0. 4207076	0. 562873 9. 7504103	0. 558853 9. 7472978	0. 397442 9. 5992734	1	0. 571394 9. 7569355	0. 761868 9. 8818742
0. 7420158 9. 8704131	4. 610755 0. 6637721	0. 985088 9. 9934749	0. 978053 9. 9903624	6. 955654 0. 8423380	1. 750107 0. 2430645	1	1. 333333 0. 1249387
0. 5565118 9. 7454744	3. 458067 0. 5388334	0. 738816 9. 8685361	0. 733540 9. 8654236	5. 216740 0. 7173992	1. 312580 0. 1181258	0. 750000 9. 8750613	1

1 English or French geographical mile = $\frac{1}{60}$ of a degree of longitude at the equator = 2028.7 English yards.

		Eng. stat. miles.			Eng. stat. miles.
Modern Roman mile....	=	0. 925	Portugal league	=	3. 841
Tuscan mile	=	1. 027	Flanders league.........	=	3. 900
Old Scottish mile	=	1. 127	Spanish common league .	=	4. 214
Irish mile..............	=	1. 273	Hungarian mile.........	=	5. 178
French posting league ..	=	2. 422	Swedish mile	=	6. 648

Table for converting Metres into Toises and French and English Feet and Inches.

Metres.	Toises.	French.			English.	
		Feet.	Inches.	Lines.	Feet.	Inches.
1	0. 51307	3	0	11. 296	3	3. 3708
2	1. 02615	6	1	10. 592	6	6. 7416
3	1. 53922	9	2	9. 888	9	10. 1124
4	2. 05230	12	3	9. 184	13	1. 4832
5	2. 56537	15	4	8. 480	16	4. 8539
6	3. 07844	18	5	7. 776	19	8. 2247
7	3. 59152	21	6	7. 072	22	11. 5955
8	4. 10459	24	7	6. 368	26	2. 9663
9	4. 61767	27	8	5. 664	29	6. 3371
10	5. 13074	30	9	4. 960	32	9. 7079
20	10, 26148	61	6	9. 920	65	7. 4158
30	15. 39222	92	4	2. 880	98	5. 1237
40	20. 52296	123	1	7. 840	131	2. 8316
50	25. 65370	153	11	0. 800	164	0. 5395
60	30. 78444	184	8	5. 760	196	10. 2474
70	35. 91519	215	5	10. 720	229	7. 9553
80	41. 04593	246	3	3. 680	262	5. 6632
90	46. 17667	277	0	8. 640	295	3. 3711
100	51. 30741	307	10	1 600	328	1. 0790
200	102. 61481	615	8	3. 200	656	2. 1580
300	153. 92222	923	6	4. 800	984	3. 2370
400	205. 22963	1231	4	6. 400	1312	4. 3160
500	256. 53704	1539	2	8. 000	1640	5. 3950
600	307. 84444	1847	0	9. 600	1968	6. 4740
700	359. 15185	2154	10	11. 200	2296	7. 5530
800	410. 45926	2462	9	0. 800	2624	8. 6320
900	461. 76667	2770	7	2. 400	2952	9. 7110
1000	513. 07407	3078	5	4. 000	3280	10. 7900
2000	1026. 14815	6156	10	8. 000	6561	9. 5800
3000	1539. 22222	9235	4	0. 000	9842	8. 3700
4000	2052. 29630	12313	9	4. 000	13123	7. 1600
5000	2565. 37037	15392	2	8. 000	16404	5. 9500
6000	3078. 44444	18470	8	0. 000	19685	4. 7400
7000	3591 51852	21549	1	4. 000	22966	3. 5300
8000	4104. 59259	24627	6	8. 000	26247	2. 3200
9000	4617. 66667	27706	0	0. 000	29528	1. 1100
10000	5130. 74074	30784	5	4. 000	32808	11. 9000

Table for converting English Feet into French Toises, Metres, and Feet.

English feet.	Toises.	Metres.	French.		
			Feet.	Inches.	Lines.
1	0.15638	0.30479	0	11	3.114
2	0.31276	0.60959	1	10	6.228
3	0.46915	0.91438	2	9	9.343
4	0.62553	1.21918	3	9	0.457
5	0.78191	1.52397	4	8	3.571
6	0.93829	1.82877	5	7	6.685
7	1.09468	2.13356	6	6	9.799
8	1.25106	2.43836	7	6	0.913
9	1.40744	2.74315	8	5	4.028
10	1.56382	3.04794	9	4	7.142
20	3.12764	6.09589	18	9	2.284
30	4.69146	9.14383	28	1	9.425
40	6.25529	12.19178	37	6	4.567
50	7.81911	15.23972	46	10	11.709
60	9.38293	18.28767	56	3	6.851
70	10.94675	21.33561	65	8	1.993
80	12.51057	24.38536	75	0	9.134
90	14.07439	27.43150	84	5	4.276
100	15.63822	30.47945	93	9	11.418
200	31.27643	60.95850	187	7	10.836
300	46.91465	91.43835	281	5	10.254
400	62.55286	121.91780	375	3	9.672
500	78.19108	152.39725	469	1	9.090
600	93.82929	182.87670	562	11	8.508
700	109.46751	213.35615	656	9	7.926
800	125.10572	243.83559	750	7	7.344
900	140.74394	274.31504	844	5	6.762
1000	156.38215	304.79449	938	3	6.180
2000	312.76431	609.58899	1876	7	0.360
3000	469.14646	914.38348	2814	10	6.539
4000	625.52861	1219.17797	3753	2	0.719
5000	781.91076	1523.97246	4691	5	6.899
6000	938.29292	1828.76696	5629	9	1.079
7000	1094.67507	2133.56145	6568	0	7.259
8000	1251.05722	2438.35594	7506	4	1.438
9000	1407.43937	2743.15044	8444	7	7.618
10000	1563.82153	3047.94493	9382	11	1.798

XIV.—*Analytical Expressions for different Lines, Surfaces, and Solids.*

1.—Lines.

Ratio of diagonal to side of square $= \sqrt{2} = 1.414 = \frac{10}{7}$, nearly.

Log $\sqrt{2} = 0.1505149978$

Side of inscribed square : R :: $\sqrt{2}$: 1

Side of inscribed equilateral triangle : R :: $\sqrt{3}$: 1

Side of inscribed regular hexagon = R

Side of inscribed regular decagon $= \frac{1}{2}$ R $(-1 + \sqrt{5}) = 0.618$ R

Circle.

Ratio of circumference to diameter $= 3.1415926 = \frac{355}{113}$, nearly.

Length of an arc $= \frac{a \pi r}{180}$, r being the radius of the circle and a the number of degrees in the arc; or nearly $= \frac{8 c' - c}{3}$, c being the chord of the arc, and c' (the chord of half the arc) $= \sqrt{\frac{1}{4} c^2 + \text{ver sin}^2}$

Ellipse.

Circumference $= \frac{199}{200} \pi \sqrt{\frac{1}{2} (a^2 + b^2)}$ nearly; a and b being the axes.

Lengths of Circular Arcs, taking the Base of Segments as Unity.

Ver sin.	Length.	Ver sin.	Length.	Ver sin.	Length.	Ver sin.	Length.	Ver sin.	Length.
.01	1.000	.11	1.032	.21	1.114	.31	1.239	.41	1.401
.02	1.000	.12	1.038	.22	1.124	.32	1.254	.42	1.418
.03	1.000	.13	1.044	.23	1.135	.33	1.269	.43	1.437
.04	1.000	.14	1.051	.24	1.147	.34	1.284	.44	1.455
.05	1.000	.15	1.059	.25	1.159	.35	1.300	.45	1.474
.06	1.006	.16	1.067	.26	1.171	.36	1.316	.46	1.493
.07	1.014	.17	1.075	.27	1.184	.37	1.332	.47	1.512
.08	1.018	.18	1.084	.28	1.197	.38	1.349	.48	1.531
.09	1.020	.19	1.093	.29	1.212	.39	1.366	.49	1.551
.10	1.026	.20	1.103	.30	1.225	.40	1.383	.50	1.571

XIV.—*Analytical Expressions, &c.*—Continued.

2.—SURFACES.

1. Triangle in terms of—

its base and its altitude . . . . $= \frac{b\,A}{2}$

two sides and the included angle . . . . $= \frac{a\,b \sin C}{2}$

its three sides . . . . $= [s\,(s - a)\,(s - b)\,(s - c)\,]^{\frac{1}{2}}$

where

$A =$ the altitude;

$a,\ b,\ c =$ the three sides;

$C =$ the angle included between a and b; and

$s = \frac{a + b + c}{2}$

2. Parallelogram in terms of—

its base and its altitude . . . . $= b\,A$

two sides and the included angle . . . . $= a\,b \sin C$

two sides and their corresponding diagonal

$= 2\,[s\,(s - a)\,(s - b)\,(s - c)]^{\frac{1}{2}}$

where

$C =$ the angle included between two adjacent sides $a,\ b$;

$c =$ the diagonal opposite; and

$s = \frac{a + b + c}{2}$

3. Trapezium in terms of—

its two parallel bases and its altitude . . . . $= \frac{B + b}{2} A$

its two parallel bases, one of its oblique sides, and the angle between one of these bases and this side . . . $= \frac{B + b}{2} l \sin C$

where

$A =$ the distance between the two parallel bases $B,\ b$;

$l =$ the length of one of the oblique sides; and

$C =$ the angle between one of these bases and this side.

4. Any quadrilateral = half the product of its two diagonals multiplied by the sine of the included angle.

XIV.—*Analytical Expressions, &c.*—Continued.

5. Regular polygon . . . . $= \dfrac{n\left(\frac{a}{2}\right)^2}{\tan \frac{180^\circ}{n}}$

where

$n =$ the number of sides; and
$a =$ the length of one of them.

6. Circle . . . . $= \pi R^2$

7. Ellipse . . . . $= \pi a b$
a and b being the semi-axes.

8. Right cylinder, exclusive of its bases . . . . $= 2 \pi R A$

9. Sphere . . . . $= 4 \pi R^2$

10. Zone . . . . . . $= 4 \pi R^2 \sin \frac{1}{2} (L' - L) \cos \frac{1}{2} (L' + L)$

11. Right cone . . . . $= \pi R L$

12. Frustum of cone with parallel bases . . . $= \pi l (R + r)$

where

R and $r =$ the radii of the bases of these solids; and
L and $l =$ the lengths of their generating elements.

13. Spherical quadrilateral, formed by two parallels of latitude and two meridians

$$= \frac{\pi}{90^\circ} (M' - M) R^2 \sin \tfrac{1}{2} (L' - L) \cos \tfrac{1}{2} (L' + L)$$

where

R = the radius of the sphere;
L, L′ = the latitudes of the bases of the zone, + when north, — when south; and
M′, M = the longitudes of the extreme meridians of the quadrilateral, (M′ — M) being expressed in degrees and decimals.

In the place of R, the normal N, of the mean latitude $\left(\frac{L' + L}{2}\right)$, can be used.

XIV.—*Analytical Expressions, &c.*—Continued.

3.—SOLIDS.

14. Prism . . . . $= B\ A$

where

$B =$ the area of the base; and

$A =$ the altitude.

15. Rectangular parallelopepidon . . . . $= p \times q \times r$

Cube . . . . $= p^3$

where $p, q, r,$ = the lengths of the three contiguous edges.

16. Pyramid . . . . $= \frac{B\ A}{3}$

The area, B, being found from No. 5.

17. Right cylinder . . . . $= \pi R^2 A$

18. Right cone . . . . $= \frac{1}{3} \pi R^2 A$

19. Sphere . . . . $= \frac{4}{3} \pi R^3$

20. Prismoid, or solid figure, similar to that which is formed in excavations or embankments of roads, terminated by parallel cross-sections.

Solid content = area of each end, added to four times the middle area, and the sum multiplied by the length divided by 6, or

$$= \left\{ (b + r h') h' + (b + r h) h + 4 \left(b + r \times \frac{h + h'}{2} \right) \frac{h + h'}{2} \right\} \frac{l}{6}$$

where

$b =$ the breadth at the bottom of the cutting;

$h =$ the perpendicular depth of cutting at higher end;

$h' =$ the perpendicular depth of cutting at lower end;

$l =$ the length of the solid; and

$r =$ the ratio of the perpendicular height of the slope to its horizontal base.

XV.—*Progression.*

1. Arithmetical:

$$a = z - (n - 1)\,d \qquad z = a + (n - 1)\,d$$

$$d = \frac{z - a}{n - 1} \qquad n = \frac{z - a}{d} + 1 \qquad s \doteq \frac{a + z}{2}\,n$$

2. Geometrical:

$$a = \frac{z}{r^{n-1}} \qquad z = a r^{n-1} \qquad r = \left(\frac{z}{a}\right)^{\frac{1}{n-1}}$$

$$n = \log.\ \frac{rz}{a} \div \log.\ r \qquad s = \frac{rz - a}{r - 1} \qquad = \frac{r^n - 1}{r - 1}\,a$$

$$a = \frac{r - 1}{r^n - 1}\,s \qquad a = zr - (r - 1)\,s \qquad r = \frac{s - a}{s - z}$$

where

a = the least term;	r = the common ratio;
z = the greatest term;	n = the number of terms;
d = the common diff.;	s = the sum of the terms.

XVI.—*Force of Gravity.*

The velocity acquired at the end of one second by a body falling in vacuo, at the level of the sea, in the latitude of London = 32.1915 feet.

The force of gravity at the latitude of 45° = 32.17 feet per second being represented by g, for any other latitude, l,

$$g' = g\,(1 - 0.002588 \cos 2\,l)$$

If g represents the force of gravity at the height h, and r the radius of the earth, the force of gravity at the level of the sea

$$= g' = g\left(1 + \frac{5}{4}\frac{h}{r}\right)$$

Length, in inches, of a pendulum vibrating seconds at the level of the sea:

Equator	= 39.0152	London, lat. 51° 31′ . .	= 39.1393
New York, lat. 40° 43′	= 39.1017	Spitzbergen, lat. 75° 50′	= 39.2147

XVII.—*Land-Surveying with Compass and Chain.*

To calculate the Area or Content of Land.

If the sum of each adjacent pair of distances perpendicular to a meridian (*departures*) assumed without the survey be multiplied by the northing or southing between them in succession round the figure in the same order, the difference between the sum of the *north* products and the sum of the *south* products will be double the area of the tract.

The *meridian distance* of a course is the distance of the middle point of that course from an assumed meridian.

Hence, the double meridian distance of the first course is equal to its departure.

And the double meridian distance of any course is equal to the double meridian distance of the preceding course, plus its departure, plus the departure of the course itself, having regard to the algebraic sign of each.

Then, to find the area—

1. Multiply the double meridian distance of each course by its northing or southing.
2. Place all the *plus* products in one column, and all the *minus* products in another.
3. Add up each column separately, and take their difference. This difference will be *double* the area of the land.

In *balancing* the work, the error for each particular course is found by the proportion—

> As the sum of the courses is to the error of latitude, (or departure,) so is each particular course to its correction.

When a bearing is due east or west, the error of latitude is nothing, and the course must be subtracted from the sum of the courses before balancing the columns of latitude. And so with the departures.

EXAMPLE.—It is required to find the content of a piece of land, of which the following are the field-notes:

Sta.	Course.	Dist.	Sta.	Course.	Dist.
1	North 46½° west.	20. chains.	4	South 56° east ..	27.60 chains.
2	North 51¾° east.	13.80 chains.	5	South 33½° west.	18.80 chains.
3	East..........	21.25 chains.	6	North 74½° west.	30.95 chains.

XVII.—*Land-Surveying, &c.*—Continued.

Calculation.

Stations.	Courses.	Dist., chains.	Diff. lat.		Departure.		Balanced.		D. M. D. +	Area. +	Area. —
			N. +	S. —	E. +	W. —	Latitude.	Departure.			
1	N. 46½° W...	20.00	13.77	....	....	14.51	+ 13.88	— 14.56	14.56	202.0928	....
2	N. 51¾° E....	13.80	8.54	....	10.84	....	+ 8.61	+ 10.81	10.81	93.0741	....
3	East.........	21.25	....	....	21.25	....	....	+ 21.20	42.82	....	....
4	S. 56° E.....	27.60	....	15.44	22.88	....	— 15.29	+ 22.82	86.84	....	1327.7836
5	S. 33¼° W....	18.80	....	15.72	....	10.31	— 15.63	— 10.36	99.30	....	1552.0590
6	N. 74½° W...	30.95	8.27	....	....	29.83	+ 8.43	— 29.91	59.03	497.6229	....
	Sums........	132.40	30.58	31.16	54.97	54.65				792.7898	2879.8426
				30.58	54.65						792.7898
	Error in northing..........			0.58	0.32, error in westing.						2)2087.0528
	Answer 104 A. 1 R. 16 P........................										1043.5264

100000 square links of Gunter's chain = 1 acre.

1 square chain = 66 feet square = $\frac{1}{10}$ acre.

XVIII.—*Table showing Differences of Latitude and Departures.*

Minutes.	Distance.	0°		1°		2°		Distance.	Minutes.
		Lat.	Dep.	Lat.	Dep.	Lat.	Dep.		
	1	1.00000	0.00000	0.99984	0.01745	0.99939	0.03490	1	
	2	2.00000	0.00000	1.99969	0.03490	1.99878	0.06980	2	
	3	3.00000	0.00000	2.99954	0.05235	2.99817	0.10470	3	
	4	4.00000	0.00000	3.99939	0.06980	3.99756	0.13960	4	
0	5	5.00000	0.00000	4.99923	0.08726	4.99695	0.17450	5	60
	6	6.00000	0.00000	5.99908	0.10471	5.99634	0.20940	6	
	7	7.00000	0.00000	6.99893	0.12216	6.99573	0.24430	7	
	8	8.00000	0.00000	7.99878	0.13961	7.99512	0.27920	8	
	9	9.00000	0.00000	8.99862	0.15707	8.99451	0.31410	9	
	1	0.99999	0.00436	0.99976	0.02181	0.99922	0.03925	1	
	2	1.99998	0.00872	1.99952	0.04363	1.99845	0.07851	2	
	3	2.99997	0.01308	2.99928	0.06544	2.99768	0.11777	3	
	4	3.99996	0.01745	3.99904	0.08725	3.99691	0.15703	4	
15	5	4.99995	0.02181	4.99881	0.10907	4.99614	0.19629	5	45
	6	5.99994	0.02617	5.99857	0.13089	5.99537	0.23555	6	
	7	6.99993	0.03054	6.99833	0.15270	6.99460	0.27481	7	
	8	7.99992	0.03490	7.99809	0.17452	7.99383	0.31407	8	
	9	8.99991	0.03926	8.99785	0.19633	8.99306	0.35333	9	
	1	0.99996	0.00872	0.99965	0.02617	0.99904	0.04361	1	
	2	1.99992	0.01745	1.99931	0.05235	1.99809	0.08723	2	
	3	2.99988	0.02617	2.99897	0.07853	2.99714	0.13085	3	
	4	3.99984	0.03490	3.99862	0.10470	3.99619	0.17447	4	
30	5	4.99981	0.04363	4.99828	0.13088	4.99524	0.21809	5	30
	6	5.99977	0.05235	5.99794	0.15706	5.99428	0.26171	6	
	7	6.99973	0.06108	6.99760	0.18323	6.99333	0.30533	7	
	8	7.99969	0.06981	7.99725	0.20941	7.99238	0.34895	8	
	9	8.99965	0.07853	8.99691	0.23559	8.99143	0.39257	9	
	1	0.99991	0.01308	0.99953	0.03053	0.99884	0.04797	1	
	2	1.99982	0.02617	1.99906	0.06107	1.99769	0.09595	2	
	3	2.99974	0.03926	2.99860	0.09161	2.99654	0.14393	3	
	4	3.99965	0.05235	3.99813	0.12215	3.99539	0.19191	4	
45	5	4.99957	0.06544	4.99766	0.15269	4.99424	0.23989	5	15
	6	5.99948	0.07853	5.99720	0.18323	5.99309	0.28786	6	
	7	6.99940	0.09162	6.99673	0.21376	6.99193	0.33584	7	
	8	7.99931	0.10471	7.99626	0.24430	7.99078	0.38382	8	
	9	8.99922	0.11780	8.99580	0.27484	8.98963	0.43180	9	
Minutes.	Distance.	Dep.	Lat.	Dep.	Lat.	Dep.	Lat.	Distance.	Minutes.
		89°		88°		87°			

Differences of Latitude and Departures—Continued.

Minutes.	Distance.	3°		4°		5°		Distance.	Minutes.
		Lat.	Dep.	Lat.	Dep.	Lat.	Dep.		
	1	0.99863	0.05233	0.99756	0.06975	0.99619	0.08715	1	
	2	1.99726	0.10467	1.99512	0.13951	1.99238	0.17431	2	
	3	2.99589	0.15700	2.99269	0.20926	2.98858	0.26146	3	
	4	3.99452	0.20934	3.99025	0.27902	3.98477	0.34862	4	
0	5	4.99315	0.26168	4.98782	0.34878	4.98097	0.43577	5	60
	6	5.99178	0.31401	5.98538	0.41853	5.97716	0.52293	6	
	7	6.99041	0.36635	6.98294	0.48829	6.97336	0.61008	7	
	8	7.98904	0.41868	7.98051	0.55805	7.96955	0.69724	8	
	9	8.98767	0.47102	8.97807	0.62780	8.96575	0.78440	9	
	1	0.99839	0.05669	0.99725	0.07410	0.09580	0.09150	1	
	2	1.99678	0.11338	1.99450	0.14821	1.99160	0.18300	2	
	3	2.99517	0.17007	2.99175	0.22232	2.98741	0.27450	3	
	4	3.99356	0.22677	3.98900	0.29643	3.98321	0.36600	4	
15	5	4.99195	0.28346	4.98625	0.37054	4.97902	0.45750	5	45
	6	5.99035	0.34015	5.98350	0.44465	5.97482	0.54900	6	
	7	6.98874	0.39684	6.98075	0.51875	6.97063	0.64051	7	
	8	7.98713	0.45354	7.97800	0.59286	7.96643	0.73201	8	
	9	8.98552	0.51023	8.97525	0.66697	8.96224	0.82351	9	
	1	0.99813	0.06104	0.99691	0.07845	0.99539	0.09584	1	
	2	1.99626	0.12209	1.99383	0.15691	1.99079	0.19169	2	
	3	2.99440	0.18314	2.99075	0.23537	2.98618	0.28753	3	
	4	3.99253	0.24419	3.98766	0.31383	3.98158	0.38338	4	
30	5	4.99067	0.30524	4.98458	0.39229	4.97698	0.47922	5	30
	6	5.98880	0.36629	5.98150	0.47075	5.97237	0.57507	6	
	7	6.98694	0.42733	6.97842	0.54921	6.96777	0.67092	7	
	8	7.98507	0.48838	7.97533	0.62767	7.96316	0.76676	8	
	9	8.98321	0.54943	8.97225	0.70613	8.95856	0.86261	9	
	1	9.99785	0.06540	0.99656	0.08280	0.99496	0.10018	1	
	2	1.99571	0.13080	1.99313	0.16561	1.98993	0.20037	2	
	3	2.99357	0.19620	2.98969	0.24842	2.98490	0.30056	3	
	4	3.99143	0.26161	3.98626	0.33123	3.97987	0.40075	4	
45	5	4.98929	0.32701	4.98282	0.41404	4.97484	0.50094	5	15
	6	5.98715	0.39241	5.97939	0.49684	5.96981	0.60112	6	
	7	6.98501	0.45782	6.97595	0.57965	6.96477	0.70131	7	
	8	7.98287	0.52322	7.97252	0.66246	7.95974	0.80150	8	
	9	8.98073	0.58862	8.96908	0.74527	8.95471	0.90169	9	
Minutes.	Distance.	Dep.	Lat.	Dep.	Lat.	Dep.	Lat.	Distance.	Minutes.
		86°		85°		84°			

Differences of Latitude and Departures—Continued.

Minutes.	Distance.	6° Lat.	6° Dep.	7° Lat.	7° Dep.	8° Lat.	8° Dep.	Distance.	Minutes.
	1	0 99452	0.10452	0.99254	0.12186	0.99026	0.13917	1	
	2	1.98904	0.20905	1.98509	0.24373	1.98053	0.27834	2	
	3	2.98356	0.31358	2.97763	0.36560	2.97080	0.41751	3	
	4	3.97808	0.41811	3.97018	0.48747	3.96107	0.55669	4	
0	5	4.97261	0.52264	4.96273	0.60934	4.95134	0.69586	5	60
	6	5.96713	0.62717	5.95519	0.73121	5.94160	0.83503	6	
	7	6.96165	0.73169	6.94782	0.85308	6.93187	0.97421	7	
	8	7.95617	0.83622	7.94038	0.97495	7.92214	0.11338	8	
	9	8.95069	0.94075	8.93291	0.09682	8.91241	0.25255	9	
	1	0.99405	0.10886	0.99200	0.12619	0.98965	0.14349	1	
	2	1.98811	0.21773	1.98400	0.25239	1.97930	0.28698	2	
	3	2.98216	0.32660	2.97601	0.37859	2.96895	0.43047	3	
	4	3.97622	0.43546	3.96801	0.50479	3.95860	0.57397	4	
15	5	4.97028	0.54433	4.96002	0.63099	4.94825	0.71746	5	45
	6	5.96433	0.65320	5.95202	0.75719	5.93790	0.86095	6	
	7	6.95839	0.76206	6.94403	0.88339	6.92755	1.00444	7	
	8	7.95245	0.87093	7.93603	1.00959	7.91721	1.14794	8	
	9	8.94650	0.97980	8.92804	1.13579	8.90686	1.29143	9	
	1	0.99357	0.11320	0.99144	0.13052	0.98901	0.14780	1	
	2	1.98714	0.22640	1.98288	0.26105	1.97803	0.29561	2	
	3	2.98071	0.33960	2.97433	0.39157	2.96704	0.44342	3	
	4	3.97428	0.45281	3.96577	0.52210	3.95606	0.59123	4	
30	5	4.96786	0.56601	4.95722	0.65263	4.94508	0.73904	5	30
	6	5.96143	0.67921	5.94866	0.78315	5.93409	0.88685	6	
	7	6.95500	0.79242	6.94011	0.91368	6.92311	1.03466	7	
	8	7.94857	0.90562	7.93155	1.04420	7.91212	1.18247	8	
	9	8.94214	1.01882	8.92300	1.17473	8.90114	1.33028	9	
	1	0.99306	0.11753	0.99086	0.13485	0.98836	0.15212	1	
	2	1.98613	0.23507	1.98173	0.26970	1.97672	0.30424	2	
	3	2.97920	0.35261	2.97259	0.40455	2.96508	0.45637	3	
	4	3.97227	0.47014	3.96346	0.53940	3.95344	0.60849	4	
45	5	4.96534	0.58768	4.95432	0.67425	4.94180	0.76061	5	15
	6	5.95841	0.70522	5.94519	0.80910	5.93016	0.91274	6	
	7	6.95147	0.82276	6.93606	0.94395	6.91853	1.06486	7	
	8	7.94454	0.94029	7.92692	1.07880	7.90689	1.21698	8	
	9	8.93761	1.05783	8.91779	1.21365	8.89525	1.36911	9	
Minutes.	Distance.	Dep.	Lat.	Dep.	Lat.	Dep.	Lat.	Distance.	Minutes.
		83°		82°		81°			

Differences of Latitude and Departures—Continued.

Minutes.	Distance.	9° Lat.	9° Dep.	10° Lat.	10° Dep.	11° Lat.	11° Dep.	Distance.	Minutes.
	1	0.98768	0.15643	0.98480	0.17364	0.98162	0.19081	1	
	2	1.97537	0.31286	1.96961	0.34729	1.96325	0.38162	2	
	3	2.96306	0.46930	2.95442	0.52094	2.94488	0.57243	3	
	4	3.95075	0.62573	3.93923	0.69459	3.92650	0.76324	4	
0	5	4.93844	0.78217	4.92403	0.86824	4.90813	0.95405	5	60
	6	5.92612	0.93860	5.90884	1.04188	5.88976	1.14486	6	
	7	6.91381	1.09504	6.89365	1.21553	6.87139	1.33566	7	
	8	7.90150	1.25147	7.87846	1.38918	7.85301	1.52648	8	
	9	8.88919	1.40791	8.86327	1.56283	8.83464	1.71729	9	
	1	0.98699	0.16074	0.98404	0.17794	0.98078	0.19509	1	
	2	1.97399	0.32148	1.96808	0.35588	1.96157	0.39018	2	
	3	2.96098	0.48222	2.95212	0.53383	2.94235	0.58527	3	
	4	3.94798	0.64297	3.93616	0.71177	3.92314	0.78036	4	
15	5	4.93498	0.80371	4.92020	0.88971	4.90392	0.97545	5	45
	6	5.92197	0.96445	5.90424	1.06766	5.88471	1.17054	6	
	7	6.90897	1.12519	6.88828	1.24560	6.86549	1.36563	7	
	8	7.89597	1.28594	7.87232	1.42354	7.84628	1.56072	8	
	9	8.88296	1.44668	8.85636	1.60149	8.82706	1.75581	9	
	1	0.98628	0.16504	0.98325	0.18223	0.97992	0.19936	1	
	2	1.97257	0.33009	1.96650	0.36447	1.95984	0.39873	2	
	3	2.95885	0.49514	2.94976	0.54670	2.93977	0.59810	3	
	4	3.94514	0.66019	3.93301	0.72894	3.91969	0.79747	4	
30	5	4.93142	0.82523	4.91627	0.91117	4.89962	0.99683	5	30
	6	5.91771	0.99028	5.89952	1.09341	5.87954	1.19620	6	
	7	6.90399	1.15533	6.88278	1.27564	6.85947	1.39557	7	
	8	7.89028	1.32038	7.86603	1.45788	7.83939	1.59494	8	
	9	8.87657	1.48542	8.84929	1.64011	8.81932	1.79431	9	
	1	0.98555	0.16935	0.98245	0.18652	0.97904	0.20364	1	
	2	1.97111	0.33870	1.96490	0.37304	1.95809	0.40728	2	
	3	2.95666	0.50805	2.94735	0.55957	2.93713	0.61092	3	
	4	3.94222	0.67740	3.92980	0.74609	3.91618	0.81456	4	
45	5	4.92778	0.84675	4.91225	0.93262	4.89522	1.01820	5	15
	6	5.91333	1.01610	5.89470	1.11914	5.87427	1.22185	6	
	7	6.89889	1.18545	6.87715	1.30566	6.85331	1.42549	7	
	8	7.88444	1.35480	7.85960	1.49219	7.83236	1.62913	8	
	9	8.87000	1.52415	8.84205	1.67871	8.81140	1.83277	9	
Minutes.	Distance.	Dep.	Lat.	Dep.	Lat.	Dep.	Lat.	Distance.	Minutes.
		80°		79°		78°			

Differences of Latitude and Departures—Continued.

Minutes.	Distance.	12° Lat.	12° Dep.	13° Lat.	13° Dep.	14° Lat.	14° Dep.	Distance.	Minutes.
	1	0.97814	0.20791	0.97437	0.22495	0.97029	0.24192	1	
	2	1.95629	0.41582	1.94874	0.44990	1.94059	0.48384	2	
	3	2.93444	0.62373	2.92311	0.67485	2.91088	0.72576	3	
	4	3.91259	0.83164	3.89748	0.89980	3.88118	0.96768	4	
0	5	4.89073	1.03955	4.87185	1.12475	4.85147	1.20961	5	60
	6	5.86888	1.24747	5.84622	1.34970	5.82177	1.45153	6	
	7	6.84703	1.45538	6.82059	1.57465	6.79206	1.69345	7	
	8	7.82518	1.66329	7.79496	1.79960	7.76236	1.93537	8	
	9	8.80332	1.87120	8.76933	2.02455	8.73266	2.17729	9	
	1	0.97723	0.21217	0.97337	0.22920	0.96923	0.24615	1	
	2	1.95446	0.42435	1.94675	0.45840	1.93846	0.49230	2	
	3	2.93169	0.63653	2.92013	0.68760	2.90769	0.73845	3	
	4	3.90892	0.84871	3.89351	0.91680	3.87692	0.98461	4	
15	5	4.88615	1.06088	4.86689	1.14600	4.84615	1.23076	5	45
	6	5.86338	1.27306	5.84027	1.37520	5.81538	1.47691	6	
	7	6.84061	1.48524	6.81365	1.60440	6.78461	1.72307	7	
	8	7.81784	1.69742	7.78703	1.83360	7.75384	1.96922	8	
	9	8.79507	1.90959	8.76041	2.06280	8.72307	2.21537	9	
	1	0.97629	0.21644	0.97237	0.23344	0.96814	0.25038	1	
	2	1.95259	0.43288	1.94474	0.46689	1.93629	0.50076	2	
	3	2.92888	0.64932	2.91711	0.70033	2.90444	0.75114	3	
	4	3.90518	0.86576	3.88948	0.93378	3.87259	1.00152	4	
30	5	4.88148	1.08220	4.86185	1.16722	4.84073	1.25190	5	30
	6	5.85777	1.29864	5.83422	1.40067	5.80888	1.50228	6	
	7	6.83407	1.51508	6.80659	1.63411	6.77703	1.75266	7	
	8	7.81036	1.73152	7.77896	1.86756	7.74518	2.00304	8	
	9	8.78666	1.94796	8.75133	2.10100	8.71332	2.25342	9	
	1	0.97534	0.22069	0.97134	0.23768	0.96704	0.25460	1	
	2	1.95068	0.44139	1.94268	0.47537	1.93409	0.50920	2	
	3	2.92602	0.66209	2.91402	0.71305	2.90113	0.76380	3	
	4	3.90136	0.88278	3.88536	0.95074	3.86818	1.01840	4	
45	5	4.87671	1.10348	4.85671	1.18843	4.83523	1.27301	5	15
	6	5.85205	1.32418	5.82805	1.42611	5.80227	1.52761	6	
	7	6.82739	1.54488	6.79939	1.66380	6.76932	1.78221	7	
	8	7.80273	1.76557	7.77073	1.90148	7.73636	2.03681	8	
	9	8.77808	1.98627	8.74207	2.13917	8.70341	2.29141	9	
Minutes.	Distance.	Dep.	Lat.	Dep.	Lat.	Dep.	Lat.	Distance.	Minutes.
		77°		76°		75°			

Differences of Latitude and Departures—Continued.

Minutes.	Distance.	15°		16°		17°		Distance.	Minutes.
		Lat.	Dep.	Lat.	Dep.	Lat.	Dep.		
	1	0.96592	0.25881	0.96126	0.27563	0.95630	0.29237	1	
	2	1.93185	0.51763	1.92252	0.55127	1.91260	0.58474	2	
	3	2.89777	0.77645	2.88378	0.82691	2.86891	0.87711	3	
	4	3.86370	1.03527	3.84504	1.10254	3.82521	1.16948	4	
0	5	4.82962	1.29409	4.80630	1.37818	4.78152	1.46185	5	60
	6	5.79555	1.55291	5.76757	1.65382	5.73782	1.75423	6	
	7	6.76148	1.81173	6.72883	1.92946	6.69413	2.04660	7	
	8	7.72740	2.07055	7.69009	2.20509	7.65043	2.33897	8	
	9	8.69333	2.32937	8.65135	2.48073	8.60674	2.63134	9	
	1	0.96478	0.26303	0.96005	0.27982	0.95502	0.29654	1	
	2	1.92957	0.52606	1.92010	0.55965	1.91004	0.59308	2	
	3	2.89436	0.78909	2.88015	0.83948	2.86506	0.88962	3	
	4	3.85914	1.05212	3.84020	1.11931	3.82008	1.18616	4	
15	5	4.82393	1.31515	4.80025	1.39914	4.77510	1.48270	5	45
	6	5.78872	1.57818	5.76030	1.67897	5.73012	1.77924	6	
	7	6.75351	1.84121	6.72035	1.95880	6.68514	2.07579	7	
	8	7.71829	2.10424	7.68040	2.23863	7.64016	2.37233	8	
	9	8.68308	2.36728	8.64045	2.51846	8.59518	2.66887	9	
	1	0.96363	0.26723	0.95882	0.28401	0.95371	0.30070	1	
	2	1.92726	0.53447	1.91764	0.56803	1.90743	0.60141	2	
	3	2.89089	0.80171	2.87646	0.85204	2.86115	0.90211	3	
	4	3.85452	1.06895	3.83528	1.13606	3.81486	1.20282	4	
30	5	4.81815	1.33619	4.79410	1.42007	4.76858	1.50352	5	30
	6	5.78178	1.60343	5.75292	1.70409	5.72230	1.80423	6	
	7	6.74541	1.87066	6.71174	1.98810	6.67601	2.10494	7	
	8	7.70904	2.13790	7.67056	2.27212	7.62973	2.40564	8	
	9	8.67267	2.40514	8.62938	2.55613	8.58345	2.70635	9	
	1	0.96245	0.27144	0.95757	0.28819	0.95239	0.30486	1	
	2	1.92491	0.54288	1.91514	0.57639	1.90479	0.60972	2	
	3	2.88736	0.81432	2.87271	0.86458	2.85718	0.91459	3	
	4	3.84982	1.08576	3.83028	1.15278	3.80958	1.21945	4	
45	5	4.81227	1.35720	4.78785	1.44098	4.76197	1.52432	5	15
	6	5.77473	1.62864	5.74542	1.72917	5.71437	1.82918	6	
	7	6.73718	1.90008	6.70299	2.01737	6.66677	2.13405	7	
	8	7.69964	2.17152	7.66057	2.30557	7.61916	2.43891	8	
	9	8.66209	2.44296	8.61814	2.59376	8.57156	2.74377	9	
Minutes.	Distance.	Dep.	Lat.	Dep.	Lat.	Dep.	Lat.	Distance.	Minutes.
		74°		73°		72°			

Differences of Latitude and Departures—Continued.

Minutes.	Distance.	18° Lat.	18° Dep.	19° Lat.	19° Dep.	20° Lat.	20° Dep.	Distance.	Minutes.
	1	0.95105	0.30901	0.94551	0.32556	0.93969	0.34202	1	
	2	1.90211	0.61803	1.89103	0.65113	1.87938	0.68404	2	
	3	2.85316	0.92705	2.83655	0.97670	2.81907	1.02606	3	
	4	3.80422	1.23606	3.78207	1.30227	3.75877	1.36808	4	
0	5	4.75528	1.54508	4.72759	1.62784	4.69846	1.71010	5	60
	6	5.70633	1.85410	5.67311	1.95340	5.63815	2.05212	6	
	7	6.65739	2.16311	6.61863	2.27897	6.57784	2.39414	7	
	8	7.60845	2.47213	7.56414	2.60454	7.51754	2.73616	8	
	9	8.55950	2.78115	8.50966	2.93011	8.45723	3.07818	9	
	1	0.94969	0.31316	0.94408	0.32969	0.93819	0.34611	1	
	2	1.89939	0.62632	1.88817	0.65938	1.87638	0.69223	2	
	3	2.84909	0.93949	2.83226	0.98907	2.81457	1.03835	3	
	4	3.79879	1.25265	3.77635	1.31876	3.75276	1.38446	4	
15	5	4.74849	1.56581	4.72044	1.64845	4.69095	1.73058	5	45
	6	5.69819	1.87898	5.66453	1.97814	5.62914	2.07670	6	
	7	6.64789	2.19214	6.60862	2.30783	6.56733	2.44281	7	
	8	7.59759	2.50531	7.55271	2.63752	7.50553	2.76893	8	
	9	8.54729	2.81847	8.49680	2.96721	8.44372	3.11505	9	
	1	0.94832	0.31730	0.94264	0.33380	0.93667	0.35020	1	
	2	1.89664	0.63460	1.88528	0.66761	1.87334	0.70041	2	
	3	2.84497	0.95191	2.82792	1.00142	2.81001	1.05062	3	
	4	3.79329	1.26921	3.77056	1.33522	3.74668	1.40082	4	
30	5	4.74161	1.58652	4.71320	1.66903	4.68336	1.75103	5	30
	6	5.68994	1.90382	5.65584	2.00284	5.62003	2.10124	6	
	7	6.63826	2.22113	6.59849	2.33664	6.55670	2.45145	7	
	8	7.58658	2.53843	7.54113	2.67045	7.49337	2.80165	8	
	9	8.53491	2.85574	8.48377	3.00426	8.43004	3.15186	9	
	1	0.94693	0.32143	0.94117	0.33791	0.93513	0.35429	1	
	2	1.89386	0.64287	1.88235	0.67583	1.87027	0.70858	2	
	3	2.84079	0.96431	2.82352	1.01375	2.80540	1.06287	3	
	4	3.78772	1.28575	3.76470	1.35166	3.74054	1.41716	4	
45	5	4.73465	1.60719	4.70588	1.68958	4.67567	1.77145	5	15
	6	5.68158	1.92863	5.64705	2.02750	5.61081	2.12574	6	
	7	6.62851	2.25007	6.58823	2.36541	6.54594	2.48003	7	
	8	7.57544	2.57151	7.52940	2.70333	7.48108	2.83432	8	
	9	8.52237	2.89295	8.47058	3.04125	8.41621	3.18861	9	
Minutes.	Distance.	Dep.	Lat.	Dep.	Lat.	Dep.	Lat.	Distance.	Minutes.
		71°		70°		69°			

Differences of Latitude and Departures—Continued.

Minutes.	Distance.	21°		22°		23°		Distance.	Minutes.
		Lat.	Dep.	Lat.	Dep.	Lat.	Dep.		
	1	0.93358	0.35836	0.92718	0.37460	0.92050	0.39073	1	
	2	1.86716	0.71673	1.85436	0.74921	1.84100	0.78146	2	
	3	2.80074	1.07510	2.78155	1.12381	2.76151	1.17219	3	
	4	3.73432	1.43347	3.70873	1.49842	3.68201	1.56292	4	
0	5	4.66790	1.79183	4.63591	1.87303	4.60252	1.95365	5	60
	6	5.60148	2.15020	5.56310	2.24763	5.52302	2.34438	6	
	7	6.53506	2.50857	6.49028	2.62224	6.44353	2.73511	7	
	8	7.46864	2.86694	7.41747	2.99685	7.36403	3.12584	8	
	9	8.40222	3.22531	8.34465	3.37145	8.28454	3.51657	9	
	1	0.93200	0.36243	0.92554	0.37864	0.91879	0.39474	1	
	2	1.86401	0.72487	1.85108	0.75729	1.83758	0.78948	2	
	3	2.79602	1.08731	2.77662	1.13594	2.75637	1.18423	3	
	4	3.72803	1.44975	3.70216	1.51459	3.67516	1.57897	4	
15	5	4.66004	1.81219	4.62770	1.89324	4.59395	1.97372	5	45
	6	5.59204	2.17462	5.55324	2.27189	5.51274	2.36846	6	
	7	6.52405	2.53706	6.47878	2.65054	6.43153	2.76320	7	
	8	7.45606	2.89950	7.40432	3.02918	7.35032	3.15795	8	
	9	8.38807	3.26194	8.32986	3.40783	8.26912	3.55269	9	
	1	0.93041	0.36650	0.92388	0.38268	0.91706	0.39874	1	
	2	1.86083	0.73300	1.84776	0.76536	1.83412	0.79749	2	
	3	2.79125	1.09950	2.77164	1.14805	2.75118	1.19624	3	
	4	3.72167	1.46600	3.69552	1.53073	3.66824	1.59499	4	
30	5	4.65208	1.83250	4.61940	1.91341	4.58530	1.99374	5	30
	6	5.58250	2.19900	5.54328	2.29610	5.50236	2.39249	6	
	7	6.51292	2.56550	6.46716	2.67878	6.41942	2.79124	7	
	8	7.44334	2.93200	7.39104	3.06146	7.33648	3.18999	8	
	9	8.37375	3.29851	8.31492	3.44415	8.25354	3.58874	9	
	1	0.92881	0.37055	0.92220	0.38671	0.91531	0.40274	1	
	2	1.85762	0.74111	1.84440	0.77342	1.83062	0.80549	2	
	3	2.78643	1.11167	2.76660	1.16013	2.74593	1.20824	3	
	4	3.71524	1.48222	3.68880	1.54684	3.66124	1.61098	4	
45	5	4.64405	1.85278	4.61100	1.93355	4.57655	2.01373	5	15
	6	5.57286	2.22334	5.53320	2.32026	5.49186	2.41648	6	
	7	6.50167	2.59390	6.45540	2.70697	6.40718	2.81922	7	
	8	7.43048	2.96445	7.37760	3.09368	7.32249	3.22197	8	
	9	8.35929	3.33501	8.29980	3.48039	8.23780	3.62472	9	
		Dep.	Lat.	Dep.	Lat.	Dep.	Lat.		
Minutes.	Distance.	68°		67°		66°		Distance.	Minutes.

Differences of Latitude and Departures—Continued.

Minutes.	Distance.	24°		25°		26°		Distance.	Minutes.
		Lat.	Dep.	Lat.	Dep.	Lat.	Dep.		
	1	0.91354	0.40673	0.90630	0.42261	0.89879	0.43837	1	
	2	1.82709	0.81347	1.81261	0.84523	1.79758	0.87674	2	
	3	2.74063	1.22020	2.71892	1.26785	2.69638	1.31511	3	
	4	3.65418	1.62694	3.62523	1.69047	3.59517	1.75348	4	
0	5	4.56772	2.03368	4.53153	2.11309	4.49397	2.19185	5	60
	6	5.48127	2.44041	5.43784	2.53570	5.39276	2.63022	6	
	7	6.39481	2.84715	6.34415	2.95832	6.29155	3.06859	7	
	8	7.30836	3.25389	7.25046	3.38094	7.19035	3.50696	8	
	9	8.22190	3.66062	8.15677	3.80356	8.08914	3.94533	9	
	1	0.91176	0.41071	0.90445	0.42656	0.89687	0.44228	1	
	2	1.82352	0.82143	1.80891	0.85313	1.79374	0.88457	2	
	3	2.73528	1.23215	2.71336	1.27970	2.69061	1.32686	3	
	4	3.64704	1.64287	3.61782	1.70627	3.58749	1.76915	4	
15	5	4.55881	2.05359	4.52227	2.13284	4.48436	2.21144	5	45
	6	5.47057	2.46431	5.42673	2.55941	5.38123	2.65373	6	
	7	6.38233	2.87503	6.33118	2.98598	6.27810	3.09602	7	
	8	7.29409	3.28575	7.23564	3.41254	7.17498	3.53830	8	
	9	8.20585	3.69647	8.14009	3.83911	8.07185	3.98059	9	
	1	0.90996	0.41469	0.90258	0.43051	0.89493	0.44619	1	
	2	1.81992	0.82938	1.80517	0.86102	1.78986	0.89239	2	
	3	2.72988	1.24407	2.70775	1.29153	2.68480	1.33859	3	
	4	3.63984	1.65877	3.61034	1.72204	3.57973	1.78479	4	
30	5	4.54980	2.07346	4.51292	2.15255	4.47467	2.23098	5	30
	6	5.45976	2.48815	5.41551	2.58306	5.36960	2.67718	6	
	7	6.36972	2.90285	6.31809	3.01357	6.26454	3.12338	7	
	8	7.27969	3.31754	7.22068	3.44408	7.15947	3.56958	8	
	9	8.18965	3.73223	8.12326	3.87459	8.05440	4.01578	9	
	1	0.90814	0.41866	0.90069	0.43444	0.89297	0.45009	1	
	2	1.81628	0.83732	1.80139	0.86889	1.78595	0.90019	2	
	3	2.72442	1.25598	2.70209	1.30333	2.67893	1.35029	3	
	4	3.63257	1.67464	3.60279	1.73778	3.57191	1.80039	4	
45	5	4.54071	2.09330	4.50349	2.17222	4.46489	2.25049	5	15
	6	5.44885	2.51196	5.40418	2.60667	5.35787	2.70059	6	
	7	6.35700	2.93062	6.30488	3.04111	6.25085	3.15068	7	
	8	7.26514	3.34928	7.20558	3.47556	7.14383	3.60078	8	
	9	8.17328	3.76794	8.10628	3.91000	8.03681	4.05088	9	
Minutes.	Distance.	Dep.	Lat.	Dep.	Lat.	Dep.	Lat.	Distance.	Minutes.
		65°		64°		63°			

Differences of Latitude and Departures—Continued.

Minutes.	Distance.	27°		28°		29°		Distance.	Minutes.
		Lat.	Dep.	Lat.	Dep.	Lat.	Dep.		
	1	0.89100	0.45399	0.88294	0.46947	0.87462	0.48481	1	
	2	1.78201	0.90798	1.76589	0.93894	1.74924	0.96962	2	
	3	2.67301	1.36197	2.64884	1.40841	2.62386	1.45443	3	
	4	3.56402	1.81596	3.53179	1.87788	3.49848	1.93924	4	
0	5	4.45503	2.26995	4.41473	2.34735	4.37310	2.42405	5	60
	6	5.34603	2.72394	5.29768	2.81682	5.24772	2.90886	6	
	7	6.23704	3.17793	6.18063	3.28630	6.12234	3.39367	7	
	8	7.12805	3.63193	7.06358	3.75577	6.99696	3.87848	8	
	9	8.01905	4.08591	7.94652	4.22524	7.87156	4.36329	9	
	1	0.88901	0.45787	0.88089	0.47332	0.87249	0.48862	1	
	2	1.77803	0.91574	1.76178	0.94664	1.74499	0.97724	2	
	3	2.66705	1.37362	2.64267	1.41996	2.61748	1.46566	3	
	4	3.55606	1.83149	3.52356	1.89328	3.48998	1.95448	4	
15	5	4.44508	2.28937	4.40445	2.36660	4.36248	2.44310	5	45
	6	5.33410	2.74724	5.28534	2.83992	5.23497	2.93172	6	
	7	6.22311	3.20511	6.16623	3.31324	6.10747	3.42034	7	
	8	7.11213	3.66299	7.04712	3.78656	6.97996	3.90896	8	
	9	8.00115	4.12086	7.92801	4.25988	7.85246	4.39759	9	
	1	0.88701	0.46174	0.87881	0.47715	0.87035	0.49242	1	
	2	1.77402	0.92349	1.75763	0.95431	1.74071	0.98484	2	
	3	2.66103	1.38524	2.63645	1.43147	2.61106	1.47727	3	
	4	3.54804	1.84699	3.51526	1.90863	3.48142	1.96969	4	
30	5	4.43505	2.30874	4.39408	2.38579	4.35177	2.46211	5	30
	6	5.32206	2.77049	5.27290	2.86295	5.22213	2.95454	6	
	7	6.20907	3.23224	6.15171	3.34011	6.09248	3.44696	7	
	8	7.09608	3.69398	7.03053	3.81727	6.96284	3.93938	8	
	9	7.98309	4.15573	7.90935	4.29442	7.83320	4.43181	9	
	1	0.88498	0.46561	0.87672	0.48098	0.86819	0.49621	1	
	2	1.76997	0.93122	1.75345	0.96197	1.73639	0.99243	2	
	3	2.65496	1.39684	2.63018	1.44296	2.60459	1.48864	3	
	4	3.53995	1.86245	3.50690	1.92395	3.47279	1.98486	4	
45	5	4.42493	2.32807	4.38363	2.40494	4.34099	2.48108	5	15
	6	5.30992	2.79368	5.26036	2.88593	5.20919	2.97729	6	
	7	6.19491	3.25930	6.13708	3.36692	6.07739	3.47351	7	
	8	7.07990	3.72491	7.01381	3.84791	6.94559	3.96973	8	
	9	7.96488	4.19053	7.89054	4.32889	7.81378	4.46594	9	
Minutes.	Distance.	Dep.	Lat.	Dep.	Lat.	Dep.	Lat.	Distance.	Minutes.
		62°		61°		60°			

Differences of Latitude and Departures—Continued.

Minutes.	Distance.	30° Lat.	30° Dep.	31° Lat.	31° Dep.	32° Lat.	32° Dep.	Distance.	Minutes.
	1	0.86602	0.50000	0.85716	0.51503	0.84804	0.52991	1	
	2	1.73205	1.00000	1.71433	1.03007	1.69609	1.05983	2	
	3	2.59807	1.50000	2.57150	1.54511	2.54414	1.58975	3	
	4	3.46410	2.00000	3.42866	2.06015	3.39219	2.11967	4	
0	5	4.33012	2.50000	4.28583	2.57519	4.24024	2.64959	5	60
	6	5.19615	3.00000	5.14300	3.09022	5.08828	3.17951	6	
	7	6.06217	3.50000	6.00017	3.60526	5.93633	3.70943	7	
	8	6.92820	4.00000	6.85733	4.12030	6.78438	4.23935	8	
	9	7.79422	4.50000	7.71450	4.63534	7.63243	4.76927	9	
	1	0.86383	0.50377	0.85491	0.51877	0.84572	0.53361	1	
	2	1.72767	1.00754	1.70982	1.03754	1.69145	1.06722	2	
	3	2.59150	1.51132	2.56473	1.55631	2.53718	1.60084	3	
	4	3.45534	2.01509	3.41964	2.07509	3.38291	2.13445	4	
15	5	4.31917	2.51887	4.27456	2.59386	4.22863	2.66807	5	45
	6	5.18301	3.02264	5.12947	3.11263	5.07436	3.20168	6	
	7	6.04684	3.52641	5.98438	3.63141	5.92009	3.73530	7	
	8	6.91068	4.03019	6.83929	4.15018	6.76582	4.26891	8	
	9	7.77451	4.53396	7.69420	4.66895	7.61155	4.80253	9	
	1	0.86162	0.50753	0.85264	0.52249	0.84339	0.53730	1	
	2	1.72325	1.01507	1.70528	1.04499	1.68678	1.07460	2	
	3	2.58488	1.52261	2.55792	1.56749	2.53017	1.61190	3	
	4	3.44651	2.03015	3.41056	2.08999	3.37356	2.14920	4	
30	5	4.30814	2.53769	4.26320	2.61249	4.21695	2.68650	5	30
	6	5.16977	3.04523	5.11584	3.13499	5.06034	3.22380	6	
	7	6.03140	3.55276	5.96948	3.65749	5.90373	3.76110	7	
	8	6.89303	4.06030	6.82112	4.17998	6.74713	4.29840	8	
	9	7.75466	4.56784	7.67376	4.70248	7.59052	4.83570	9	
	1	0.85940	0.51129	0.85035	0.52621	0.84103	0.54097	1	
	2	1.71881	1.02258	1.70070	1.05242	1.68207	1.08194	2	
	3	2.57821	1.53387	2.55105	1.57864	2.52311	1.62292	3	
	4	3.43762	2.04517	3.40140	2.10485	3.36415	2.16389	4	
45	5	4.29703	2.55646	4.25176	2.63107	4.20519	2.70487	5	15
	6	5.15643	3.06775	5.10211	3.15728	5.04623	3.24584	6	
	7	6.01584	3.57905	5.95246	3.68349	5.88827	3.78682	7	
	8	6.87525	4.09034	6.80281	4.20971	6.72831	4.32779	8	
	9	7.73465	4.60163	7.65316	4.73592	7.56935	4.86877	9	
Minutes.	Distance.	Dep.	Lat.	Dep.	Lat.	Dep.	Lat.	Distance.	Minutes.
		59°		58°		57°			

Differences of Latitude and Departures—Continued.

Minutes.	Distance.	33°		34°		35°		Distance.	Minutes.
		Lat.	Dep.	Lat.	Dep.	Lat.	Dep.		
	1	0.83867	0.54463	0.82903	0.55919	0.81915	0.57357	1	
	2	1.67734	1.08927	1.65807	1.11838	1.63830	1.14715	2	
	3	2.51601	1.63391	2.48711	1.67757	2.45745	1.72072	3	
	4	3.35468	2.17855	3.31615	2.23677	3.27660	2.29430	4	
0	5	4.19335	2.72319	4.14518	2.79596	4.09576	2.86788	5	60
	6	5.03202	3.26783	4.97422	3.35515	4.91491	3.44145	6	
	7	5.87069	3.81247	5.80326	3.91435	5.73406	4.01503	7	
	8	6.70936	4.35711	6.63230	4.47354	6.55321	4.58861	8	
	9	7.54803	4.90175	7.46133	5.03273	7.37236	5.16218	9	
	1	0.83628	0.54829	0.82659	0.56280	0.81664	0.57714	1	
	2	1.67257	1.09658	1.65318	1.12560	1.63328	1.15429	2	
	3	2.50885	1.64487	2.47977	1.68841	2.44992	1.73143	3	
	4	3.34514	2.19317	3.30636	2.25121	3.26656	2.30858	4	
15	5	4.18143	2.74146	4.13295	2.81402	4.08320	2.88572	5	45
	6	5.01771	3.28975	4.95954	3.37682	4.89984	3.46287	6	
	7	5.85400	3.83805	5.78613	3.93963	5.71649	4.04001	7	
	8	6.69028	4.38634	6.61272	4.50243	6.53313	4.61716	8	
	9	7.52657	4.93463	7.43931	5.06524	7.34977	5.19430	9	
	1	0.83388	0.55193	0.82412	0.56640	0.81411	0.58070	1	
	2	1.66777	1.10387	1.64825	1.13281	1.62823	1.16140	2	
	3	2.50165	1.65581	2.47237	1.69921	2.44234	1.74210	3	
	4	3.33554	2.20774	3.29650	2.26562	3.25646	2.32281	4	
30	5	4.16942	2.75968	4.12063	2.83203	4.07057	2.90351	5	30
	6	5.00331	3.31162	4.94475	3.39843	4.88469	3.48421	6	
	7	5.83720	3.86355	5.76888	3.96484	5.69880	4.06492	7	
	8	6.67108	4.41549	6.59300	4.53124	6.51292	4.64562	8	
	9	7.50497	4.96743	7.41713	5.09765	7.32703	5.22632	9	
	1	0.83147	0.55557	0.82164	0.56999	0.81157	0.58425	1	
	2	1.66294	1.11114	1.64329	1.13999	1.62314	1.16850	2	
	3	2.49441	1.66671	2.46494	1.70999	2.43472	1.75275	3	
	4	3.32588	2.22228	3.28658	2.27998	3.24629	2.33700	4	
45	5	4.15735	2.77785	4.10823	2.84998	4.05787	2.92125	5	15
	6	4.98882	3.33342	4.92988	3.41998	4.86944	3.50550	6	
	7	5.82029	3.88899	5.75152	3.98997	5.68101	4.08975	7	
	8	6.65176	4.44456	6.57317	4.55997	6.49260	4.67400	8	
	9	7.48323	5.00013	7.39482	5.12997	7.30416	5.25825	9	
Minutes.	Distance.	Dep.	Lat.	Dep.	Lat.	Dep.	Lat.	Distance.	Minutes.
		56°		55°		54°			

Differences of Latitude and Departures—Continued.

Minutes.	Distance.	36° Lat.	36° Dep.	37° Lat.	37° Dep.	38° Lat.	38° Dep.	Distance.	Minutes.
	1	0.80901	0.58778	0.79863	0.60181	0.78801	0.61566	1	
	2	1.61803	1.17557	1.59727	1.20363	1.57602	1.23132	2	
	3	2.42705	1.76335	2.39590	1.80544	2.36403	1.84698	3	
	4	3.23606	2.35114	3.19454	2.40726	3.15204	2.46264	4	
0	5	4.04508	2.93892	3.99317	3.00907	3.94005	3.07830	5	60
	6	4.85410	3.52671	4.79181	3.61089	4.72806	3.69396	6	
	7	5.66311	4.11449	5.59044	4.21270	5.51607	4.30963	7	
	8	6.47213	4.70228	6.38908	4.81452	6.30408	4.92529	8	
	9	7.28115	5.29006	7.18771	5.41633	7.09209	5.54095	9	
	1	0.80644	0.59130	0.79600	0.60529	1.78531	0.61909	1	
	2	1.61288	1.18261	1.59200	1.21058	1.57063	1.23818	2	
	3	2.41933	1.77392	2.38800	1.81588	2.35595	1.85728	3	
	4	3.22577	2.36523	3.18400	2.42117	3.14126	2.47637	4	
15	5	4.03222	2.95654	3.98001	3.02647	3.92658	3.09547	5	45
	6	4.83866	3.54785	4.77601	3.63176	4.71190	3.71456	6	
	7	5.64511	4.13916	5.57201	4.23705	5.49721	4.33365	7	
	8	6.45155	4.73047	6.36801	4.84235	6.28253	4.95275	8	
	9	7.25800	5.32178	7.16401	5.44764	7.06785	5.57184	9	
	1	0.80385	0.59482	0.79335	0.60876	0.78260	0.62251	1	
	2	1.60771	1.18964	1.58670	1.21752	1.56521	1.24502	2	
	3	2.41157	1.78446	2.38005	1.82628	2.34782	1.86754	3	
	4	3.21542	2.37929	3.17341	2.43504	3.13043	2.49005	4	
30	5	4.01928	2.97411	3.96676	3.04380	3.91304	3.11257	5	30
	6	4.82314	3.56893	4.76011	3.65256	4.69564	3.73508	6	
	7	5.62699	4.16375	5.55347	4.26132	5.47825	4.35760	7	
	8	6.43085	4.75858	6.34682	4.87009	6.26086	4.98011	8	
	9	7.23471	5.35340	7.14017	5.47885	7.04347	5.60263	9	
	1	0.80125	0.59832	0.79068	0.61221	0.77988	0.62592	1	
	2	1.60250	1.19664	1.58137	1.22443	1.55946	1.25184	2	
	3	2.40376	1.79497	2.37206	1.83665	2.33965	1.87777	3	
	4	3.20501	2.39329	3.16275	2.44886	3.11953	2.50369	4	
45	5	4.00626	2.99162	3.95344	3.06108	3.89942	3.12961	5	15
	6	4.80752	3.58994	4.74413	3.67330	4.67930	3.75554	6	
	7	5.60877	4.18827	5.53482	4.28552	5.45919	4.38146	7	
	8	6.41003	4.78659	6.32551	4.89773	6.23907	5.00738	8	
	9	7.21128	5.38492	7.11620	5.50995	7.01896	5.63331	9	
Minutes.	Distance.	Dep.	Lat.	Dep.	Lat.	Dep.	Lat.	Distance.	Minutes.
		53°		52°		51°			

Differences of Latitude and Departures—Continued.

Minutes.	Distance.	39° Lat.	39° Dep.	40° Lat.	40° Dep.	41° Lat.	41° Dep.	Distance.	Minutes.
	1	0.77714	0.62932	0.76604	0.64278	0.75470	0.65605	1	
	2	1.55429	1.25864	1.53208	1.28557	1.50941	1.31211	2	
	3	2.33143	1.88796	2.29813	1.92836	2.26412	1.96817	3	
	4	3.10858	2.51728	3.06417	2.57115	3.01883	2.62423	4	
0	5	3.88573	3.14660	3.83022	3.21393	3.77354	3.28029	5	60
	6	4.66287	3.77592	4.59626	3.85672	4.52825	3.93635	6	
	7	5.44002	4.40524	5.36231	4.49951	5.28296	4.59241	7	
	8	6.21716	5.03456	6.12835	5.14230	6.03767	5.24847	8	
	9	6.99431	5.66388	6.89439	5.78508	6.79238	5.90453	9	
	1	0.77439	0.63270	0.76323	0.64612	0.75184	0.65934	1	
	2	1.54878	1.26541	1.52646	1.29224	1.50368	1.31869	2	
	3	2.32317	1.89811	2.28969	1.93837	2.25552	1.97803	3	
	4	3.09757	2.53082	3.05293	2.58449	3.00736	2.63738	4	
15	5	3.87196	3.16352	3.81616	3.23062	3.75920	3.29672	5	45
	6	4.64635	3.79623	4.57939	3.87674	4.51104	3.95607	6	
	7	5.42074	4.42893	5.34262	4.52286	5.26288	4.61542	7	
	8	6.19514	5.06164	6.10586	5.16899	6.01472	5.27476	8	
	9	6.96953	5.69434	6.86909	5.81511	6.76656	5.93411	9	
	1	0.77162	0.63607	0.76040	0.64944	0.74895	0.66262	1	
	2	1.54324	1.27215	1.52081	1.29889	1.49791	1.32524	2	
	3	2.31487	1.90823	2.28121	1.94834	2.24686	1.98786	3	
	4	3.08649	2.54431	3.04162	2.59779	2.99582	2.65048	4	
30	5	3.85812	3.18039	3.80203	3.24724	3.74477	3.31310	5	30
	6	4.62974	3.81646	4.56243	3.89668	4.49373	3.97572	6	
	7	5.40137	4.45254	5.32284	4.54613	5.24268	4.63834	7	
	8	6.17299	5.08862	6.08324	5.19558	5.99164	5.30096	8	
	9	6.94462	5.72470	6.84365	5.84503	6.74060	5.96358	9	
	1	0.76884	0.63943	0.75756	0.65276	0.74605	0.66588	1	
	2	1.53768	1.27887	1.51513	1.30552	1.49211	1.33176	2	
	3	2.30652	1.91831	2.27269	1.95828	2.23817	1.99764	3	
	4	3.07536	2.55775	3.03026	2.61104	2.98422	2.66352	4	
45	5	3.84420	3.19719	3.78782	2.26380	3.73028	3.32940	5	15
	6	4.61305	3.83663	4.54539	3.91656	4.47634	3.99529	6	
	7	5.38189	4.47607	5.30295	4.56932	5.22240	4.66117	7	
	8	6.15073	5.11551	6.06052	5.22208	5.96845	5.32705	8	
	9	6.91957	5.75495	6.81808	5.87484	6.71451	5.99293	9	
Minutes.	Distance.	Dep.	Lat.	Dep.	Lat.	Dep.	Lat.	Distance.	Minutes.
		50°		49°		48°			

Differences of Latitude and Departures—Continued.

Minutes.	Distance.	42°		43°		44°		Distance.	Minutes.
		Lat.	Dep.	Lat.	Dep.	Lat.	Dep.		
	1	0.74314	0.66913	0.73135	0.68199	0.71933	0.69465	1	
	2	1.48628	1.33826	1.46270	1.36399	1.43867	1.38931	2	
	3	2.22943	2.00739	2.19406	2.04599	2.15801	2.08397	3	
	4	2.97257	2.67652	2.92541	2.72799	2.87735	2.77863	4	
0	5	3.71572	3.34565	3.65676	3.40999	3.59669	3.47329	5	60
	6	4.45886	4.01478	4.38812	4.09199	4.31603	4.16795	6	
	7	5.20201	4.68391	5.11947	4.77398	5.03537	4.86260	7	
	8	5.94515	5.35304	5.85082	5.45598	5.75471	5.55726	8	
	9	6.68830	6.02217	6.58218	6.13798	6.47405	6.25192	9	
	1	0.74021	0.67236	0.72837	0.68518	0.71630	0.69779	1	
	2	1.48043	1.34473	1.45674	1.37036	1.43260	1.39558	2	
	3	2.22065	2.01710	2.18511	2.05554	2.14890	2.09337	3	
	4	2.96087	2.68946	2.91348	2.74073	2.86520	2.79116	4	
15	5	3.70109	3.36183	3.64185	3.42591	3.58151	3.48895	5	45
	6	4.44130	4.03420	4.37022	4.11109	4.29781	4.18674	6	
	7	5.18152	4.70656	5.09859	4.79628	5.01411	4.88453	7	
	8	5.92174	5.37893	5.82696	5.48146	5.73041	5.58232	8	
	9	6.66196	6.05130	6.55533	6.16664	6.44671	6.28011	9	
	1	0.73727	0.67559	0.72537	0.68835	0.71325	0.70090	1	
	2	1.47455	1.35118	1.45074	1.37670	1.42650	1.40181	2	
	3	2.21183	2.02677	2.17612	2.06506	2.13975	2.10272	3	
	4	2.94910	2.70236	2.90149	2.75341	2.85300	2.80363	4	
30	5	3.68638	3.37795	3.62687	3.44177	3.56625	3.50454	5	30
	6	4.42366	4.05354	4.35224	4.13012	4.27950	4.20545	6	
	7	5.16094	4.72913	5.07762	4.81848	4.99275	4.90636	7	
	8	5.89821	5.40472	5.80299	5.50683	5.70600	5.60727	8	
	9	6.63549	6.08031	6.52836	6.19519	6.41925	6.30818	9	
	1	0.73432	0.67880	0.72236	0.69151	0.71018	0.70401	1	
	2	1.46864	1.35760	1.44472	1.38302	1.42037	1.40802	2	
	3	2.20296	2.03640	2.16709	2.07453	2.13055	2.11204	3	
	4	2.93729	2.71520	2.88945	2.76605	2.84074	2.81605	4	
45	5	3.67161	3.39400	3.61182	3.45756	3.55092	3.52007	5	15
	6	4.40593	4.07280	4.33418	4.14907	4.26111	4.22408	6	
	7	5.14025	4.75160	5.05654	4.84059	4.97129	4.92810	7	
	8	5.87458	5.43040	5.77891	5.53210	5.68148	5.63211	8	
	9	6.60890	6.10920	6.50127	6.22361	6.39166	6.33613	9	
Minutes.	Distance.	Dep.	Lat.	Dep.	Lat.	Dep.	Lat.	Distance.	Minutes.
		47°		46°		45°			

Differences of Latitude and Departures—Continued.

	45°		
	Lat.	Dep.	
1	0. 70710	0. 70710	1
2	1. 41421	1. 41421	2
3	2. 12132	2. 12132	3
4	2. 82842	2. 82842	4
5	3. 53553	3. 53553	5
6	4. 24264	4, 24264	6
7	4. 94974	4. 94974	7
8	5. 65685	5. 65685	8
9	6. 36396	6. 36396	9
	Dep.	Lat.	
	45°		

Chains, Yards, and Feet,

WITH THEIR RECIPROCAL EQUIVALENTS.

Link = 7. 92 inches. Chain = 66 feet = 792 inches.

CHAINS INTO FEET.				FEET INTO LINKS.		
Chains.	Links.	Yards.	Feet.	Feet.	Yards.	Links.
0	1	0. 22	0. 66	0. 10	. 033	0. 15
0	2	0. 44	1. 32	0. 20	. 066	0. 30
0	3	0. 66	1. 98	0. 25	. 082	0. 38
0	4	0. 88	2. 64	0. 30	. 010	0. 45
0	5	1. 10	3. 30	0. 40	. 133	0. 60
0	6	1. 32	3. 96	0. 50	. 166	0. 76
0	7	1. 54	4. 62	0. 60	. 200	0. 91
0	8	1. 76	5. 28	0. 70	. 233	1. 06
0	9	1. 98	5. 94	0. 75	. 250	1. 13
0	10	2. 20	6. 60	0. 80	. 266	1. 21

Chains, Yards, and Feet—Continued.

CHAINS INTO FEET.				FEET INTO LINKS.		
Chains.	Links.	Yards.	Feet.	Feet.	Yards.	Links.
0	20	4.40	13.20	0.9	.30	1.36
0	30	6.60	19.80	1.0	.33	1.51
0	40	8.80	26.40	2.0	.66	3.0
0	50	11.00	33.00	3.0	1.00	4.5
0	60	13.20	39.60	4.0	1.33	6.0
0	70	15.40	46.20	5.0	1.66	7.5
0	80	17.60	52.80	6.0	2.00	9.1
0	90	19.80	59.40	7.0	2.33	10.6
1	00	22.00	66.00	8.0	2.66	12.1
2	00	44.00	132	9.0	3.00	13.6
3		66	198	10	3.33	15.1
4		88	264	15	5.00	22.7
5		110	330	20	6.66	30.3
6		132	396	24	8.00	36.3
7		154	462	27	9.00	40.9
8		176	528	30	10.00	45.4
9		198	594	33	11.00	50.0
10		220	660	36	12.00	54.5
20		440	1320	39	13.00	59.1
30		660	1980	40	13.33	60.6
35		770	2310	42	14.00	63.3
40		880	2640	45	15.00	68.2
45		990	2970	48	16.00	72.7
50		1100	3300	50	16.66	75.7
55		1210	3630	51	17.00	77.3
60		1320	3960	54	18.00	81.8
65		1430	4290	57	19.00	86.3
70		1540	4620	60	20.00	90.9
75		1650	4950	63	21.00	95.4
80		1760	5280	66	22.00	100

XIX.—*To trace Railroad Curves by means of Deflections.*

GENERAL PROPOSITIONS.

1. The angle formed by a tangent and a chord is equal to half the angle at the center of the circle subtended by the chord.

2. The angle of deflection formed by any two equal chords meeting at the circumference is equal to the angle at the center, subtended by either cord.

3. A line bisecting the angle of deflection formed by any two equal chords is a tangent to the arc at the point where the two chords meet.

4. If an arc of a circle be subdivided into any number of equal parts, and lines be drawn from the several points of subdivision so as to meet at any point in the circumference, these several lines will form equal angles at the point of meeting, and the angles thus formed will be respectively measured by one-half the subdivided arc.

METHOD BY DEFLECTION-ANGLES.

The *degree* of a curve is determined by the angle subtended at its center by a chord of 100 feet.

The *deflection-angle* of a curve is the acute angle formed at any point between a tangent and a chord. It is, therefore, half the degree of the curve.

In order to unite two straight lines by a curve, the angle of intersection is measured, and then a radius for the curve may be assumed and the tangent calculated, or the tangent may be assumed of a certain length and the radius calculated.

Let I = angle of intersection of the two lines;
R = radius of circle;
T = length of tangent, or distance from point of intersection to point where the curvature is to commence; and
D = angle of deflection.

Then

$$T = R \tan \tfrac{1}{2} I$$

$$R = T \cot \tfrac{1}{2} I$$

$$\sin D = \frac{50}{R} = \frac{50 \tan \frac{1}{2} I}{T}$$

XIX.—*To trace Railroad Curves, &c.*—Continued.

To lay out a curve, set the instrument at the point at which the curvature is to commence, lay off the given deflection-angle, and the *first* point in the curve will be at the end of 100 feet measured on this new direction.

Then lay off another deflection-angle equal to the first; attach the 100-foot chain to the point last found, and swing it, stretched, until its extremity intersects the new direction, which will be the *second* point; and so on. Should it be found necessary to remove the instrument from its first position, either on account of the length of the curve or of some obstruction to the sight, the first deflection at the new position of the instrument will be equal to the total deflection from the preceding position.

METHOD BY TANGENT AND CHORD DEFLECTION.

Tangent-deflection is the distance between the extremities of a tangent and a chord, each 100 feet long.

Chord-deflection is the distance from the extremity of the first chord, produced an additional 100 feet, to the extremity of the next, and is, therefore, double the tangent-deflection.

To lay out a curve, stretch the 100-foot chain from the point of beginning in the direction of the tangent, and mark its extremity; swing the chain toward the direction of the curve, keeping the initial point fixed, until it has diverged a distance equal to the *tangent-deflection*, which will be the *first* point of the curve.

Produce the first chord an additional 100 feet, and swing the chain (round the extremity of the first chord as a pivot) until it has diverged a distance equal to the *chord-deflection*, which will be the *second* point of the curve.

Continue to lay off the chord-deflection from the preceding chord produced until the curve is finished.

LENGTH OF CIRCULAR ARCS IN PARTS OF RADIUS.

°			′			″		
1	.01745	32925	1	.00029	08882	1	.00000	48481
2	.03490	65850	2	.00058	17764	2	.00000	96962
3	.05235	98775	3	.00087	26646	3	.00001	45444
4	.06981	31700	4	.00116	35528	4	.00001	93925
5	.08726	64625	5	.00145	44410	5	.00002	42406
6	.10471	97551	6	.00174	53292	6	.00002	90888
7	.12217	30476	7	.00203	62174	7	.00003	39369
8	.13962	63401	8	.00232	71056	8	.00003	87850
9	.15707	96326	9	.00261	79938	9	.00004	36332

XIX.—*To trace Railroad Curves, &c.*—Continued.

Degree.	Radii.	ORDINATES. To circular arcs on a chord of 100 feet. 12½	25	37½	50	Tangent-deflection.	Chord-deflection.
° ′	*Feet.*						
0 5	68754.94	.008	.014	.017	.018	.073	.145
10	34377.48	.016	.027	.034	.036	.145	.291
15	22918.33	.024	.041	.051	.055	.218	.436
20	17188.76	.032	.055	.068	.073	.291	.582
25	13751.02	.040	.068	.085	.091	.364	.727
30	11459.19	.048	.082	.102	.109	.436	.873
35	9822.18	.056	.095	.119	.127	.509	1.018
40	8594.41	.064	.109	.136	.145	.582	1.164
45	7639.49	.072	.123	.153	.164	.654	1.309
50	6875.55	.080	.136	.170	.182	.727	1.454
55	6250.51	.087	.150	.187	.200	.800	1.600
1 0	5729.65	.095	.164	.205	.218	.873	1.745
5	5288.92	.103	.177	.222	.236	.945	1.891
10	4911.15	.111	.191	.239	.255	1.018	2.036
15	4583.75	.119	.205	.256	.273	1.091	2.182
20	4297.28	.127	.218	.273	.291	1.164	2.327
25	4044.51	.135	.232	.290	.309	1.236	2.472
30	3819.83	.143	.245	.307	.327	1.309	2.618
35	3618.80	.151	.259	.324	.345	1.382	2.763
40	3437.87	.159	.273	.341	.364	1.454	2.909
45	3274.17	.167	.286	.358	.382	1.527	3.054
50	3125.36	.175	.300	.375	.400	1.600	3.200
55	2989.48	.183	.314	.392	.418	1.673	3.345
2 0	2864.93	.191	.327	.409	.436	1.745	3.490
5	2750.35	.199	.341	.426	.455	1.818	3.636
10	2644.58	.207	.355	.443	.473	1.891	3.781
15	2546.64	.215	.368	.460	.491	1.963	3.927
20	2455.70	.223	.382	.477	.509	2.036	4.072
25	2371.04	.231	.395	.494	.527	2.109	4.218
30	2292.01	.239	.409	.511	.545	2.181	4.363
35	2218.09	.247	.423	.528	.564	2.254	4.508
40	2148.79	.255	.436	.545	.582	2.327	4.654
45	2083.68	.263	.450	.562	.600	2.400	4.799
50	2022.41	.270	.464	.580	.618	2.472	4.945
55	1964.64	.278	.477	.597	.636	2.545	5.090

XIX.—*To trace Railroad Curves, &c.*—Continued.

Degree.	Radii.	ORDINATES. To circular arcs on a chord of 100 feet.				Tangent-deflection.	Chord-deflection.
		12½	25	37½	50		
° ′							
3 0	1910.08	.286	.491	.614	.655	2.618	5.235
5	1858.47	.294	.505	.631	.673	2.690	5.381
10	1809.57	.302	.518	.648	.691	2.763	5.526
15	1763.18	.310	.532	.665	.709	2.836	5.672
20	1719.12	.318	.545	.682	.727	2.908	5.817
25	1677.20	.326	.559	.699	.745	2.981	5.962
30	1637.28	.334	.573	.716	.764	3.054	6.108
35	1599.21	.342	.586	.733	.782	3.127	6.253
40	1562.88	.350	.600	.750	.800	3.199	6.398
45	1528.16	.358	.614	.767	.818	3.272	6.544
50	1494.95	.366	.627	.784	.836	3.345	6.689
55	1463.16	.374	.641	.801	.855	3.417	6.835
4 0	1432.69	.382	.655	.818	.873	3.490	6.980
5	1403.46	.390	.668	.835	.891	3.563	7.125
10	1375.40	.398	.682	.852	.909	3.635	7.271
15	1348.45	.406	.695	.869	.927	3.708	7.416
20	1322.53	.414	.709	.886	.945	3.781	7.561
25	1297.58	.422	.723	.903	.964	3.853	7.707
30	1273.57	.430	.736	.921	.982	3.926	7.852
35	1250.42	.438	.750	.938	1.000	3.999	7.997
40	1228.11	.446	.764	.955	1.018	4.071	8.143
45	1206.57	.454	.777	.972	1.036	4.144	8.288
50	1185.78	.462	.791	.989	1.055	4.217	8.433
55	1165.70	.469	.805	1.006	1.073	4.289	8.579
5 0	1146.28	.477	.818	1.023	1.091	4.362	8.724
5	1127.50	.485	.832	1.040	1.109	4.435	8.869
10	1109.33	.493	.846	1.057	1.127	4.507	9.014
15	1091.73	.501	.859	1.074	1.146	4.580	9.160
20	1074.68	.509	.873	1.091	1.164	4.653	9.305
25	1058.16	.517	.887	1.108	1.182	4.725	9.450
30	1042.14	.525	.900	1.125	1.200	4.798	9.596
35	1026.60	.533	.914	1.142	1.218	4.870	9.741
40	1011.51	.541	.928	1.159	1.237	4.943	9.886
45	996.87	.549	.941	1.176	1.255	5.016	10.031

XIX.—*To trace Railroad Curves, &c.*—Continued.

Degree.	Radii.	ORDINATES. To circular arcs on a chord of 100 feet. 12½	25	37½	50	Tangent-deflection.	Chord-deflection.
° ′							
5 50	982.64	.557	.955	1.193	1.273	5.088	10.177
55	968.81	.565	.968	1.210	1.291	5.161	10.322
6 0	955.37	.573	.982	1.228	1.309	5.234	10.467
5	942.29	.581	.996	1.245	1.327	5.306	10.612
10	929.57	.589	1.009	1.262	1.346	5.379	10.758
15	917.19	.597	1.023	1.279	1.364	5.451	10.903
20	905.13	.605	1.037	1.296	1.382	5.524	11.048
25	893.39	.613	1.050	1.313	1.400	5.597	11.193
30	881.95	.621	1.064	1.330	1.418	5.669	11.339
35	870.79	.629	1.078	1.347	1.437	5.742	11.484
40	859.92	.637	1.091	1.364	1.455	5.814	11.629
45	849.32	.645	1.105	1.381	1.473	5.887	11.774
50	838.97	.653	1.118	1.398	1.491	5.960	11.919
55	828.88	.661	1.132	1.415	1.510	6.032	12.065
7 0	819.02	.669	1.146	1.432	1.528	6.105	12.210
5	809.40	.677	1.159	1.449	1.546	6.177	12.355
10	800.00	.685	1.173	1.466	1.564	6.250	12.500
15	790.81	.693	1.187	1.483	1.582	6.323	12.645
20	781.84	.701	1.200	1.501	1.600	6.395	12.790
25	773.07	.709	1.214	1.517	1.619	6.468	12.936
30	764.49	.717	1.228	1.535	1.637	6.540	13.081
35	756.10	.725	1.242	1.552	1.655	6.613	13.226
40	747.89	.733	1.255	1.569	1.673	6.685	13.371
45	739.86	.740	1.269	1.586	1.691	6.758	13.516
50	732.01	.748	1.283	1.603	1.710	6.831	13.661
55	724.31	.756	1.296	1.620	1.728	6.903	13.806
8 0	716.78	.764	1.310	1.637	1.746	6.976	13.951
5	709.40	.772	1.324	1.654	1.764	7.048	14.096
10	702.18	.780	1.337	1.671	1.782	7.121	14.241
15	695.09	.788	1.351	1.688	1.801	7.193	14.387
20	688.16	.796	1.365	1.705	1.819	7.266	14.532
25	681.35	.804	1.378	1.722	1.837	7.338	14.677
30	674.69	.812	1.392	1.739	1.855	7.411	14.822
35	668.15	.820	1.406	1.757	1.873	7.483	14.967

XIX.—*To trace Railroad Curves, &c.*—Continued.

Degree.	Radii.	ORDINATES. To circular arcs on a chord of 100 feet.				Tangent-deflection.	Chord-deflection.
		12½	25	37½	50		
° ′							
8 40	661.74	.828	1.419	1.774	1.892	7.556	15.112
45	655.45	.836	1.433	1.791	1.910	7.628	15.257
50	649.27	.844	1.447	1.808	1.928	7.701	15.402
55	643.22	.852	1.460	1.825	1.946	7.773	15.547
9 0	637.27	.860	1.474	1.842	1.965	7.846	15.692
5	631.44	.868	1.488	1.859	1.983	7.918	15.837
10	625.71	.876	1.501	1.876	2.001	7.991	15.982
15	620.09	.884	1.515	1.893	2.019	8.063	16.127
20	614.56	.892	1.529	1.910	2.037	8.136	16.272
25	609.14	.900	1.542	1.927	2.056	8.208	16.417
30	603.80	.908	1.556	1.944	2.074	8.281	16.562
35	598.57	.916	1.570	1.961	2.092	8.353	16.707
40	593.42	.924	1.583	1.979	2.110	8.426	16.852
45	588.36	.932	1.597	1.996	2.128	8.498	16.996
50	583.38	.940	1.611	2.013	2.147	8.571	17.141
55	578.49	.948	1.624	2.030	2.165	8.643	17.286
10 0	573.69	.956	1.638	2.047	2.183	8.716	17.431
10	564.31	.972	1.665	2.081	2.219	8.860	17.721
20	555.23	.988	1.693	2.115	2.256	9.005	18.011
30	546.44	1.004	1.720	2.149	2.292	9.150	18.300
40	537.92	1.020	1.748	2.184	2.329	9.295	18.590
50	529.67	1.036	1.775	2.218	2.365	9.440	18.880
11 0	521.67	1.052	1.802	2.252	2.402	9.585	19.169
10	513.91	1.068	1.830	2.286	2.438	9.729	19.459
20	506.38	1.084	1.857	2.320	2.475	9.874	19.748
30	499.06	1.100	1.884	2.354	2.511	10.019	20.038
40	491.96	1.116	1.912	2.389	2.547	10.164	20.327
50	485.05	1.132	1.938	2.423	2.584	10.308	20.616
12 0	478.34	1.148	1.967	2.457	2.620	10.453	20.906
10	471.81	1.164	1.994	2.491	2.657	10.597	21.195
20	465.46	1.180	2.021	2.525	2.693	10.742	21.484
30	459.28	1.196	2.049	2.560	2.730	10.887	21.773
40	453.26	1.212	2.076	2.594	2.766	11.031	22.063
50	447.40	1.228	2.104	2.628	2.803	11.176	22.352

XIX.—*To trace Railroad Curves, &c.*—Continued.

Degree.	Radii.	ORDINATES. To circular arcs on a chord of 100 feet.				Tangent-deflection.	Chord-deflection.
		12½	25	37½	50		
° ′							
13 0	441.68	1.244	2.131	2.662	2.839	11.320	22.641
10	436.12	1.260	2.159	2.697	2.876	11.465	22.930
20	430.69	1.277	2.186	2.731	2.912	11.609	23.219
30	425.40	1.293	2.213	2.765	2.949	11.754	23.507
40	420.23	1.309	2.241	2.799	2.985	11.898	23.796
50	415.19	1.325	2.268	2.833	3.022	12.043	24.085
14 0	410.28	1.341	2.296	2.868	3.058	12.187	24.374
10	405.47	1.357	2.323	2.902	3.095	12.331	24.663
20	400.78	1.373	2.351	2.936	3.131	12.476	24.951
30	396.20	1.389	2.378	2.970	3.168	12.620	25.240
40	391.72	1.405	2.406	3.005	3.204	12.764	25.528
50	387.34	1.421	2.433	3.039	3.241	12.908	25.817
15 0	383.06	1.437	2.461	3.073	3.277	13.053	26.105
10	378.88	1.453	2.488	3.107	3.314	13.197	26.394
20	374.79	1.469	2.515	3.142	3.350	13.341	26.682
30	370.78	1.486	2.543	3.176	3.387	13.485	26.970
40	366.86	1.502	2.570	3.210	3.423	13.629	27.258
50	363.02	1.518	2.598	3.245	3.460	13.773	27.547
16 0	359.26	1.534	2.625	3.279	3.496	13.917	27.835
10	355.59	1.550	2.653	3.313	3.533	14.061	28.123
20	351.98	1.566	2.680	3.347	3.569	14.205	28.411
30	348.45	1.582	2.708	3.382	3.606	14.349	28.699
40	344.99	1.598	2.736	3.416	3.643	14.493	28.986
50	341.60	1.615	2.763	3.450	3.679	14.637	29.274
17 0	338.27	1.631	2.791	3.485	3.716	14.781	29.562
10	335.01	1.647	2.818	3.519	3.752	14.925	29.850
20	331.82	1.663	2.846	3.553	3.789	15.069	30.137
30	328.68	1.679	2.873	3.588	3.825	15.212	30.425
40	325.60	1.695	2.901	3.622	3.862	15.356	30.712
50	322.59	1.711	2.928	3.656	3.898	15.500	31.000
18 0	319.62	1.728	2.956	3.691	3.935	15.643	31.287
10	316.71	1.744	2.983	3.725	3.972	15.787	31.574
20	313.86	1.760	3.011	3.759	4.008	15.931	31.861

XIX.—*To trace Railroad Curves, &c.*—Continued.

Degree.	Radii.	ORDINATES. To circular arcs on a chord of 100 feet. 12½	25	37½	50	Tangent-deflection.	Chord-deflection.
° ′							
18 30	311.06	1.776	3.039	3.794	4.045	16.074	32.149
40	308.30	1.792	3.066	3.828	4.081	16.218	32.436
50	305.60	1.809	3.094	3.862	4.118	16.361	32.723
19 0	302.94	1.825	3.121	3.897	4.155	16.505	33.010
10	300.33	1.841	3.149	3.931	4.191	16.648	33.296
20	297.77	1.857	3.177	3.965	4.228	16.792	33.583
30	295.25	1.873	3.204	4.000	4.265	16.935	33.870
40	292.77	1.890	3.232	4.034	4.301	17.078	34.157
50	290.33	1.906	3.259	4.069	4.338	17.222	34.443

LONG CHORDS.

Degree of curve.	2 stations.	3 stations.	4 stations.	5 stations.	6 stations.
° ′					
0 10	200.000	299.999	399.998	499.996	599.993
20	199.999	.997	.992	.983	.970
30	.998	.992	.981	.962	.933
40	.997	.986	.966	.932	.882
50	.995	.979	.947	.894	.815
1 0	199.992	299.970	399.924	499.848	599.733
10	.990	.959	.896	.793	.637
20	.986	.946	.865	.729	.526
30	.983	.932	.829	.657	.401
40	.979	.915	.789	.577	.260
50	.974	.898	.744	.488	.105
2 0	199.970	299.878	399.695	499.391	598.934
10	.964	.857	.643	.285	.750
20	.959	.834	.586	.171	.550
30	.952	.810	.524	.049	.336
40	.946	.783	.459	498.918	.106
50	.939	.756	.389	.778	597.862

XIX.—*To trace Railroad Curves, &c.*—Continued.

LONG CHORDS—Continued.

Degree of curve.	2 stations.	3 stations.	4 stations.	5 stations.	6 stations.
° ′					
3 0	199.931	299.726	399.315	498.630	597.604
10	.924	.695	.237	.474	.331
20	.915	.662	.154	.309	.043
30	.907	.627	.068	.136	596.740
40	.898	.591	398.977	497.955	.423
50	.888	.553	.882	.765	.091
4 0	199.878	299.513	398.782	497.566	595.744
10	.868	.471	.679	.360	.383
20	.857	.428	.571	.145	.007
30	.846	.383	.459	496.921	594.617
40	.834	.337	.343	.689	.212
50	.822	.289	.223	.449	593.792
5 0	199.810	299.239	398.099	496.200	593.358
10	.797	.187	397.970	495.944	592.909
20	.783	.134	.837	.678	.446
30	.770	.079	.700	.405	591.968
40	.756	.023	.559	.123	.476
50	.741	298.964	.413	494.832	590.970
6 0	199.726	298.904	397.264	494.534	590.449
10	.710	.843	.110	.227	589.913
20	.695	.779	396.952	493.912	.364
30	.678	.714	.790	.588	588.800
40	.662	.648	.623	.257	.221
50	.644	.579	.453	492.917	587.628
7 0	199.627	298.509	396.278	492.568	587.021
10	.609	.438	.099	.212	586.400
20	.591	.364	395.916	491.847	585.765
30	.572	.289	.729	.474	.115
40	.553	.212	.538	.093	584.451
50	.533	.134	.342	490.704	583.773
8 0	199.513	298.054	395.142	490.306	583.081

XX.—*To ascertain the Discharge of Water in any Stream.*

1. For practically gauging large rivers a locality is selected in a straight portion of the stream where the water flows smoothly and without obstruction. A base-line about 200 feet long is laid out parallel to the current, and the exact cross-section in front of this base is determined by careful sounding.

To obtain the discharge, two theodolites are established, and the angular distance from, and the times of transit past, each end of the base, of numerous floats, well distributed between the banks, are noted.

The floats should be made double, the surface-float being a minute tin ellipsoid, a piece of cork, or some other small light body, bearing a small flag. The lower float may be a large box or keg without top or bottom, kept upright by lead ballasting; or better, because lighter, two sheets of tin bent at right angles, and soldered together at the bend, so as to make all the angles between the four faces right angles; the essential conditions being that the lower float shall so greatly preponderate in area over the upper, and shall be connected by so fine a wire or cord, that its rate of movement will govern the whole combination.

The center of the lower float should be placed at the *mid-depth of the stream*, in each vertical plane of transit, because the rate of movement will then be unaffected by wind.

As it is sometimes troublesome to adjust to mid-depth in the different planes of transit, when there is a tolerably uniform and symmetrical cross-section, the average mid-depth of the river may be adopted for all the floats without sensible error.

If floats passing near the surface are used, errors in the computed discharge may be caused by an ordinary breeze, and as these errors are positive or negative according to the direction of the wind, discrepancies may result in the measurements of different days when there is no real variation in the discharge. All this uncertainty is avoided by using mid-depth floats.

The exact level of the water-surface on a permanent gauge-rod should be carefully noted when the observations begin and terminate.

XX.—*To ascertain the Discharge of Water, &c.*—Continued.

Upon a sheet of section-paper the base-line and the two perpendiculars across which the times of transit were noted are then laid down, and, from the recorded angles and a table of natural tangents, the distances from the base-line to the points at which each float passed both lines are plotted. These points, being connected, indicate the paths of the floats. Upon each path the difference between the two recorded times of transit is written in seconds. These seconds of transit are next examined, and the total width of the river is marked off into as many "divisions" as it seems proper to assume are traversed by water moving with sensibly unvarying veolcity, say, for instance, that each division is about $\frac{1}{10}$ of the width of the stream. A mean of the seconds of transit of all the floats in each "division" is next taken, and, when reduced to velocity in feet per second, is adopted as the mid-depth velocity in that "division."

A mean of all these mean mid-depth velocities—interpolations being made if any are missing—closely approximates the mean velocity of the river, provided the "divisions" are equal in width.

This method involves two errors, which nearly balance each other, viz: the inequality in area of the divisions, and the difference between the mid-depth velocity and the mean velocity in any vertical plane, giving a resulting mean velocity of about 0.95 times its true value.

The mean velocity in each plane is obtained from the mid-depth velocities, $V \frac{1}{2} D$, by the formula—

$$v = 1.075\, V \tfrac{1}{2} D + 0.004\, b - 0.093\, (V \tfrac{1}{2} D \,.\, b)^{\frac{1}{2}}$$

and—

$$b = \frac{1.69}{(D + 1.5)^{\frac{1}{2}}}$$

where—

D = the depth of the stream at any point of the surface.

If a is the area of cross-section, and a', a'', etc., the partial division areas, the discharge may be found by—

$$Q = v\, a = \left[a' \, (V \tfrac{1}{2} D) - \tfrac{1}{12} \, (b \,.\, v)^{\frac{1}{2}} \right]$$

where $[\;\;]$ denotes the sum of similar quantities.

XX.—*To ascertain the Discharge of Water, &c.*—Continued.

2. Determination of the mean velocity in terms of the dimensions of the cross-section and the slope.

Humphreys-Abbot Formula.

(Not applicable to water flowing in smooth artificial channels.)

$$v = \left(\sqrt{0.0081\, b + (225\, r_1\, s^{\frac{1}{2}})^{\frac{1}{2}}} - 0.09\,(b)^{\frac{1}{2}} \right)^2 - \frac{2.4\,(v')^{\frac{1}{2}}}{1 + p}$$

where the symbols have the following signification, all expressed in English feet:

v = mean velocity $= \frac{Q}{a}$;

a = area of cross-section;

W = width;

r = mean radius, or $\frac{a}{p}$;

v' = value of first term in expression for v;

p = wetted perimeter;

Q = discharge in cubic feet per second;

$r_1 = \frac{a}{p + W}$;

s = sine of slope of water surface corrected for bends;

b = function of the depth, for small streams $= \frac{1.69}{(r + 1.5)^{\frac{1}{2}}}$

For rivers whose mean radius exceeds 12 or 15 feet, b may be assumed to be 0.1856, which will make the numerical value of the term involving b so small that it may be generally neglected, reducing the above equation to—

$$v = ([\,225\, r_1\, s^{\frac{1}{2}}\,]^{\frac{1}{4}} - 0.0388)^2$$

The following formulæ give the value of each variable in terms of the others and known quantities:

$$z = 0.93\, v + 0.167\,(b\, v)^{\frac{1}{2}};$$

and when p is not known by measurement it may, for ordinary natural channels, be assumed to be 1.015 W.

$$s = \left(\frac{(p + W)\, z^2}{195\, a}\right)^2$$

$$a = \frac{(p + W)\, z^2}{195\, (s)^{\frac{1}{2}}}$$

$$p + W = \frac{195\, a\, (s)^{\frac{1}{2}}}{z^2}$$

XX.—*To ascertain the Discharge of Water, &c.*—Continued.

APPLICATION.—The variables which enter these formulæ require a knowledge of the mean cross-section of the stream, and a map of the course of the channel between two selected points of the water-surface, whose difference of level should be exactly known.

Whenever practicable, the two points should be located on a straight and regular portion of the river to eliminate the effects of bends. As this is not always possible, the general case is considered in the above formulæ, and bends are assumed to exist between the points selected.

The field-operations consist in a survey of the channel, with numerous soundings between permanent bench-marks placed near the water, and in running a line of levels between those marks, so as to give their relative level with the most extreme accuracy.

These points should be located with care, as far apart as practicable, distant from any eddy, and placed where the current on the banks flows with *equal velocity*. This latter condition is necessary, because, as water in motion exerts less pressure than when at rest, if it moved rapidly past one bench-mark and was nearly stationary at the other, a difference of level, which has nothing to do with the motive power of the stream, would vitiate the observation.

In determining the mean dimensions of cross-sections, care must be taken to extend the soundings throughout the entire distance between the bench-marks, and it must be borne in mind that measured fall in water-surface between two stations corresponds to *the mean channel between them*.

When the soundings are made, the water-level should be referred to the bench-marks in order to determine the area corresponding to any subsequent stand of the river.

These soundings completed, frequent gauging of the river can be made by referring at any time, by accurate levels, the water-surface at the two points to their respective bench-marks, thus determining the fall and corresponding cross-section of the river from which the discharge is computed.

The observations must be simultaneous in order to avoid the effect of any oscillation in the river, and *calm days* should be selected because waves render it difficult to determine the exact level of the water-surface; and, also, changes of level result from the general piling up or lowering of the water under the influence of winds.

XX.—*To ascertain the Discharge of Water, &c.*—Continued.

Correction for Bends.—A line following the mid-channel is drawn on the map, composed of straight lines with angular changes, wherever necessary, of 30°. A mean velocity is assumed, to be corrected subsequently if required, and the value of h is computed in the following formulæ, in which N represents the number of deflections:

$$h = \frac{v^2 \text{ N} \sin^2 30^\circ}{134}$$

The deduced value of h is next subtracted from the total fall in the water-surface between the two stations; the remainder divided by the distance in feet between these stations, measured on the middle line of the river, is the true value of s in the formula for mean velocity.

If any material error has been made in assuming v, the computation should be repeated until the requisite approximation has been made.

By expressing the formula in the form—

$$v = \left(\sqrt{\text{M} + \left(\frac{225\, a \sqrt{s}}{p + \text{W}}\right)^{\frac{1}{2}}} - \sqrt{\text{M}}\right)^2 - \text{M}' \sqrt{v}$$

The following table will facilitate its application:

r	M	$\sqrt{\text{M}}$	p	M′	Log M′
1	0.0087	0.0930	5	0.400	9.602060
2	73	855	6	.343	9.535294
3	65	803	7	.300	9.477121
4	58	764	8	.267	9.426511
5	54	733	9	.240	9.380211
6	50	707	10	.218	9.338456
7	47	685	12	.185	9.267172
8	44	666	14	.160	9.204120
9	42	649	16	.141	9.149219
10	40	634	18	.126	9.100371
12	37	610	20	.114	9.056905
14	35	590	22	.104	9.017033
16	33	573	24	.096	8.982271
18	31	558	26	.089	8.949390
20	29	544	28	.083	8.919078
30	24	494	30	.078	8.892095
50	0.0019	0.0437	50	0.047	8.672098

For streams larger than 50 or 100 feet in cross-section the term involving M′ may be dropped, and for larger rivers, exceeding 12 or 20 feet in mean radius, M, but not $\sqrt{\text{M}}$, may be neglected.

XX.—*To ascertain the Discharge of Water, &c.*—Continued.

EXAMPLES OF RESULTING MEAN VELOCITIES FROM GIVEN CROSS-SECTIONS AND SLOPES.

Stream.	Locality.	Date.	Dimensions of cross-section.				Slope.			Mean velocity.		Authority.
			Area.	Width.	Peri-meter.	Max. Depth.	s	log s	log $s^{\frac{1}{2}}$	Observed.	Computed.	
			Sq. feet.	*Feet.*	*Feet.*	*Feet.*						
1. Mississippi River..	Carrolton..........	H. W. of 1851	193968	2658	2693	136	0.00002051	5.3119657	7.6559828	5.9288	5.8903	Delta survey.
2. Bayou Plaquemine.	Near upper mouth	Jan. 16, 1859	4259	268	278	24	0.00014372	6.1575172	8.0787586	3.9589	4.3460	Do.
3. Chesapeake & Ohio Canal feeder.	Near Georgetown, D. C.	Nov. 26, 1859	121	23	32.7	7.6	0.00069851	6.8441726	8.4220863	3.0323	3.1032	Do.

XX.—*To ascertain the Discharge of Water, &c.*—Continued.

3. Formulæ for the mean velocity, from other authorities:

Chezy...	Downing's and others' co-efficient	$v = 100.0\,(rs)^{\frac{1}{2}}$
	Eytelwein's co-efficient	$v = 93.4\,(rs)^{\frac{1}{2}}$
	Young's co-efficient	$v = 84.3\,(rs)^{\frac{1}{2}}$
De Prony	For canals	$v = (0.0556 + 10593\,rs)^{\frac{1}{2}} - 0.2357$
	For canals and pipes	$v = (0.0237 + 9966\,rs)^{\frac{1}{2}} - 0.1542$
	Eytelwein's co-efficient	$v = (0.0119 + 8963\,rs)^{\frac{1}{2}} - 0.1089$
	Weisbach's co-efficient	$v = (0.00024 + 8675\,rs)^{\frac{1}{2}} - 0.0154$
Darcy-Bazin		$v = r\left(\dfrac{1000\,s}{0.08534\,r + 0.35}\right)^{\frac{1}{2}}$

XXI.—*Motion of Water in Conduit-Pipes.*

Discharge through pipes of uniform dimensions, and having no sudden changes of direction:

For ordinary cases:

$$Q = 38.436\sqrt{\frac{H\,D^5}{L}} - 0.070862\,D^2$$

In great velocities:

$$Q = 36.769\sqrt{\frac{H\,D^5}{L}}$$

If the velocity be required, divide the discharge by the area of the section ($0.7854\,D^2$).

To find the diameter of a conduit-pipe for a given discharge under a given head:

$$D = 0.2323\sqrt[5]{\frac{L\,Q^2}{H}}$$

Where Q = discharge in cubic feet per second; H, the head, and D and L the diameter and length of the pipe in feet.

The resistance of curves is proportional to the square of the velocity of the fluid, to the number of angles of reflextion, and to the square of their sine,

$$\text{or, in function of } Q, = 0.006079\,\frac{Q^2}{D^4}\cdot s^2$$

s^2 being the sum of the squares of all the sines of the angles of reflexion.

XXII.—*Logarithms of Numbers.*

Nat. Nos.	0	1	2	3	4	5	6	7	8	9	Proportional parts.								
											1	2	3	4	5	6	7	8	9
10	.0000	.0043	.0086	.0128	.0170	.0212	.0253	.0294	.0334	.0374	4	8	12	17	21	25	29	33	37
11	.0414	.0453	.0492	.0531	.0569	.0607	.0645	.0682	.0719	.0755	4	8	11	15	19	23	26	30	34
12	.0792	.0828	.0864	.0899	.0934	.0969	.1004	.1038	.1072	.1106	3	7	10	14	17	21	24	28	31
13	.1139	.1173	.1206	.1239	.1271	.1303	.1335	.1367	.1399	.1430	3	6	10	13	16	19	23	26	29
14	.1461	.1492	.1523	.1553	.1584	.1614	.1644	.1673	.1703	.1732	3	6	9	12	15	18	21	24	27
15	.1761	.1790	.1818	.1847	.1875	.1903	.1931	.1959	.1987	.2014	3	6	8	11	14	17	20	22	25
16	.2041	.2068	.2095	.2122	.2148	.2175	.2201	.2227	.2253	.2279	3	5	8	11	13	16	18	21	24
17	.2304	.2330	.2355	.2380	.2405	.2430	.2455	.2480	.2504	.2529	2	5	7	10	12	15	17	20	22
18	.2553	.2577	.2601	.2625	.2648	.2672	.2695	.2718	.2742	.2765	2	5	7	9	12	14	16	19	21
19	.2788	.2810	.2833	.2856	.2878	.2900	.2923	.2945	.2967	.2989	2	4	7	9	11	13	16	18	20
20	.3010	.3032	.3054	.3075	.3096	.3118	.3139	.3160	.3181	.3201	2	4	6	8	11	13	15	17	19
21	.3222	.3243	.3263	.3284	.3304	.3324	.3345	.3365	.3385	.3404	2	4	6	8	10	12	14	16	18
22	.3424	.3444	.3464	.3483	.3502	.3522	.3541	.3560	.3579	.3598	2	4	6	8	10	12	14	15	17
23	.3617	.3636	.3655	.3674	.3692	.3711	.3729	.3747	.3766	.3784	2	4	6	7	9	11	13	15	17
24	.3802	.3820	.3838	.3856	.3874	.3892	.3909	.3927	.3945	.3962	2	4	5	7	9	11	12	14	16
25	.3979	.3997	.4014	.4031	.4048	.4065	.4082	.4099	.4116	.4133	2	3	5	7	9	10	12	14	15
26	.4150	.4166	.4183	.4200	.4216	.4232	.4249	.4265	.4281	.4298	2	3	5	7	8	10	11	13	15
27	.4314	.4330	.4346	.4362	.4378	.4393	.4409	.4425	.4440	.4456	2	3	5	6	8	9	11	13	14
28	.4472	.4487	.4502	.4518	.4533	.4548	.4564	.4579	.4594	.4609	2	3	5	6	8	9	11	12	14
29	.4624	.4639	.4654	.4669	.4683	.4698	.4713	.4728	.4742	.4757	1	3	4	6	7	9	10	12	13
30	.4771	.4786	.4800	.4814	.4829	.4843	.4857	.4871	.4886	.4900	1	3	4	6	7	9	10	11	13
31	.4914	.4928	.4942	.4955	.4969	.4983	.4997	.5011	.5024	.5038	1	3	4	6	7	8	10	11	12
32	.5051	.5065	.5079	.5092	.5105	.5119	.5132	.5145	.5159	.5172	1	3	4	5	7	8	9	11	12
33	.5185	.5198	.5211	.5224	.5237	.5250	.5263	.5276	.5289	.5302	1	3	4	5	6	8	9	10	12
34	.5315	.5328	.5340	.5353	.5366	.5378	.5391	.5403	.5416	.5428	1	3	4	5	6	8	9	10	11
35	.5441	.5453	.5465	.5478	.5490	.5502	.5514	.5527	.5539	.5551	1	2	4	5	6	7	9	10	11
36	.5563	.5575	.5587	.5599	.5611	.5623	.5635	.5647	.5658	.5670	1	2	4	5	6	7	8	10	11
37	.5682	.5694	.5705	.5717	.5729	.5740	.5752	.5763	.5775	.5786	1	2	3	5	6	7	8	9	10
38	.5798	.5809	.5821	.5832	.5843	.5855	.5866	.5877	.5888	.5899	1	2	3	5	6	7	8	9	10
39	.5911	.5922	.5933	.5944	.5955	.5966	.5977	.5988	.5999	.6010	1	2	3	4	5	7	8	9	10
40	.6021	.6031	.6042	.6053	.6064	.6075	.6085	.6096	.6107	.6117	1	2	3	4	5	6	8	9	10
41	.6128	.6138	.6149	.6160	.6170	.6180	.6191	.6201	.6212	.6222	1	2	3	4	5	6	7	8	9
42	.6232	.6243	.6253	.6263	.6274	.6284	.6294	.6304	.6314	.6325	1	2	3	4	5	6	7	8	9
43	.6335	.6345	.6355	.6365	.6375	.6385	.6395	.6405	.6415	.6425	1	2	3	4	5	6	7	8	9
44	.6435	.6444	.6454	.6464	.6474	.6484	.6493	.6503	.6513	.6522	1	2	3	4	5	6	7	8	9
45	.6532	.6542	.6551	.6561	.6571	.6580	.6590	.6599	.6609	.6618	1	2	3	4	5	6	7	8	9
46	.6628	.6637	.6646	.6656	.6665	.6675	.6684	.6693	.6702	.6712	1	2	3	4	5	6	7	7	8
47	.6721	.6730	.6739	.6749	.6758	.6767	.6776	.6785	.6794	.6803	1	2	3	4	5	5	6	7	8
48	.6812	.6821	.6830	.6839	.6848	.6857	.6866	.6875	.6884	.6893	1	2	3	4	4	5	6	7	8
49	.6902	.6911	.6920	.6928	.6937	.6946	.6955	.6964	.6972	.6981	1	2	3	4	4	5	6	7	8
50	.6990	.6998	.7007	.7016	.7024	.7033	.7042	.7050	.7059	.7067	1	2	3	3	4	5	6	7	8
51	.7076	.7084	.7093	.7101	.7110	.7118	.7126	.7135	.7143	.7152	1	2	3	3	4	5	6	7	8
52	.7160	.7168	.7177	.7185	.7193	.7202	.7210	.7218	.7226	.7235	1	2	2	3	4	5	6	7	7
53	.7243	.7251	.7259	.7267	.7275	.7284	.7292	.7300	.7308	.7316	1	2	2	3	4	5	6	6	7
54	.7324	.7332	.7340	.7348	.7356	.7364	.7372	.7380	.7388	.7396	1	2	2	3	4	5	6	6	7

XXII.—*Logarithms of Numbers*—Continued.

Nat. Nos.	0	1	2	3	4	5	6	7	8	9	Proportional parts.								
											1	2	3	4	5	6	7	8	9
55	.7404	.7412	.7419	.7427	.7435	.7443	.7451	.7459	.7466	.7474	1	2	2	3	4	5	5	6	7
56	.7482	.7490	.7497	.7505	.7513	.7520	.7528	.7536	.7543	.7551	1	2	2	3	4	5	5	6	7
57	.7559	.7566	.7574	.7582	.7589	.7597	.7604	.7612	.7619	.7627	1	2	2	3	4	5	5	6	7
58	.7634	.7642	.7649	.7657	.7664	.7672	.7679	.7686	.7694	.7701	1	1	2	3	4	4	5	6	7
59	.7709	.7716	.7723	.7731	.7738	.7745	.7752	.7760	.7767	.7774	1	1	2	3	4	4	5	6	7
60	.7782	.7789	.7796	.7803	.7810	.7818	.7825	.7832	.7839	.7846	1	1	2	3	4	4	5	6	6
61	.7853	.7860	.7868	.7875	.7882	.7889	.7896	.7903	.7910	.7917	1	1	2	3	4	4	5	6	6
62	.7924	.7931	.7938	.7945	.7952	.7959	.7966	.7973	.7980	.7987	1	1	2	3	3	4	5	6	6
63	.7993	.8000	.8007	.8014	.8021	.8028	.8035	.8041	.8048	.8055	1	1	2	3	3	4	5	5	6
64	.8062	.8069	.8075	.8082	.8089	.8096	.8102	.8109	.8116	.8122	1	1	2	3	3	4	5	5	6
65	.8129	.8136	.8142	.8149	.8156	.8162	.8169	.8176	.8182	.8189	1	1	2	3	3	4	5	5	6
66	.8195	.8202	.8209	.8215	.8222	.8228	.8235	.8241	.8248	.8254	1	1	2	3	3	4	5	5	6
67	.8261	.8267	.8274	.8280	.8287	.8293	.8299	.8306	.8312	.8319	1	1	2	3	3	4	5	5	6
68	.8325	.8331	.8338	.8344	.8351	.8357	.8363	.8370	.8376	.8382	1	1	2	3	3	4	4	5	6
69	.8388	.8395	.8401	.8407	.8414	.8420	.8426	.8432	.8439	.8445	1	1	2	2	3	4	4	5	6
70	.8451	.8457	.8463	.8470	.8476	.8482	.8488	.8494	.8500	.8506	1	1	2	2	3	4	4	5	6
71	.8513	.8519	.8525	.8531	.8537	.8543	.8549	.8555	.8561	.8567	1	1	2	2	3	4	4	5	5
72	.8573	.8579	.8585	.8591	.8597	.8603	.8609	.8615	.8621	.8627	1	1	2	2	3	4	4	5	5
73	.8633	.8639	.8645	.8651	.8657	.8663	.8669	.8675	.8681	.8686	1	1	2	2	3	4	4	5	5
74	.8692	.8698	.8704	.8710	.8716	.8722	.8727	.8733	.8739	.8745	1	1	2	2	3	4	4	5	5
75	.8751	.8756	.8762	.8768	.8774	.8779	.8785	.8791	.8797	.8802	1	1	2	2	3	3	4	5	5
76	.8808	.8814	.8820	.8825	.8831	.8837	.8842	.8848	.8854	.8859	1	1	2	2	3	3	4	5	5
77	.8865	.8871	.8876	.8882	.8887	.8893	.8899	.8904	.8910	.8915	1	1	2	2	3	3	4	4	5
78	.8921	.8927	.8932	.8938	.8943	.8949	.8954	.8960	.8965	.8971	1	1	2	2	3	3	4	4	5
79	.8976	.8982	.8987	.8993	.8998	.9004	.9009	.9015	.9020	.9025	1	1	2	2	3	3	4	4	5
80	.9031	.9036	.9042	.9047	.9053	.9058	.9063	.9069	.9074	.9079	1	1	2	2	3	3	4	4	5
81	.9085	.9090	.9096	.9101	.9106	.9112	.9117	.9122	.9128	.9133	1	1	2	2	3	3	4	4	5
82	.9138	.9143	.9149	.9154	.9159	.9165	.9170	.9175	.9180	.9186	1	1	2	2	3	3	4	4	5
83	.9191	.9196	.9201	.9206	.9212	.9217	.9222	.9227	.9232	.9238	1	1	2	2	3	3	4	4	5
84	.9243	.9248	.9253	.9258	.9263	.9269	.9274	.9279	.9284	.9289	1	1	2	2	3	3	4	4	5
85	.9294	.9299	.9304	.9309	.9315	.9320	.9325	.9330	.9335	.9340	1	1	2	2	3	3	4	4	5
86	.9345	.9350	.9355	.9360	.9365	.9370	.9375	.9380	.9385	.9390	1	1	2	2	3	3	4	4	5
87	.9395	.9400	.9405	.9410	.9415	.9420	.9425	.9430	.9435	.9440	0	1	1	2	2	3	3	4	4
88	.9445	.9450	.9455	.9460	.9465	.9469	.9474	.9479	.9484	.9489	0	1	1	2	2	3	3	4	4
89	.9494	.9499	.9504	.9509	.9513	.9518	.9523	.9528	.9533	.9538	0	1	1	2	2	3	3	4	4
90	.9542	.9547	.9552	.9557	.9562	.9566	.9571	.9576	.9581	.9586	0	1	1	2	2	3	3	4	4
91	.9590	.9595	.9600	.9605	.9609	.9614	.9619	.9624	.9628	.9633	0	1	1	2	2	3	3	4	4
92	.9638	.9643	.9647	.9652	.9657	.9661	.9666	.9671	.9675	.9680	0	1	1	2	2	3	3	4	4
93	.9685	.9689	.9694	.9699	.9703	.9708	.9713	.9717	.9722	.9727	0	1	1	2	2	3	3	4	4
94	.9731	.9736	.9741	.9745	.9750	.9754	.9759	.9763	.9768	.9773	0	1	1	2	2	3	3	4	4
95	.9777	.9782	.9786	.9791	.9795	.9800	.9805	.9809	.9814	.9818	0	1	1	2	2	3	3	4	4
96	.9823	.9827	.9832	.9836	.9841	.9845	.9850	.9854	.9859	.9863	0	1	1	2	2	3	3	4	4
97	.9868	.9872	.9877	.9881	.9886	.9890	.9894	.9899	.9903	.9908	0	1	1	2	2	3	3	4	4
98	.9912	.9917	.9921	.9926	.9930	.9934	.9939	.9943	.9948	.9952	0	1	1	2	2	3	3	4	4
99	.9956	.9961	.9965	.9969	.9974	.9978	.9983	.9987	.9991	.9996	0	1	1	2	2	3	3	3	4

XXIII.—*Logarithms of Sines and Tangents.*

	0°				1°				
	Sin.	Cos.	Tan.	Cot.	Sin.	Cos.	Tan.	Cot.	
0′		0.0000			8.2419	9.9999	8.2419	1.7581	60′
1	6.4637	.0000	6.4637	3.5363	.2490	.9999	.2491	.7509	59
2	.7648	.0000	.7648	.2352	.2561	.9999	.2562	.7438	58
3	6.9408	.0000	6.9408	3.0592	.2630	.9999	.2631	.7369	57
4	7.0658	.0000	7.0658	2.9342	.2699	.9999	.2700	.7300	56
5	.1627	.0000	.1627	.8373	.2766	.9999	.2767	.7233	55
6	.2419	.0000	.2419	.7581	.2832	.9999	.2833	.7167	54
7	.3088	.0000	.3088	.6912	.2898	.9999	.2899	.7101	53
8	.3668	.0000	.3668	.6332	.2962	.9999	.2963	.7037	52
9	.4180	.0000	.4180	.5820	.3025	.9999	.3026	.6974	51
10	.4637	.0000	.4637	.5363	.3088	.9999	.3089	.6911	50
11	.5051	.0000	.5051	.4949	.3150	.9999	.3150	.6850	49
12	.5429	.0000	.5429	.4571	.3210	.9999	.3211	.6789	48
13	.5777	.0000	.5777	.4223	.3270	.9999	.3271	.6729	47
14	.6099	.0000	.6099	.3901	.3329	.9999	.3330	.6670	46
15	.6398	.0000	.6398	.3602	.3388	.9999	.3389	.6611	45
16	.6678	.0000	.6678	.3322	.3445	.9999	.3446	.6554	44
17	.6942	.0000	.6942	.3058	.3502	.9999	.3503	.6497	43
18	.7190	.0000	.7190	.2810	.3558	.9999	.3559	.6441	42
19	.7425	.0000	.7425	.2575	.3613	.9999	.3614	.6386	41
20	.7648	.0000	.7648	.2352	.3668	.9999	.3669	.6331	40
21	.7859	.0000	.7860	.2140	.3722	.9999	.3723	.6277	39
22	.8061	.0000	.8062	.1938	.3775	.9999	.3776	.6224	38
23	.8255	.0000	.8255	.1745	.3828	.9999	.3829	.6171	37
24	.8439	.0000	.8439	.1561	.3880	.9999	.3881	.6119	36
25	.8617	.0000	.8617	.1383	.3931	.9999	.3932	.6068	35
26	.8787	.0000	.8787	.1213	.3982	.9999	.3983	.6017	34
27	.8951	.0000	.8951	.1049	.4032	.9999	.4033	.5967	33
28	.9109	.0000	.9109	.0891	.4082	.9999	.4083	.5917	32
29	.9261	.0000	.9261	.0739	.4131	.9999	.4132	.5868	31
30	.9408	.0000	.9409	.0591	.4179	.9999	.4181	.5819	30
31	.9551	.0000	.9551	.0449	.4227	.9998	.4229	.5771	29
32	.9689	.0000	.9689	.0311	.4275	.9998	.4276	.5724	28
33	.9822	.0000	.9823	.0177	.4322	.9998	.4323	.5677	27
34	7.9952	.0000	7.9952	2.0048	.4368	.9998	.4370	.5630	26
35	8.0078	.0000	8.0078	1.9922	.4414	.9998	.4416	.5584	25
36	.0200	.0000	.0200	.9800	.4459	.9998	.4461	.5539	24
37	.0319	.0000	.0319	.9681	.4504	.9998	.4506	.5494	23
38	.0435	.0000	.0435	.9565	.4549	.9998	.4551	.5449	22
39	.0548	.0000	.0548	.9452	.4593	.9998	.4595	.5405	21
40	.0658	.0000	.0658	.9342	.4637	.9998	.4638	.5362	20
41	.0765	.0000	.0765	.9235	.4680	.9998	.4682	.5318	19
42	.0870	.0000	.0870	.9130	.4723	.9998	.4725	.5275	18
43	.0972	.0000	.0972	.9028	.4765	.9998	.4767	.5233	17
44	.1072	.0000	.1072	.8928	.4807	.9998	.4809	.5191	16
45	.1169	.0000	.1170	.8830	.4848	.9998	.4851	.5149	15
46	.1265	.0000	.1265	.8735	.4890	.9998	.4892	.5108	14
47	.1358	.0000	.1359	.8641	.4930	.9998	.4933	.5067	13
48	.1450	.0000	.1450	.8550	.4971	.9998	.4973	.5027	12
49	.1539	.0000	.1540	.8460	.5011	.9998	.5013	.4987	11
50	.1627	.0000	.1627	.8373	.5050	.9998	.5053	.4947	10
51	.1713	.0000	.1713	.8287	.5090	.9998	.5092	.4908	9
52	.1797	0.0000	.1798	.8202	.5129	.9998	.5131	.4869	8
53	.1880	9.9999	.1880	.8120	.5167	.9998	.5170	.4830	7
54	.1961	.9999	.1962	.8038	.5206	.9998	.5208	.4792	6
55	.2041	.9999	.2041	.7959	.5243	.9998	.5246	.4754	5
56	.2119	.9999	.2120	.7880	.5281	.9998	.5283	.4717	4
57	.2196	.9999	.2196	.7804	.5318	.9997	.5321	.4679	3
58	.2271	.9999	.2272	.7728	.5355	.9997	.5358	.4642	2
59	.2346	.9999	.2346	.7654	.5392	.9997	.5394	.4606	1
60	8.2419	9.9999	8.2419	1.7581	8.5428	9.9997	8.5431	1.4569	0
	Cos.	Sin.	Cot.	Tan.	Cos.	Sin.	Cot.	Tan.	
	89°				88°				

XXIII.—*Logarithms of Sines and Tangents*—Continued.

	2°				3°				4°				
	Sin.	Cos.	Tan.	Cot.	Sin.	Cos.	Tan.	Cot.	Sin.	Cos.	Tan.	Cot.	
0′	8.5428	9.9997	8.5431	1.4569	8.7188	9.9994	8.7194	1.2806	8.8436	9.9989	8.8446	1.1554	60′
1	.5464	.9997	.5467	.4533	.7212	.9994	.7218	.2782	.8454	.9989	.8465	.1535	59
2	.5500	.9997	.5503	.4497	.7236	.9994	.7242	.2758	.8472	.9989	.8483	.1517	58
3	.5535	.9997	.5538	.4462	.7260	.9994	.7266	.2734	.8490	.9989	.8501	.1499	57
4	.5571	.9997	.5573	.4427	.7283	.9994	.7290	.2710	.8508	.9989	.8518	.1482	56
5	.5605	.9997	.5608	.4392	.7307	.9994	.7313	.2687	.8525	.9989	.8536	.1464	55
6	.5640	.9997	.5643	.4357	.7330	.9994	.7337	.2663	.8543	.9989	.8554	.1446	54
7	.5674	.9997	.5677	.4323	.7354	.9994	.7360	.2640	.8560	.9989	.8572	.1428	53
8	.5708	.9997	.5711	.4289	.7377	.9994	.7383	.2617	.8578	.9989	.8589	.1411	52
9	.5742	.9997	.5745	.4255	.7400	.9993	.7406	.2594	.8595	.9989	.8607	.1393	51
10	.5776	.9997	.5779	.4221	.7423	.9993	.7429	.2571	.8613	.9989	.8624	.1376	50
11	.5809	.9997	.5812	.4188	.7445	.9993	.7452	.2548	.8630	.9988	.8642	.1358	49
12	.5842	.9997	.5845	.4155	.7468	.9993	.7475	.2525	.8647	.9988	.8659	.1341	48
13	.5875	.9997	.5878	.4122	.7491	.9993	.7497	.2503	.8665	.9988	.8676	.1324	47
14	.5907	.9997	.5911	.4089	.7513	.9993	.7520	.2480	.8682	.9988	.8694	.1306	46
15	.5939	.9997	.5943	.4057	.7535	.9993	.7542	.2458	.8699	.9988	.8711	.1289	45
16	.5972	.9997	.5975	.4025	.7557	.9993	.7565	.2435	.8716	.9988	.8728	.1272	44
17	.6003	.9997	.6007	.3993	.7580	.9993	.7587	.2413	.8733	.9988	.8745	.1255	43
18	.6035	.9996	.6038	.3962	.7602	.9993	.7609	.2391	.8749	.9988	.8762	.1238	42
19	.6066	.9996	.6070	.3930	.7623	.9993	.7631	.2369	.8766	.9988	.8778	.1222	41
20	.6097	.9996	.6101	.3899	.7645	.9993	.7652	.2348	.8783	.9988	.8795	.1205	40
21	.6128	.9996	.6132	.3868	.7667	.9993	.7674	.2326	.8799	.9987	.8812	.1188	39
22	.6159	.9996	.6163	.3837	.7688	.9992	.7696	.2304	.8816	.9987	.8829	.1171	38
23	.6189	.9996	.6193	.3807	.7710	.9992	.7717	.2283	.8833	.9987	.8845	.1155	37
24	.6220	.9996	.6223	.3777	.7731	.9992	.7739	.2261	.8849	.9987	.8862	.1138	36
25	.6250	.9996	.6254	.3746	.7752	.9992	.7760	.2240	.8865	.9987	.8878	.1122	35
26	.6279	.9996	.6283	.3717	.7773	.9992	.7781	.2219	.8882	.9987	.8895	.1105	34
27	.6309	.9996	.6313	.3687	.7794	.9992	.7802	.2198	.8898	.9987	.8911	.1089	33
28	.6339	.9996	.6343	.3657	.7815	.9992	.7823	.2177	.8914	.9987	.8927	.1073	32
29	.6368	.9996	.6372	.3628	.7836	.9992	.7844	.2156	.8930	.9987	.8944	.1056	31
30	.6397	.9996	.6401	.3599	.7857	.9992	.7865	.2135	.8946	.0987	.8960	.1040	30
31	.6426	.9996	.6430	.3570	.7877	.9992	.7886	.2114	.8962	.9986	.8976	.1024	29
32	.6454	.9996	.6459	.3541	.7898	.9992	.7906	.2094	.8978	.9986	.8992	.1008	28
33	.6483	.9996	.6487	.3513	.7918	.9992	.7927	.2073	.8994	.9986	.9008	.0992	27
34	.6511	.9996	.6515	.3485	.7939	.9992	.7947	.2053	.9010	.9986	.9024	.0976	26
35	.6539	.9996	.6544	.3456	.7959	.9992	.7967	.2033	.9026	.9986	.9040	.0960	25
36	.6567	.9996	.6571	.3429	.7979	.9991	.7988	.2012	.9342	.9986	.9056	.0944	24
37	.6595	.9995	.6599	.3401	.7999	.9991	.8008	.1992	.9057	.9986	.9071	.0929	23
38	.6622	.9995	.6627	.3373	.8019	.9991	.8028	.1972	.9073	.9986	.9087	.0913	22
39	.6650	.9995	.6654	.3346	.8039	.9991	.8048	.1952	.9089	.9986	.9103	.0897	21
40	.6677	.9995	.6682	.3318	.8059	.9991	.8067	.1933	.9104	.9986	.9118	.0882	20
41	.6704	.9995	.6709	.3291	.8078	.9991	.8087	.1913	.9119	.9985	.9134	.0866	19
42	.6731	.9995	.6736	.3264	.8098	.9991	.8107	.1893	.9135	.9985	.9150	.0850	18
43	.6758	.9995	.6762	.3238	.8117	.9991	.8126	.1874	.9150	.9985	.9165	.0835	17
44	.6784	.9995	.6789	.3211	.8137	.9991	.8146	.1854	.9166	.9985	.9180	.0820	16
45	.6810	.9995	.6815	.3185	.8156	.9991	.8165	.1835	.9181	.9985	.9196	.0804	15
46	.6837	.9995	.6842	.3158	.8175	.9991	.8185	.1815	.9196	.9985	.9211	.0789	14
47	.6863	.9995	.6868	.3132	.8194	.9991	.8204	.1796	.9211	.9985	.9226	.0774	13
48	.6889	.9995	.6894	.3106	.8213	.9990	.8223	.1777	.9226	.9985	.9241	.0759	12
49	.6914	.9995	.6920	.3080	.8232	.9990	.8242	.1758	.9241	.9985	.9256	.0744	11
50	.6940	.9995	.6945	.3055	.8251	.9990	.8261	.1739	.9256	.9985	.9272	.0728	10
51	.6965	.9995	.6971	.3029	.8270	.9990	.8280	.1720	.9271	.9984	.9287	.0713	9
52	.6991	.9995	.6996	.3004	.8289	.9990	.8299	.1701	.9286	.9984	.9302	.0698	8
53	.7016	.9994	.7021	.2979	.8307	.9990	.8317	.1683	.9301	.9984	.9316	.0684	7
54	.7041	.9994	.7046	.2954	.8326	.9990	.8336	.1664	.9315	.9984	.9331	.0669	6
55	.7066	.9994	.7071	.2929	.8345	.9990	.8355	.1645	.9330	.9984	.9346	.0654	5
56	.7090	.9994	.7096	.2904	.8363	.9990	.8373	.1627	.9345	.9984	.9361	.0639	4
57	.7115	.9994	.7121	.2879	.8381	.9990	.8392	.1608	.9359	.9984	.9376	.0624	3
58	.7140	.9994	.7145	.2855	.8400	.9990	.8410	.1590	.9374	.9984	.9390	.0610	2
59	.7164	.9994	.7170	.2830	.8418	.9989	.8428	.1572	.9388	.9984	.9405	.0595	1
60	8.7188	9.9994	8.7194	1.2806	8.8436	9.9989	8.8446	1.1554	8.9403	9.9983	8.9420	1.0580	0
	Cos.	Sin.	Cot.	Tan.	Cos.	Sin.	Cot.	Tan.	Cos.	Sin.	Cot.	Tan.	
	87°				86°				85°				

XXIII.—*Logarithms of Sines and Tangents*—Continued.

Arc	Sin.	Df.	Cos.	Df.	Tan.	Df.	Cot.	Arc	Arc	Sin.	Df.	Cos.	Df.	Tan.	Df.	Cot.	Arc
° ′								° ′	° ′								° ′
5 0	8.9403	142	9.9983	1	8.9420	143	1.0580	85 0	15 0	9.4130	47	9.9849	3	9.4281	50	0.5719	75 0
10	.9545	137	.9982	1	.9563	138	.0437	50	10	.4177	46	.9846	3	.4331	50	.5669	50
20	.9682	134	.9981	1	.9701	135	.0299	40	20	.4223	46	.9843	4	.4381	49	.5619	40
30	.9816	129	.9980	1	.9836	130	.0164	30	30	.4269	45	.9839	3	.4430	49	.5570	30
40	8.9945	125	.9979	2	8.9966	127	1.0034	20	40	.4314	45	.9836	4	.4479	48	.5521	20
50	9.0070	122	.9977	1	9.0093	123	0.9907	10	50	.4359	44	.9832	4	.4527	48	.5473	10
6 0	.0192	119	.9976	1	.0216	120	.9784	84 0	16 0	.4403	44	.9828	3	.4575	47	.5425	74 0
10	.0311	115	.9975	2	.0336	117	.9664	50	10	.4447	44	.9825	4	.4622	47	.5378	50
20	.0426	113	.9973	1	.0453	114	.9547	40	20	.4491	42	.9821	4	.4669	47	.5331	40
30	.0539	109	.9972	1	.0567	111	.9433	30	30	.4533	43	.9817	3	.4716	46	.5284	30
40	.0648	107	.9971	2	.0678	108	.9322	20	40	.4576	42	.9814	4	.4762	46	.5238	20
50	.0755	104	.9969	1	.0786	105	.9214	10	50	.4618	41	.9810	4	.4808	45	.5192	10
7 0	.0859	102	.9968	2	.0891	104	.9109	83 0	17 0	.4659	41	.9806	4	.4853	45	.5147	73 0
10	.0961	99	.9966	2	.0995	101	.9005	50	10	.4700	41	.9802	4	.4898	45	.5102	50
20	.1060	97	.9964	1	.1096	98	.8904	40	20	.4741	40	.9798	4	.4943	44	.5057	40
30	.1157	95	.9963	2	.1194	97	.8806	30	30	.4781	40	.9794	4	.4987	44	.5013	30
40	.1252	93	.9961	2	.1291	94	.8709	20	40	.4821	40	.9790	4	.5031	44	.4969	20
50	.1345	91	.9959	1	.1385	93	.8615	10	50	.4861	39	.9786	4	.5075	43	.4925	10
8 0	.1436	89	.9958	2	.1478	91	.8522	82 0	18 0	.4900	39	.9782	4	.5118	43	.4882	72 0
10	.1525	87	.9956	2	.1569	89	.8431	50	10	.4939	38	.9778	4	.5161	42	.4839	50
20	.1612	85	.9954	2	.1658	87	.8342	40	20	.4977	38	.9774	4	.5203	42	.4797	40
30	.1697	84	.9952	2	.1745	86	.8255	30	30	.5015	37	.9770	5	.5245	42	.4755	30
40	.1781	82	.9950	2	.1831	84	.8169	20	40	.5052	38	.9765	4	.5287	42	.4713	20
50	.1863	80	.9948	2	.1915	82	.8085	10	50	.5090	36	.9761	4	.5329	41	.4671	10
9 0	.1943	79	.9946	2	.1997	81	.8003	81 0	19 0	.5126	37	.9757	5	.5370	41	.4630	71 0
10	.2022	78	.9944	2	.2078	80	.7922	50	10	.5163	36	.9752	4	.5411	40	.4589	50
20	.2100	76	.9942	2	.2158	78	.7842	40	20	.5199	36	.9748	5	.5451	40	.4549	40
30	.2176	75	.9940	2	.2236	77	.7764	30	30	.5235	35	.9743	4	.5491	40	.4509	30
40	.2251	73	.9938	2	.2313	76	.7687	20	40	.5270	36	.9739	5	.5531	40	.4469	20
50	.2324	73	.9936	2	.2389	74	.7611	10	50	.5306	35	.9734	4	.5571	40	.4429	10
10 0	.2397	71	.9934	3	.2463	73	.7537	80 0	20 0	.5341	34	.9730	5	.5611	39	.4389	70 0
10	.2468	70	.9931	2	.2536	73	.7464	50	10	.5375	34	.9725	4	.5650	39	.4350	50
20	.2538	68	.9929	2	.2609	71	.7391	40	20	.5409	34	.9721	5	.5689	38	.4311	40
30	.2606	68	.9927	3	.2680	70	.7320	30	30	.5443	34	.9716	5	.5727	39	.4273	30
40	.2674	66	.9924	2	.2750	69	.7250	20	40	.5477	33	.9711	5	.5766	38	.4234	20
50	.2740	66	.9922	3	.2819	68	.7181	10	60	.5510	33	.9706	4	.5804	38	.4196	10
11 0	.2806	64	.9919	2	.2887	66	.7113	79 0	21 0	.5543	33	.9702	5	.5842	37	.4158	69 0
10	.2870	64	.9917	3	.2953	67	.7047	50	10	.5576	33	.9697	5	.5879	38	.4121	50
20	.2934	63	.9914	2	.3020	65	.6980	40	20	.5609	32	.9692	5	.5917	37	.4083	40
30	.2997	61	.9912	3	.3085	64	.6915	30	30	.5641	32	.9687	5	.5954	37	.4046	30
40	.3058	61	.9909	2	.3149	63	.6851	20	40	.5673	31	.9682	5	.5991	37	.4009	20
50	.3119	60	.9907	3	.3212	63	.6788	10	50	.5704	32	.9677	5	.6028	36	.3972	10
12 0	.3179	59	.9904	3	.3275	61	.6725	78 0	22 0	.5736	31	.9672	5	.6064	36	.3936	68 0
10	.3238	58	.9901	2	.3336	61	.6664	50	10	.5767	31	.9667	6	.6100	36	.3900	50
20	.3296	57	.9899	3	.3397	61	.6603	40	20	.5798	30	.9661	5	.6136	36	.3864	40
30	.3353	57	.9896	3	.3458	59	.6542	30	30	.5828	31	.9656	5	.6172	36	.3828	30
40	.3410	56	.9893	3	.3517	59	.6483	20	40	.5859	30	.9651	5	.6208	35	.3792	20
50	.3466	55	.9890	3	.3576	58	.6424	10	50	.5889	30	.9646	6	.6243	36	.3757	10
13 0	.3521	54	.9887	3	.3634	57	.6366	77 0	23 0	.5919	29	.9640	5	.6279	35	.3721	67 0
10	.3575	54	.9884	3	.3691	57	.6309	50	10	.5948	30	.9635	6	.6314	34	.3686	50
20	.3629	53	.9881	3	.3748	56	.6252	40	20	.5978	29	.9629	5	.6348	35	.3652	40
30	.3682	52	.9878	3	.3804	55	.6196	30	30	.6007	29	.9624	6	.6383	34	.3617	30
40	.3734	52	.9875	3	.3859	55	.6141	20	40	.6036	29	.9618	5	.6417	35	.3583	20
50	.3786	51	.9872	3	.3914	54	.6086	10	50	.6065	28	.9613	6	.6452	34	.3548	10
14 0	.3837	50	.9869	3	.3968	53	.6032	76 0	24 0	.6093	28	.9607	5	.6486	34	.3514	66 0
10	.3887	50	.9866	3	.4021	53	.5979	50	10	.6121	28	.9602	6	.6520	33	.3480	50
20	.3937	49	.9863	4	.4074	53	.5926	40	20	.6149	28	.9596	6	.6553	34	.3447	40
30	.3986	49	.9859	3	.4127	52	.5873	30	30	.6177	28	.9590	6	.6587	33	.3413	30
40	.4035	48	.9856	3	.4178	51	.5822	20	40	.6205	27	.9584	5	.6620	34	.3380	20
50	.4083	47	.9853	4	.4230	51	.5770	10	50	.6232	27	.9579	6	.6654	33	.3346	10
15 0	9.4130	47	9.9849	3	9.4281	50	0.5719	75 0	25 0	0.6259	27	9.9573	7	9.6687	33	0.3313	65 0
Arc	Cos.	Df.	Sin.	Df.	Cot.	Df.	Tan.	Arc	Arc	Cos.	Df.	Sin.	Df.	Cot.	Df.	Tan.	Arc

XXIII.—*Logarithms of Sines and Tangents*—Continued.

Arc	Sin.	Df.	Cos.	Df.	Tan.	Df.	Cot.	Arc	Arc	Sin.	Df.	Cos.	Df.	Tan.	Df.	Cot.	Arc
° ′								° ′	° ′								° ′
25 0	9.6259	27	9.9573	6	9.6687	33	0.3313	65 0	35 0	9.7586	18	9.9134	9	9.8452	27	0.1548	55 0
10	.6286	27	.9567	6	.6720	32	.3280	50	10	.7604	18	.9125	9	.8479	27	.1521	50
20	.6313	27	.9561	6	.6752	33	.3248	40	20	.7622	18	.9116	9	.8506	27	.1494	40
30	.6340	26	.9555	6	.6785	32	.3215	30	30	.7640	17	.9107	9	.8533	26	.1467	30
40	.6366	26	.9549	6	.6817	33	.3183	20	40	.7657	18	.9098	9	.8559	27	.1441	20
50	.6392	26	.9543	6	.6850	32	.3150	10	50	.7675	17	.9089	9	.8586	27	.1414	10
26 0	.6418	26	.9537	7	.6882	32	.3118	64 0	36 0	.7692	18	.9080	10	.8613	26	.1387	54 0
10	.6444	26	.9530	6	.6914	32	.3086	50	10	.7710	17	.9070	9	.8639	27	.1361	50
20	.6470	25	.9524	6	.6946	31	.3054	40	20	.7727	17	.9061	9	.8666	26	.1334	40
30	.6495	26	.9518	6	.6977	32	.3023	30	30	.7744	17	.9052	10	.8692	26	.1308	30
40	.6521	25	.9512	7	.7009	31	.2991	20	40	.7761	17	.9042	9	.8718	27	.1282	20
50	.6546	24	.9505	6	.7040	32	.2960	10	50	.7778	17	.9033	10	.8745	26	.1255	10
27 0	.6570	25	.9499	7	.7072	31	.2928	63 0	37 0	.7795	16	.9023	9	.8771	26	.1229	53 0
10	.6595	25	.9492	6	.7103	31	.2897	50	10	.7811	17	.9014	10	.8797	27	.1203	50
20	.6620	24	.9486	7	.7134	31	.2866	40	20	.7828	16	.9004	9	.8824	26	.1176	40
30	.6644	24	.9479	6	.7165	31	.2835	30	30	.7844	17	.8995	10	.8850	26	.1150	30
40	.6668	24	.9473	7	.7196	30	.2804	20	40	.7861	16	.8985	10	.8876	26	.1124	20
50	.6692	24	.9466	7	.7226	31	.2774	10	50	.7877	16	.8975	10	.8902	26	.1098	10
28 0	.6716	24	.9459	6	.7257	30	.2743	62 0	38 0	.7893	17	.8965	10	.8928	26	.1072	52 0
10	.6740	23	.9453	7	.7287	30	.2713	50	10	.7910	16	.8955	10	.8954	26	.1046	50
20	.6763	24	.9446	7	.7317	31	.2683	40	20	.7926	15	.8945	10	.8980	26	.1020	40
30	.6787	23	.9439	7	.7348	30	.2652	30	30	.7941	16	.8935	10	.9006	26	.0994	30
40	.6810	23	.9432	7	.7378	30	.2622	20	40	.7957	16	.8925	10	.9032	26	.0968	20
50	.6833	23	.9425	7	.7408	30	.2592	10	50	.7973	16	.8915	10	.9058	26	.0942	10
29 0	.6856	22	.9418	7	.7438	29	.2562	61 0	39 0	.7989	15	.8905	10	.9084	26	.0916	51 0
10	.6878	23	.9411	7	.7467	30	.2533	50	10	.8004	16	.8895	11	.9110	25	.0890	50
20	.6901	22	.9404	7	.7497	29	.2503	40	20	.8020	15	.8884	10	.9135	26	.0865	40
30	.6923	23	.9397	7	.7526	30	.2474	30	30	.8035	15	.8874	10	.9161	26	.0839	30
40	.6946	22	.9390	7	.7556	29	.2444	20	40	.8050	16	.8864	11	.9187	25	.0813	20
50	.6968	22	.9383	8	.7585	29	.2415	10	50	.8066	15	.8853	10	.9212	26	.0788	10
30 0	.6990	22	.9375	7	.7614	30	.2386	60 0	40 0	.8081	15	.8843	11	.9238	26	.0762	50 0
10	.7012	21	.9368	7	.7644	29	.2356	50	10	.8096	15	.8832	11	.9264	25	.0736	50
20	.7033	22	.9361	8	.7673	28	.2327	40	20	.8111	14	.8821	11	.9289	26	.0711	40
30	.7055	21	.9353	7	.7701	29	.2299	30	30	.8125	15	.8810	10	.9315	26	.0685	30
40	.7076	21	.9346	8	.7730	29	.2270	20	40	.8140	15	.8800	11	.9341	25	.0659	20
50	.7097	21	.9338	7	.7759	29	.2241	10	50	.8155	14	.8789	11	.9366	26	.0634	10
31 0	.7118	21	.9331	8	.7788	28	.2212	59 0	41 0	.8169	15	.8778	11	.9392	25	.0608	49 0
10	.7139	21	.9323	8	.7816	29	.2184	50	10	.8184	14	.8767	11	.9417	26	.0583	50
20	.7160	21	.9315	7	.7845	28	.2155	40	20	.8198	15	.8756	11	.9443	25	.0557	40
30	.7181	20	.9308	8	.7873	29	.2127	30	30	.8213	14	.8745	12	.9468	26	.0532	30
40	.7201	21	.9300	8	.7902	28	.2098	20	40	.8227	14	.8733	11	.9494	25	.0506	20
50	.7222	20	.9292	8	.7930	28	.2070	10	50	.8241	14	.8722	11	.9519	25	.0481	10
32 0	.7242	20	.9284	8	.7958	28	.2042	58 0	42 0	.8255	14	.8711	12	.9544	26	.0456	48 0
10	.7262	20	.9276	8	.7986	28	.2014	50	10	.8269	14	.8699	11	.9570	25	.0430	50
20	.7282	20	.9268	8	.8014	28	.1986	40	20	.8283	14	.8688	12	.9595	26	.0405	40
30	.7302	20	.9260	8	.8042	28	.1958	30	30	.8297	14	.8676	11	.9621	25	.0379	30
40	.7322	20	.9252	8	.8070	27	.1930	20	40	.8311	13	.8665	12	.9646	25	.0354	20
50	.7342	19	.9244	8	.8097	28	.1903	10	50	.8324	14	.8653	12	.9671	26	.0329	10
33 0	.7361	19	.9236	8	.8125	28	.1875	57 0	43 0	.8338	13	.8641	12	.9697	25	.0303	47 0
10	.7380	20	.9228	9	.8153	27	.1847	50	10	.8351	14	.8629	11	.9722	25	.0278	50
20	.7400	19	.9219	8	.8180	28	.1820	40	20	.8365	13	.8618	12	.9747	25	.0253	40
30	.7419	19	.9211	8	.8208	27	.1792	30	30	.8378	13	.8606	12	.9772	26	.0228	30
40	.7438	19	.9203	9	.8235	28	.1765	20	40	.8391	14	.8594	12	.9798	25	.0202	20
50	.7457	19	.9194	8	.8263	27	.1737	10	50	.8405	13	.8582	13	.9823	25	.0177	10
34 0	.7476	18	.9186	9	.8390	27	.1710	56 0	44 0	.8418	13	.8569	12	.9848	26	.0152	46 0
10	.7494	19	.9177	8	.8317	27	.1683	50	10	.8431	13	.8557	12	.9874	25	.0126	50
20	.7513	18	.9169	9	.8344	27	.1656	40	20	.8444	13	.8545	13	.9899	25	.0101	40
30	.7531	19	.9160	9	.8371	27	.1629	30	30	.8457	12	.8532	12	.9924	25	.0076	30
40	.7550	18	.9151	9	.8398	27	.1602	20	40	.8469	13	.8520	13	.9949	26	.0051	20
50	.7568	18	.9142	8	.8425	27	.1575	10	50	.8482	13	.8507	12	9.9975	25	.0025	10
35 0	9.7586	18	9.9134	9	9.8452	27	0.1548	55 0	45 0	9.8495		9.8495		0.0000		0.0000	45 0
Arc	Cos.	Df.	Sin.	Df.	Cot.	Df.	Tan.	Arc	Arc	Cos.	Df.	Sin.	Df.	Cot.	Df.	Tan.	Arc

XXIV.—*Squares and Square Roots.*

No.	Square.	Square root.	No.	Square.	Square root.
1	1	1.000	51	2601	7.141
2	4	1.414	52	2704	7.211
3	9	1.732	53	2809	7.280
4	16	2.000	54	2916	7.348
5	25	2.236	55	3025	7.416
6	36	2.449	56	3136	7.483
7	49	2.646	57	3249	7.550
8	64	2.828	58	3364	7.616
9	81	3.000	59	3481	7.681
10	100	3.162	60	3600	7.746
11	121	3.317	61	3721	7.810
12	144	3.464	62	3844	7.874
13	169	3.606	63	3969	7.937
14	196	3.742	64	4096	8.000
15	225	3.873	65	4225	8.062
16	256	4.000	66	4356	8.124
17	289	4.123	67	4489	8.185
18	324	4.243	68	4624	8.246
19	361	4.359	69	4761	8.307
20	400	4.472	70	4900	8.367
21	441	4.583	71	5041	8.426
22	484	4.690	72	5184	8.485
23	529	4.796	73	5329	8.544
24	576	4.899	74	5476	8.602
25	625	5.000	75	5625	8.660
26	676	5.099	76	5776	8.718
27	729	5.196	77	5929	8.775
28	784	5.292	78	6084	8.832
29	841	5.385	79	6241	8.888
30	900	5.477	80	6400	8.944
31	961	5.568	81	6561	9.000
32	1024	5.657	82	6724	9.055
33	1089	5.745	83	6889	9.110
34	1156	5.831	84	7056	9.165
35	1225	5.916	85	7225	9.220
36	1296	6.000	86	7396	9.274
37	1369	6.083	87	7569	9.327
38	1444	6.164	88	7744	9.381
39	1521	6.245	89	7921	9.434
40	1600	6.325	90	8100	9.487
41	1681	6.403	91	8281	9.539
42	1764	6.481	92	8464	9.592
43	1849	6.557	93	8649	9.644
44	1936	6.633	94	8836	9.695
45	2025	6.708	95	9025	9.747
46	2116	6.782	96	9216	9.798
47	2209	6.856	97	9409	9.849
48	2304	6.928	98	9604	9.899
49	2401	7.000	99	9801	9.950
50	2500	7.071	100	10000	10.000

XXIV.—*Squares and Square Roots*—Continued.

No.	Square.	Square root.	No.	Square.	Square root.
101	10201	10.050	151	22801	12.288
102	10404	10.100	152	23104	12.329
103	10609	10.149	153	23409	12.369
104	10816	10.198	154	23716	12.410
105	11025	10.247	155	24025	12.450
106	11236	10.296	156	24336	12.490
107	11449	10.344	157	24649	12.530
108	11664	10.392	158	24964	12.570
109	11881	10.440	159	25281	12.610
110	12100	10.488	160	25600	12.650
111	12321	10.536	161	25921	12.689
112	12544	10.583	162	26244	12.728
113	12769	10.630	163	26569	12.767
114	12996	10.677	164	26896	12.806
115	13225	10.724	165	27225	12.845
116	13456	10.771	166	27556	12.884
117	13689	10.817	167	27889	12.923
118	13924	10.863	168	28224	12.961
119	14161	10.909	169	28561	13.000
120	14400	10.954	170	28900	13.038
121	14641	11.000	171	29241	13.077
122	14884	11.045	172	29584	13.115
123	15129	11.091	173	29929	13.153
124	15376	11.136	174	30276	13.191
125	15625	11.180	175	30625	13.229
126	15876	11.225	176	30976	13.266
127	16129	11.269	177	31329	13.304
128	16384	11.314	178	31684	13.342
129	16641	11.358	179	32041	13.379
130	16900	11.402	180	32400	13.416
131	17161	11.446	181	32761	13.454
132	17424	11.489	182	33124	13.491
133	17689	11.533	183	33489	13.528
134	17956	11.576	184	33856	13.565
135	18225	11.619	185	34225	13.601
136	18496	11.662	186	34596	13.638
137	18769	11.705	187	34969	13.675
138	19044	11.747	188	35344	13.711
139	19321	11.790	189	35721	13.748
140	19600	11.832	190	36100	13.784
141	19881	11.874	191	36481	13.820
142	20164	11.916	192	36864	13.856
143	20449	11.958	193	37249	13.892
144	20736	12.000	194	37636	13.928
145	21025	12.042	195	38025	13.964
146	21316	12.083	196	38416	14.000
147	21609	12.124	197	38809	14.036
148	21904	12.166	198	39204	14.071
149	22201	12.207	199	39601	14.107
150	22500	12.247	200	40000	14.142

XXIV.—*Squares and Square Roots*—Continued.

No.	Square.	Square root.	No.	Square.	Square root.
201	40401	14.177	251	63001	15.843
202	40804	14.213	252	63504	15.875
203	41209	14.248	253	64009	15.906
204	41616	14.283	254	64516	15.937
205	42025	14.318	255	65025	15.969
206	42436	14.353	256	65536	16.000
207	42849	14.387	257	66049	16.031
208	43264	14.422	258	66564	16.062
209	43681	14.457	259	67081	16.093
210	44100	14.491	260	67600	16.125
211	44521	14.526	261	68121	16.155
212	44944	14.560	262	68644	16.186
213	45369	14.595	263	69169	16.217
214	45796	14.629	264	69696	16.248
215	46225	14.663	265	70225	16.279
216	46656	14.697	266	70756	16.310
217	47089	14.731	267	71289	16.340
218	47524	14.765	268	71824	16.371
219	47961	14.799	269	72361	16.401
220	48400	14.832	270	72900	16.432
221	48841	14.866	271	73441	16.462
222	49284	14.900	272	73984	16.492
223	49729	14.933	273	74529	16.523
224	50176	14.967	274	75076	16.553
225	50625	15.000	275	75625	16.583
226	51076	15.033	276	76176	16.613
227	51529	15.067	277	76729	16.643
228	51984	15.100	278	77284	16.673
229	52441	15.133	279	77841	16.703
230	52900	15.166	280	78400	16.733
231	53361	15.199	281	78961	16.763
232	53824	15.232	282	79524	16.793
233	54289	15.264	283	80089	16.823
234	54756	15.297	284	80656	16.852
235	55225	15.330	285	81225	16.882
236	55696	15.362	286	81796	16.912
237	56169	15.395	287	82369	16.941
238	56644	15.427	288	82944	16.971
239	57121	15.460	289	83521	17.000
240	57600	15.492	290	84100	17.029
241	58081	15.524	291	84681	17.059
242	58564	15.556	292	85264	17.088
243	59049	15.588	293	85849	17.117
244	59536	15.620	294	86436	17.146
245	60025	15.652	295	87025	17.176
246	60516	15.684	296	87616	17.205
247	61009	15.716	297	88209	17.234
248	61504	15.748	298	88804	17.263
249	62001	15.780	299	89401	17.292
250	62500	15.811	300	90000	17.321

XXIV.—*Squares and Square Roots*—Continued.

No.	Square.	Square root.	No.	Square.	Square root.
301	90601	17.349	351	123201	18.735
302	91204	17.378	352	123904	18.762
303	91809	17.407	353	124609	18.788
304	92416	17.436	354	125316	18.815
305	93025	17.464	355	126025	18.841
306	93636	17.493	356	126736	18.868
307	94249	17.521	357	127449	18.894
308	94864	17.550	358	128164	18.921
309	95481	17.578	359	128881	18.947
310	96100	17.607	360	129600	18.974
311	96721	17.635	361	130321	19.000
312	97344	17.664	362	131044	19.026
313	97969	17.692	363	131769	19.053
314	98596	17.720	364	132496	19.079
315	99225	17.748	365	133225	19.105
316	99856	17.776	366	133956	19.131
317	100489	17.804	367	134689	19.157
318	101124	17.833	368	135424	19.183
319	101761	17.861	369	136161	19.209
320	102400	17.889	370	136900	19.235
321	103041	17.916	371	137641	19.261
322	103684	17.944	372	138384	19.287
323	104329	17.972	373	139129	19.313
324	104976	18.000	374	139876	19.339
325	105625	18.028	375	140625	19.365
326	106276	18.055	376	141376	19.391
327	106929	18.083	377	142129	19.416
328	107584	18.111	378	142884	19.442
329	108241	18.138	379	143641	19.468
330	108900	18.166	380	144400	19.494
331	109561	18.193	381	145161	19.519
332	110224	18.221	382	145924	19.545
333	110889	18.248	383	146689	19.570
334	111556	18.276	384	147456	19.596
335	112225	18.303	385	148225	19.621
336	112896	18.330	386	148996	19.647
337	113569	18.358	387	149769	19.672
338	114244	18.385	388	150544	19.698
339	114921	18.412	389	151321	19.723
340	115600	18.439	390	152100	19.748
341	116281	18.466	391	152881	19.774
342	116964	18.493	392	153664	19.799
343	117649	18.520	393	154449	19.824
344	118336	18.547	394	155236	19.849
345	119025	18.574	395	156025	19.875
346	119716	18.601	396	156816	19.900
347	120409	18.628	397	157609	19.925
348	121104	18.655	398	158404	19.950
349	121801	18.682	399	159201	19.975
350	122500	18.708	400	160000	20.000

XXIV.—*Squares and Square Roots*—Continued.

No.	Square.	Square root.	No.	Square.	Square root.
401	160801	20.025	451	203401	21.237
402	161604	20.050	452	204304	21.260
403	162409	20.075	453	205209	21.284
404	163216	20.100	454	206116	21.307
405	164025	20.125	455	207025	21.331
406	164836	20.149	456	207936	21.354
407	165649	20.174	457	208849	21.378
408	166464	20.199	458	209764	21.401
409	167281	20.224	459	210681	21.424
410	168100	20.248	460	211600	21.448
411	168921	20.273	461	212521	21.471
412	169744	20.298	462	213444	21.494
413	170569	20.322	463	214369	21.517
414	171396	20.347	464	215296	21.541
415	172225	20.372	465	216225	21.564
416	173056	20.396	466	217156	21.587
417	173889	20.421	467	218089	21.610
418	174724	20.445	468	219024	21.633
419	175561	20.469	469	219961	21.656
420	176400	20.494	470	220900	21.679
421	177241	20.518	471	221841	21.703
422	178084	20.543	472	222784	21.726
423	178929	20.567	473	223729	21.749
424	179776	20.591	474	224676	21.772
425	180625	20.616	475	225625	21.794
426	181476	20.640	476	226576	21.817
427	182329	20.664	477	227529	21.840
428	183184	20.688	478	228484	21.863
429	184041	20.712	479	229441	21.886
430	184900	20.736	480	230400	21.909
431	185761	20.761	481	231361	21.932
432	186624	20.785	482	232324	21.954
433	187489	20.809	483	233289	21.977
434	188356	20.833	484	234256	22.000
435	189225	20.857	485	235225	22.023
436	190096	20.881	486	236196	22.045
437	190969	20.905	487	237169	22.068
438	191844	20.928	488	238144	22.091
439	192721	20.952	489	239121	22.113
440	193600	20.976	490	240100	22.136
441	194481	21.000	491	241081	22.159
442	195364	21.024	492	242064	22.181
443	196249	21.048	493	243049	22.204
444	197136	21.071	494	244036	22.226
445	198025	21.095	495	245025	22.249
446	198916	21.119	496	246016	22.271
447	199809	21.142	497	247009	22.293
448	200704	21.166	498	248004	22.316
449	201601	21.190	499	249001	22.338
450	202500	21.213	500	250000	22.361

XXIV.—*Squares and Square Roots*—Continued.

No.	Square.	Square root.	No.	Square.	Square root.
501	251001	22.383	551	303601	23.473
502	252004	22.405	552	304704	23.495
503	253009	22.428	553	305809	23.516
504	254016	22.450	554	306916	23.537
505	255025	22.472	555	308025	23.558
506	256036	22.494	556	309136	23.580
507	257049	22.517	557	310249	23.601
508	258064	22.539	558	311364	23.622
509	259081	22.561	559	312481	23.643
510	260100	22.583	560	313600	23.664
511	261121	22.605	561	314721	23.685
512	262144	22.627	562	315844	23.707
513	263169	22.650	563	316969	23.728
514	264196	22.672	564	318096	23.749
515	265225	22.694	565	319225	23.770
516	266256	22.716	566	320356	23.791
517	267289	22.738	567	321489	23.812
518	268324	22.760	568	322624	23.833
519	269361	22.782	569	323761	23.854
520	270400	22.804	570	324900	23.875
521	271441	22.825	571	326041	23.896
522	272484	22.847	572	327184	23.917
523	273529	22.869	573	328329	23.937
524	274576	22.891	574	329476	23.958
525	275625	22.913	575	330625	23.979
526	276676	22.935	576	331776	24.000
527	277729	22.956	577	332929	24.021
528	278784	22.978	578	334084	24.042
529	279841	23.000	579	335241	24.062
530	280900	23.022	580	336400	24.083
531	281961	23.043	581	337561	24.104
532	283024	23.065	582	338724	24.125
533	284089	23.087	583	339889	24.145
534	285156	23.108	584	341056	24.166
535	286225	23.130	585	342225	24.187
536	287296	23.152	586	343396	24.207
537	288369	23.173	587	344569	24.228
538	289444	23.195	588	345744	24.249
539	290521	23.216	589	346921	24.269
540	291600	23.238	590	348100	24.290
541	292681	23.259	591	349281	24.310
542	293764	23.281	592	350464	24.331
543	294849	23.302	593	351649	24.352
544	295936	23.324	594	352836	24.372
545	297025	23.345	595	354025	24.393
546	298116	23.367	596	355216	24.413
547	299209	23.388	597	356409	24.434
548	300304	23.409	598	357604	24.454
549	301401	23.431	599	358801	24.474
550	302500	23.452	600	360000	24.495

XXIV.—*Squares and Square Roots*—Continued.

No.	Square.	Square root.	No.	Square.	Square root.
601	361201	24.515	651	423801	25.515
602	362404	24.536	652	425104	25.534
603	363609	24.556	653	426409	25.554
604	364816	24.576	654	427716	25.573
605	366025	24.597	655	429025	25.593
606	367236	24.617	656	430336	25.612
607	368449	24.637	657	431649	25.632
608	369664	24.658	658	432964	25.652
609	370881	24.678	659	434281	25.671
610	372100	24.698	660	435600	25.690
611	373321	24.718	661	436921	25.710
612	374544	24.739	662	438244	25.720
613	375769	24.759	663	439569	25.749
614	376996	24.779	664	440896	25.768
615	378225	24.799	665	442225	25.788
616	379456	24.819	666	443556	25.807
617	380689	24.839	667	444889	25.826
618	381924	24.860	668	446224	25.846
619	383161	24.880	669	447561	25.865
620	384400	24.900	670	448900	25.884
621	385641	24.920	671	450241	25.904
622	386884	24.940	672	451584	25.923
623	388129	24.960	673	452929	25.942
624	389376	24.980	674	454276	25.962
625	390625	25.000	675	455625	25.981
626	391876	25.020	676	456976	26.000
627	393129	25.040	677	458329	26.019
628	394384	25.060	678	459684	26.038
629	395641	25.080	679	461041	26.038
630	396900	25.100	680	462400	26.077
631	398161	25.120	681	463761	26.096
632	399424	25.140	682	465124	26.115
633	400689	25.160	683	466489	26.134
634	401956	25.180	684	467856	26.153
635	403225	25.200	685	469225	26.173
636	404496	25.220	686	470596	26.192
637	405769	25.239	687	471969	26.211
638	407044	25.259	688	473344	26.230
639	408321	25.278	689	474721	26.249
640	409600	25.298	690	476100	26.268
641	410881	25.318	691	477481	26.287
642	412164	25.338	692	478864	26.306
643	413449	25.357	693	480249	26.325
644	414736	25.377	694	481636	26.344
645	416025	25.397	695	483025	26.363
646	417316	25.417	696	484416	26.382
647	418609	25.436	697	485809	26.401
648	419904	25.456	698	487204	26.420
649	421201	25.475	699	488601	26.439
650	422500	25.495	700	490000	26.458

XXIV.—*Squares and Square Roots*—Continued.

No.	Square.	Square root.	No.	Square.	Square root.
701	491401	26.476	751	564001	27.404
702	492804	26.495	752	565504	27.423
703	494209	26.514	753	567009	27.441
704	495616	26.532	754	568516	27.459
705	497025	26.552	755	570025	27.477
706	498436	26.571	756	571536	27.495
707	499849	26.589	757	573049	27.514
708	501264	26.608	758	574564	27.532
709	502681	26.627	759	576081	27.550
710	504100	26.646	760	577600	27.568
711	505521	26.665	761	579121	27.586
712	506944	26.683	762	580644	27.604
713	508369	26.702	763	582169	27.622
714	509796	26.721	764	583696	27.641
715	511225	26.739	765	585225	27.659
716	512656	26.758	766	586756	27.677
717	514089	26.777	767	588289	27.695
718	515524	26.796	768	289824	27.713
719	516961	26.814	769	591361	27.731
720	518400	26.833	770	592900	27.749
721	519841	26.851	771	594441	27.767
722	521284	26.870	772	595984	27.785
723	522729	26.889	773	597529	27.803
724	524176	26.907	774	599076	27.821
725	525625	26.926	775	600625	27.839
726	527076	26.944	776	602176	27.857
727	528529	26.963	777	603729	27.875
728	529984	26.981	778	605284	27.893
729	531441	27.000	779	606841	27.911
730	532900	27.019	780	608400	27.928
731	534361	27.037	781	609961	27.946
732	535824	27.055	782	611524	27.964
733	537289	27.074	783	613089	27.982
734	538756	27.092	784	614656	28.000
735	540225	27.111	785	616225	28.018
736	541696	27.129	786	617796	28.036
737	543169	27.148	787	619369	28.054
738	544644	27.166	788	620944	28.071
739	546121	27.185	789	622521	28.089
740	547600	27.203	790	624100	28.107
741	549081	27.221	791	625681	28.125
742	550564	27.240	792	627264	28.142
743	552049	27.258	793	628849	28.160
744	553536	27.276	794	630436	28.178
745	555025	27.295	795	632025	28.196
746	556516	27.313	796	633616	28.213
747	558009	27.331	797	635209	28.231
748	559504	27.350	798	636804	28.249
749	561001	27.368	799	638401	28.267
750	562500	27.386	800	640000	28.284

XXIV.—*Squares and Square Roots*—Continued.

No.	Square.	Square root.	No.	Square.	Square root.
801	641601	28.302	851	724201	29.172
802	643204	28.320	852	725904	29.189
803	644809	28.337	853	727609	29.206
804	646416	28.555	854	729316	29.223
805	648025	28.373	855	731025	29.240
806	649636	28.390	856	732736	29.257
807	651249	28.408	857	734449	29.275
808	652864	28.425	858	736164	29.292
809	654481	28.443	859	737881	29.309
810	656100	28.460	860	739600	29.326
811	657721	28.478	861	741321	29.343
812	659344	28.496	862	743044	29.360
813	660969	28.513	863	744769	29.377
814	662596	28.531	864	746496	29.394
815	664225	28.548	865	748225	29.411
816	665856	28.566	866	749956	29.428
817	667489	28.583	867	751689	29.445
818	669124	28.601	868	753424	29.462
819	670761	28.618	869	755161	29.479
820	672400	28.636	870	756900	29.496
821	674041	28.653	871	758641	29.513
822	675684	28.671	872	760384	29.530
823	677329	28.688	873	762129	29.547
824	678976	28.705	874	763876	29.563
825	680625	28.723	875	765625	29.580
826	682276	28.740	876	767376	29.597
827	683929	28.758	877	769129	29.614
828	685584	28.775	878	770884	29.631
829	687241	28.792	879	772641	29.648
830	688900	28.810	880	774400	29.665
831	690561	28.827	881	776161	29.682
832	692224	28.844	882	777924	29.698
833	693889	28.862	883	779689	29.715
834	695556	28.879	884	781456	29.732
835	697225	28.896	885	783225	29.749
836	698896	28.914	886	784996	29.766
837	700569	28.931	887	786769	29.783
838	702244	28.948	888	788544	29.799
839	703921	28.965	889	790321	29.816
840	705600	28.983	890	792100	29.833
841	707281	29.000	891	793881	29.850
842	708964	29.017	892	795664	29.866
843	710649	29.034	893	797449	29.883
844	712336	29.052	894	799236	29.900
845	714025	29.069	895	801025	29.917
846	715716	29.086	896	802816	29.933
847	717409	29.103	897	804609	29.950
848	719104	29.120	898	806404	29.967
849	720801	29.138	899	808201	29.983
850	722500	29.155	900	810000	30.000

XXIV.—*Squares and Square Roots*—Continued.

No.	Square.	Square root.	No.	Square.	Square root.
901	811801	30.017	951	904401	30.838
902	813604	30.033	952	906304	30.854
903	815409	30.050	953	908209	30.871
904	817216	30.067	954	910116	30.887
905	819025	30.083	955	912025	30.903
906	820836	30.100	956	913936	30.919
907	822649	30.116	957	915849	30.935
908	824464	30.133	958	917764	30.952
909	826281	30.150	959	919681	30.968
910	828100	30.166	960	921600	30.984
911	829921	30.183	961	923521	31.000
912	831744	30.199	962	925444	31.016
913	833569	30.216	963	927369	31.032
914	835396	30.232	964	929296	31.048
915	837225	30.249	965	931225	31.064
916	839056	30.265	966	933156	31.081
917	840889	30.282	967	935089	31.097
918	842724	30.299	968	937024	31.113
919	844561	30.315	969	938961	31.129
920	846400	30.332	970	940900	31.145
921	848241	30.348	971	942841	31.161
922	850084	30.364	972	944784	31.177
923	851929	30.381	973	946729	31.193
924	853776	30.397	974	948676	31.209
925	855625	30.414	975	950625	31.225
926	857476	30.430	976	952576	31.241
927	859329	30.447	977	954529	31.257
928	861184	30.463	978	956484	31.273
929	863041	30.480	979	958441	31.289
930	864900	30.496	980	960400	31.305
931	866761	30.512	981	962361	31.321
932	868624	30.529	982	964324	31.337
933	870489	30.545	983	966289	31.353
934	872356	30.561	984	968256	31.369
935	874225	30.578	985	970225	31.385
936	876096	30.594	986	972196	31.401
937	877969	30.610	987	974169	31.417
938	879844	30.627	988	976144	31.432
939	881721	30.643	989	978121	31.448
940	883600	30.659	990	980100	31.464
941	885481	30.676	991	982081	31.480
942	887364	30.692	992	984064	31.496
943	889249	30.708	993	986049	31.512
944	891136	30.725	994	988036	31.528
945	893025	30.741	995	990025	31.544
946	894916	30.757	996	992016	31.560
947	896809	30.773	997	994009	31.575
948	898704	30.790	998	996004	31.591
949	900601	30.806	999	998001	31.607
950	902500	30.822	1000	1000000	31.623

Bessel's Co-efficients.

Parts of the unit of time.	2d diff.	3d diff.	4th diff.	Parts of the unit of time.	2d diff.	3d diff.	4th diff.
	$t.\frac{t-1}{2}$	$t.\frac{t-1}{2}.\frac{t-\frac{1}{2}}{3}$	$t\ldots\frac{t-2}{4}$		$t.\frac{t-1}{2}$	$t.\frac{t-1}{2}.\frac{t-\frac{1}{2}}{3}$	$t\ldots\frac{t-2}{4}$
0.01	−.00495	.00081	.00083	0.51	−.12495	−.00042	.02343
.02	.00980	.00157	.00165	.52	.12480	.00083	.02340
.03	.01455	.00228	.00246	.53	.12455	.00125	.02334
.04	.01920	.00294	.00326	.54	.12420	.00166	.02327
.05	.02375	.00356	.00405	.55	.12375	.00206	.02318
.06	.02820	.00414	.00483	.56	.12320	.00246	.02306
.07	.03255	.00467	.00560	.57	.12255	.00286	.02293
.08	.03680	.00515	.00636	.58	.12180	.00325	.02278
.09	.04095	.00560	.00711	.59	.12095	.00363	.02260
.10	.04500	.00600	.00784	.60	.12000	.00400	.02240
.11	.04895	.00636	.00856	.61	.11895	.00436	.02218
.12	.05280	.00669	.00927	.62	.11780	.00471	.02194
.13	.05655	.00697	.00996	.63	.11655	.00505	.02169
.14	.06020	.00723	.01064	.64	.11520	.00538	.02141
.15	.06375	.00744	.01130	.65	.11375	.00569	.02111
.16	.06720	.00762	.01195	.66	.11220	.00598	.02080
.17	.07055	.00776	.01259	.67	.11055	.00626	.02046
.18	.07380	.00787	.01321	.68	.10880	.00653	.02010
.19	.07695	.00795	.01381	.69	.10695	.00677	.01973
.20	.08000	.00800	.01440	.70	.10500	.00700	.01934
.21	.08295	.00802	.01497	.71	.10295	.00721	.01893
.22	.08580	.00801	.01553	.72	.10080	.00739	.01850
.23	.08855	.00797	.01606	.73	.09855	.00756	.01805
.24	.09120	.00790	.01658	.74	.09620	.00770	.01758
.25	.09375	.00781	.01709	.75	.09375	.00781	.01709
.26	.09620	.00770	.01758	.76	.09120	.00790	.01658
.27	.09855	.00756	.01805	.77	.08855	.00797	.01606
.28	.10080	.00739	.01850	.78	.08580	.00801	.01553
.29	.10295	.00721	.01893	.79	.08295	.00802	.01497
.30	.10500	.00700	.01934	.80	.08000	.00800	.01440
.31	.10695	.00677	.01973	.81	.07695	.00795	.01381
.32	.10880	.00653	.02010	.82	.07380	.00787	.01321
.33	.11055	.00626	.02046	.83	.07055	.00776	.01259
.34	.11220	.00598	.02080	.84	.06720	.00762	.01195
.35	.11375	.00569	.02111	.85	.06375	.00744	.01130
.36	.11520	.00538	.02141	.86	.06020	.00723	.01064
.37	.11655	.00505	.02169	.87	.05655	.00697	.00996
.38	.11780	.00471	.02194	.88	.05280	.00669	.00927
.39	.11895	.00436	.02218	.89	.04895	.00636	.00856
.40	.12000	.00400	.02240	.90	.04500	.00600	.00784
.41	.12095	.00363	.02260	.91	.04095	.00560	.00711
.42	.12180	.00325	.02278	.92	.03680	.00515	.00636
.43	.12255	.00286	.02293	.93	.03255	.00467	.00560
.44	.12320	.00246	.02306	.94	.02820	.00414	.00483
.45	.12375	.00206	.02318	.95	.02375	.00356	.00405
.46	.12420	.00166	.02327	.96	.01920	.00294	.00326
.47	.12455	.00125	.02334	.97	.01455	.00228	.00246
.48	.12480	.00083	.02340	.98	.00980	.00157	.00165
.49	.12495	.00042	.02343	.99	.00495	.00081	.00083
.50	−.12500	.00000	.02344	1.00	−.00000	−.00000	.00000

TABLES AND FORMULÆ.

PART II.

GEODESY.

GEODESY.

XXV.—*Reduction to Center of Station.*

Call—

P the place of the instrument;

C the center of the station;

O the angle at P, between two objects, A and B.;

y the angle at P, between C and the *left*-hand object, B;

r the distance, C P;

C the unknown angle at C;

D the distance A C; and

G the distance B C,

then—

$$C = O + \frac{r \sin (O + y)}{D \sin 1''} - \frac{r \sin y}{G \sin 1''}$$

In the use of this formula proper attention should be paid to the signs of sin $(O + y)$ and sin y; for the first term will be *positive* when $(O + y)$ is less than 180°, (the reverse with sin y;) D being the distance of the *right*-hand object, the graduation of the instrument running from left to right.

r being small, the lengths of D and G are computed with the angle O.

XXVI.—*Reduction to Center of Signal Observed, or Correction for Phase in Tin Cones Used as Signals.*

$$\text{Correctio} = \pm \frac{r \cos^2 \frac{1}{2} Z}{D \sin 1''}$$

where—

r = radius of the signal;

Z = angle at the point of observation between the sun and the signal; and

D = the distance.

XXVII.—*Spherical Excess.*

$$E = \frac{S}{r^2 \sin 1''} = \frac{a\,b \sin C}{2r^2 \sin 1''}$$

S being the area of the triangle; r, the radius of the earth.

$$S = \frac{a\,b \sin c}{2}$$

$$= \sqrt{s\,(s-a)\,(s-b)\,(s-c)}$$

s being $= \dfrac{a+b+c}{2}$

Between latitudes 45° and 25° the spherical excess amounts to about 1″ for an area of 75·5 square miles.

Hence, if the area in square miles be known, a close approximation to the spherical excess will be had by dividing the area by 75·5.

log mean radius of the earth in yards = 6.8427917

If the three angles of a triangle are assumed to have been equally well determined, the previous determination of the spherical excess is not necessary for the calculation of the sides, though it will be required for estimating the relative accuracy of the observations; for the sides of a spherical triangle may be computed as if they were rectilineal when one-third the excess of the sum of the three angles above 180° is deducted from each of the three observed angles. Then—

$$\text{side } b = \text{side } a \sin\left(B - \tfrac{1}{3}E\right) \div \sin\left(A - \tfrac{1}{3}E\right)$$

For large triangles:

$$E = \frac{a\,b \sin C\,(1 + e^2 \cos 2\,L)}{2\,A^2 \sin 1''}$$

A being the equatorial radius, and L the mean latitude of the three stations.

XXVIII.—*To Reduce the Length of an Inclined Base to Horizontal Measure.*

Let—

B be the length of the base on the inclined plane;

b that reduced to the horizontal plane; and

θ the inclination,

then—

$$b = \text{B} \cos \theta$$

But as θ is generally a small angle, and need not be known with extreme precision, it is better to compute the excess of B above b; and, supposing θ to be given in minutes,

$$\text{B} - b = \text{B}(1 - \cos \theta) = 2\,\text{B} \sin^2 \frac{\theta}{2} = \tfrac{1}{2}\,\text{B}\,\theta^2 \sin^2 1' = \frac{\sin^2 1'}{2}\,\theta^2\,\text{B}$$

or, $\text{B} - b = 0.00000004231\ \theta^2\ \text{B}$

or, by logarithms,

$$\log(\text{B} - b) = \text{const} \log \bar{2}.626422 + 2 \log \theta + \log \text{B}$$

XXIX.—*To Reduce a Broken Base to a Straight Line.*

Let—

a and b be the given sides; and

C the contained angle, very nearly 180°.

Make $\text{C} = 180° - \theta$; θ being small, and $\cos \theta = 1 - \frac{1}{2}\theta^2$,

then—

$$\text{side } c = a + b - \frac{\sin^2 1'}{2} \cdot \frac{a\,b\,\theta^2}{a+b}$$

$$= a + b - 0.00000004231 \times \frac{a\,b\,\theta}{a+b}$$

θ being expressed in minutes.

$$\log 0.00000004231 = \bar{2}.6264222$$

XXX.—*To Find the Length,* B D $= x$, *of a Portion of a Straight Line,* A H, *Knowing the Two Other Portions,* A B $= a$, D H $= b$, *and also the Angles* α, β, γ, *from any Exterior Station,* C, *between* B *and* A, D *and* A, *and* H *and* A.

The problem being intended to supply by observation any portion of a base which cannot be directly measured—

$$\tan^2 \varphi = \frac{4\,a\,b}{(a-b)^2} \times \frac{\sin\beta \sin(\gamma-\alpha)}{\sin\alpha \sin(\gamma-\beta)}$$

$$x = -\frac{a+b}{2} \pm \frac{a-b}{2\cos\varphi}$$

XXXI.—*To Reduce a Measured Base to the Level of the Sea.*

Let—

r represent the radius of the earth (or better, the normal, N,) corresponding to the base b at the level of the sea; and

$r+a$ the radius referred to the level of the measured base B,

then—

$$r+a : r :: \mathrm{B} : b = \mathrm{B} \times \frac{r}{r+a}$$

and—

$$\mathrm{B} - b = \mathrm{B} - \mathrm{B}\frac{r}{r+a} = \mathrm{B} \times \left(\frac{a}{r} - \frac{a^2}{r^2} + \text{etc.}\right)$$

But the radius of the earth being very great in comparison to the difference of level, a, we have the correction δ sufficiently accurate by retaining only the first term; hence—

$$\delta = \frac{\mathrm{B}\,a}{r}$$

XXXII.—*Correction for Temperature in Metallic Rods.*

Let—

e = the linear expansion for 1° of Fahrenheit;

l = the length of the rod before expansion;

l' = the length of the rod after expansion;

t = the number of degrees Fahrenheit,

then—

$$\text{total expansion} = e\,t$$

and—

$$l' = l\,(1 + e\,t)$$

The following expansions were adopted by Mr. Hassler in his comparisons of weights and measures, (report of 1832:)

	Expansion for 1° F. = e	For 1° in a yard's length.	
Platinum	0.0000051344	0.0001848384	English inches.
Brass bar	0.00001050903	0.00037832508	"
Iron bar	0.000006963535	0.000240687260	"

Other authorities:

	Expansion for 1° F. = e	For 1° in a yard's length.		
Brass bar	0.000010480	0.0003772800	Eng. in.	Bailey.
Brass rod	0.0000105155	0.0003785580	"	Roy.
Brass rod	0.0000106666	0.0003839976	"	Troughton.
Brass wire	0.0000107407	0.0003866652	"	Smeaton.
Iron bar	0.0000069907	0.0002516652	"	Smeaton.
Steel rod	0.0000063596	0.0002289456	"	Roy.
Glass barometer-tubes	0.0000043119	0.0001552284	"	Roy.
White Norway pine	0.0000022685	0.0000816660	"	Kater.

XXXIII.—*Measurement of Distances by Sound.*

The *velocity* of sound, in one second of time, at 32° Fahrenheit, is about 1090 English feet. For any higher temperature,

$$v = 1089^{\text{ft}}.42\ \sqrt{1 + (t - 32°) \times 0.00208}$$

t being the temperature in degrees Fahrenheit.

The velocity of sound through the air is independent of the barometric pressure, and experiments show it to be sensibly unaffected by its hygrometrical state of moisture and dryness; by the nature of the sound itself, whether produced by a blow, gunshot, the voice, or a musical instrument; by the original direction of the sound, whether, for instance, the muzzle of a gun is turned one way or the other; or by the nature and position of the ground over which the sound is conveyed.

It is affected by the wind; but, in ordinary cases, likely to be selected for experiment, its influence would be almost inappreciable.

Velocity and Force of the Wind.

Velocity in—		Pressure on 1 square foot.	Common designations of the force of the winds.
1 hour.	1 second.		
Miles.	*Feet.*	*Pounds.*	
1	1.47	0.005	Hardly perceptible.
2	2.93	0.020	Just perceptible.
3	4.40	0.044	
4	5.87	0.079	Gentle, pleasant wind.
5	7.33	0.123	
10	14.67	0.492	Pleasant, brisk breeze.
15	22.00	1.107	
20	29.34	1.968	Very brisk.
25	36.67	3.075	
30	44.01	4.429	High wind.
35	51.34	6.027	
40	58.68	7.873	Very high.
45	66.01	9.963	
50	73.35	12.300	A storm or tempest.
60	88.02	17.715	A great storm.
80	117.36	31.490	A hurricane.
100	146.70	49.200	A hurricane that tears up trees, carries buildings before it, &c.

XXXIV.—*For Reconnaissances.*

"THREE-POINT PROBLEM."

At a point, P, from whence are to be seen three points, A, C, B, forming a triangle, the elements (*i. e.*, the angles and sides) of which are known, measure the angles A P C and C P B; then, required to determine the direction and distance of the point P from each object.

Make—

$$A\ C = a;$$

$$B\ C = b;$$

$$B\,C\,A = C;$$

$$A\ P\ C = P; \text{ and}$$

$$C\ P\ B = P';$$

also, make—

$$R = 360° - P - P' - C;$$

$$x = C\ A\ P;$$

$$y = P\ B\ C.$$

Then will—

$$\cot x = \cot R \left(\frac{a \sin P'}{b \sin P \cos R} + 1 \right)$$

$$y = R - x$$

The use of these formulæ need not be embarrassing if care is taken in properly applying the signs of cos R and cot R. When R is less than 90° both cos R and cot R are plus; between 90° and 180° both are minus; between 180° and 270° cos R is minus and cot R plus; between 270° and 360° cos R is plus and cot R minus.

This problem is indeterminate when P falls upon the circumference of the circle passing through A B C. A case of this nature is of rare occurrence, however, in practice.

For the more general form of this problem, where the angles are measured from the point sought to any number of given points, to fix its position, see Coast Survey Report of 1864.

XXXV.—*For Computing the Principal Geodetic Quantities Depending on the Spheroidal Figure of the Earth at any Given Latitude.*

Eccentricity of the Earth $= e = \left(\frac{a^2 - b^2}{a^2}\right)^{\frac{1}{2}}$

$$= \left(1 - \frac{b^2}{a^2}\right)^{\frac{1}{2}} = 2\text{ E} - \text{E}^2$$

Ellipticity $= \text{E} = \frac{a - b}{a} = 1 - \frac{b}{a}$

or, very nearly—

$$e^2 = 2\text{E}\ ; \qquad \text{E} = \frac{e^2}{2}$$

Normal ending at minor axis (or radius of curvature of a section perpendicular to the meridian) $=$ N

$$= \frac{a}{(1 - e^2 \sin^2 \text{L})^{\frac{1}{2}}}$$

Normal ending at major axis..... $= \text{N}' = \text{N}\ (1 - e^2)$

$$= \frac{a\ (1 - e^2)}{(1 - e^2 \sin^2 \text{L})^{\frac{1}{2}}}$$

Tangent ending at minor axis... $= t \ = \text{N} \cot \text{L}$

Tangent ending at major axis.... $= \text{T} \ = \text{N} \tan \text{L}\ (1 - e^2)$

Radius of the parallel.......... $= \rho \ = \text{N} \cos \text{L}$

Radius of curvature of the meridian $= \text{R} = \frac{\text{N}^3}{a^2}\ (1 - e^2)$

$$= \frac{a\ (1 - e^2)}{(1 - e^2 \sin^2 \text{L})^{\frac{3}{2}}}$$

Radius of curvature of a section making an angle Z with the meridian........... $= \text{R}_z$

$$= \frac{\text{N R}}{\text{N} \cos^2 \text{Z} + \text{R} \sin^2 \text{Z}}$$

Radius of the earth.......... $= r$

$$= a \left(1 - \frac{e^2\ (1 - e^2) \sin^2 \text{L}}{1 - e^2 \sin^2 \text{L}}\right)^{\frac{1}{2}}$$

Equatorial radius.............. $= a$

Polar radius.................... $= b$

The given latitude............. $= \text{L}$

XXXVI.—*Numerical Values of Some of the Preceding Quantities, from a Discussion by* BESSEL *in the "Astronomische Nachrichten," No.* 438.

$$a = \text{equatorial radius} = 3272077.14 \text{ toises}$$
$$\log = 6.5148235337$$
$$b = \text{polar radius} = 3261139.33 \text{ toises}$$
$$\log = 6.5133693539$$

Ratio of the toise to the metre, law of France, December 10, 1799:

$$T = 1^{m}.9490363; \qquad \log = 0.2898199300$$

whence in metres—

$$a = 6377397^{m}.15; \qquad \log = 6.8046434637$$
$$b = 6356078^{m}.96; \qquad \log = 6.8031892839$$

Ratio of the axes:

$$a : b :: 299.1528 : 298.1528$$

Mean uncertainty $= \pm$ 4.667 units.

Length of the earth's quadrant $= 5131179^{t}.81 = 10000855^{m}.76$

Mean uncertainty $= \pm 498^{m}.23$

$$e = \text{eccentricity} = \left(1 - \frac{b^2}{a^2}\right)^{\frac{1}{2}} = 0.0816967$$
$$\log = 8.9122052271$$
$$E = \text{ellipticity} = \tfrac{1}{2}\, e^2 \qquad \log = 7.5233789824$$

Length, in toises, of a meridional degree whose middle latitude is φ:

$$D_m = 57013^{t}.109 - 286^{t}.337 \cos 2\varphi + 0^{t}.611 \cos 4\varphi + 0^{t}.001 \cos 6\varphi$$

Length of a degree of the parallel, in toises:

$$D_p = 57156^{t}.285 \cos \varphi - 47^{t}.825 \cos 3\varphi + 0^{t}.060 \cos 5\varphi$$

or, making $\sin \psi = e \sin \varphi$—

$$\log D_p = 4.7567009.0 + \log \cos \varphi - \log \cos \psi$$

XXXVII.—*Relative Lengths of the Yard and the Metre.*

1. From Clarke's comparisons referred to the present parliamentary standard, (*Comparison of Standards of Length made at the Ordnance Survey Office by Captain A. R. Clarke, R. E., F. R. S., published by authority*, 1866:)

Values Adopted in the Measurement, now in Progress, of an Arc of Parallel Extending from Ireland to the River Ural in Russia, as "the Exact Relative Lengths of Standards" used as the Units of Measure in the Triangulations of England, France, Belgium, Prussia, and Russia:

Standards.	Expressed in terms of the standard yard.	Expressed in inches.	Expressed in lines of the toise.	Expressed in millimetres.
The yard	1.00000000	36.000000	405.34622	914.39180
The toise	2.13151116	76.734402	864.00000	1949.03632
The metre	1.09362311	39.370432	443.29600	1000.00000

$$\log 39.370432 = 1.5951701816$$

2. From Kater's comparisons with the Shuckburg scale, (*Phil. Trans. for* 1818:)

1 metre at 32° F. = 39.370790 inches of the old imperial standard at 62° F.

$$\log = 1.5951741293$$

3. From Hassler's comparisons of the Troughton 82-inch scale, with the iron standard *committee metre* of the American Philosophical Society, (*Report of the Secretary of the Treasury on the Comparison of Weights and Measures, Twenty-Second Congress, First Session, June* 20, 1832:)

Value adopted by the United States Coast Survey.

1 metre at 32° F. = 39.36850535 inches of the Troughton 82-inch scale at 62° F.

$$\log = 1.5951489586$$

a value materially smaller than the preceding.

There is a doubt whether this discordance is to be attributed to inaccuracy in the length of the Troughton scale or in errors in the co-efficients of expansion used by Mr. Hassler.

XXXVII.—*Relative Lengths of the Yard and the Metre*—Con'd.

Logarithms to Reduce Metres to Yards.

Clarke.	Kater.	Coast Survey.
0.0388676809	0.0388716286	0.0388464579

log 3 = 0.4771212547
log 12 = 1.0791812460
log 5280 = 3.7226339225

Kater's length of the metre in English inches was adopted in the preparation of the first edition of this Collection as being at the time most generally in use. It has been retained throughout the present volume, although the results of Clarke's comparisons should now be universally adopted.

The *committee metre* of the American Philosophical Society is the unit of length to which all linear measures of the Coast Survey are referred. It was compared August 24, 1867, at Paris, directly with the standard platinum metre of the *Conservatoire des arts et métiers*, and was found (at the temperature of melting ice) = $1^{m}.00000336$ of the *platinum metre of the archives*.

The French *standard metre* has its normal length at zero centigrade, or the freezing-point. It was intended to be a natural standard, and to represent the ten-millionth part of the terrestrial arc between the equator and the pole, which was assumed to be 5130740 toises, and the length of the metre 443.29596 lines of the *toise du Pérou;* which quantity was declared by law in 1799 to be the length of the *legal* metre, and is the length of the standard *platinum metre of the archives.*

The *toise du Pérou*, made in 1735, is a bar of iron, and has its standard length at 13° Reaumur, (61°.25 Fahrenheit.) It was used by La Condamine in the measurement of an arc of the meridian in Peru in 1744. As the above determination of the length of the quarter of the meridian is now known to be erroneous, the *legal* metre becomes, in fact, but a legalized part of the *toise du Pérou*, and this last remains the primitive standard. It is the unit of length in which the greater part of the European geodetic measurements are expressed.

The standard *Klafter* of *Vienna* has its normal length at 13° Reaumur, and is = 840.76134 lines of the *toise du Pérou.*

The standard *Prussian foot* is a standard also at 13° Reaumur, and was declared by law to be = 139.13 lines of the *toise du Pérou.*

XXXVIII.—*Numerical Values of Bessel's Terrestrial Elements in English Yards, adopting Kater's Value of the Metre, viz:* 39.37079 *English Inches; log* 1.5951741293.

Log. to reduce toises to yards = 0.3286915586

Log. to reduce metres to yards = 0.0388716286

a = equatorial radius = $6974532^{y}.339$

log = 6.8435150923

b = polar radius = $6951218^{y}.059$

log = 6.8420609125

Length, in yards, of a meridional degree whose middle latitude is φ:

$$D_m = 121525^{y}.183 - 610^{y}.336 \cos 2\varphi + 1^{y}.302 \cos 4\varphi + 0^{y}.002 \cos 6\varphi$$

Length, in yards, of a degree of the parallel:

$$D_p = 121830^{y}.366 \cos\varphi - 101^{y}.941 \cos 3\varphi + 0^{y}.128 \cos 5\varphi$$

or, making $\sin\psi = e\sin\varphi$—

$$\log D_p = 5.0853925 + \log\cos\varphi - \log\cos\psi$$

or, using the logarithms of the numerical co-efficients—

$$D_m = 121525^{y}.183 - (2.7855691)\cos 2\varphi + (0.1147)\cos 4\varphi + (7.3287)\cos 6\varphi$$

$$D_p = (5.0857556)\cos\varphi - (2.00835)\cos 3\varphi + (9.1069)\cos 5\varphi$$

or—

$$D_p = \frac{(5.0853925)\cos\varphi}{\cos\psi}$$

Constant Logarithms.

$e^2 = 0.00667435$	log = 7.8244104542
$\frac{1}{2}e^2 = E$ = ellipticity = $\frac{1}{299.66}$	= 7.5233789824
$\sin 1''$	= 4.6855748668
$\frac{1}{2}\sin 1''$	= 4.3845448711
$\frac{3e^2}{2}\sin 1''$	= 2.6860751039
$(1 - e^2) = 0.99332565$	= 9.9970916404
$a(1 - e^2)$	= 6.8406067325
$a \sin 1''$	= 1.5290899591
$a \sin 1''$, (arithmetical complement)	= 8.4709100409

XXXIX.—*For Computing the Geodetic Latitudes, Longitudes, and Azimuths of Points of a Triangulation.*

1. For distances not exceeding one hundred miles:

$$-d\,L = K\,B \cos Z + K^2\,C \sin^2 Z + (\delta\,L)^2\,D - K^2\,h\,E \sin^2 Z$$

where—

$$B = \frac{1}{R \text{ arc } 1''}$$

$$C = \frac{\tan L}{2\,N\,R \text{ arc } 1''}$$

$$D = \frac{\frac{3}{2}\,e^2 \sin L \cos L \text{ arc } 1''}{(1 - e^2 \sin^2 L)^{\frac{3}{2}}}$$

$$E = \frac{1 + 3 \tan^2 L}{6\,N^2}$$

$h = K\,B \cos Z$, or first term;

$\delta\,L$ an approximate value for $-d\,L$, computed from first and second term;

$$d\,M = \frac{A'\,K \sin Z}{\cos L'}.$$

$$A' = \frac{1}{N \text{ arc } 1''} \text{ referred to second point;}$$

$$-d\,Z = \frac{d\,M \sin \frac{1}{2}\,(L + L')}{\cos \frac{1}{2}\,d\,L} + d\,M^3\,F$$

log F for latitude $25° = 7.8324$; for latitude $45° = 7.8404$.

2. For distances not exceeding twenty miles:

In terms of the sides of the triangles:

$$u'' = \frac{K}{N \sin 1''} = \frac{K\,(1 - e^2 \sin^2 L)^{\frac{1}{2}}}{a \sin 1''}$$

$$\left.\begin{aligned} L' = L &- (1 + e^2 \cos^2 L)\,u'' \cos Z \\ &- (1 + e^2 \cos^2 L)\,(u'' \sin Z)^2 \tan L \times \tfrac{1}{2} \sin 1'' \end{aligned}\right\}$$

$$M' = M + \frac{u'' \sin Z}{\cos L'}$$

$$Z' = 180° + Z - \frac{u'' \sin Z}{\cos L'} \sin \tfrac{1}{2}\,(L + L')$$

XXXIX.—*For Computing Geodetic Latitudes, &c.*—Continued.

In terms of the co-ordinates of rectangular axes referred to one of the points of the triangulation, the latitude and longitude of which are known; y being the ordinate in the direction of the meridian, and x the ordinate perpendicular to it:

$$L' = L \pm \frac{y}{R \sin 1''} - \tfrac{1}{2} \sin 1'' \left(\frac{x}{N \sin 1''} \right)^2 \times \tan \left(L \pm \frac{y}{R \sin 1''} \right)$$

$$M' = M \pm \left(\frac{x}{N \sin 1''} \right) \times \frac{1}{\cos L'}$$

$$Z' = (180 + Z) \pm \frac{x}{N \sin 1''} \tan L'$$

K = distance in yards between two stations, the latitude and longitude of one of which is known, and u'' this same distance converted to seconds of arc;

L = latitude of 1st station;

M = longitude of 1st station, + if west;

Z = azimuth of 2d station at 1st, counted from the south round by the west, from 0° to 360°; the algebraic signs of the sine and cosine of this angle must be carefully attended to;

L′, M′, Z′ the same things at 2d station, or quantities required;

a = the equatorial radius;

e = the eccentricity;

R = the radius of curvature of the meridian; and

N = the radius of curvature of a section perpendicular to the meridian.

The quantity—

$$\frac{u'' \sin Z}{\cos L'} \sin \tfrac{1}{2} (L + L')$$

or—

$$(M' - M) \sin \tfrac{1}{2} (L + L')$$

by which the azimuth at one end of a line exceeds the azimuth at the other, is called *the convergence of the meridians.*

XL.—*To Compute the Length and Direction of a Line Joining Two Points, the Latitudes and Longitudes of which are Known, or Measurement of a Base by Astronomical Observations.*

$$\frac{\beta}{2} = \frac{e^2 (L - L') \cos^2 \frac{1}{2} (L + L')}{2}$$

$$N = \frac{a}{[1 - e^2 \sin^2 \frac{1}{2} (L + L')]^{\frac{1}{2}}}$$

$$l = L - \frac{\beta}{2}$$

$$x'' = (M' - M) \cos l'$$

$$l' = L' + \frac{\beta}{2}$$

$$y'' = (l - l') - \frac{1}{2} \sin 1'' \, x''^2 \tan l$$

$$\tan Z = \frac{x''}{y''}$$

$$x = x'' \, N \sin 1''$$

$$u'' = \frac{x''}{\sin Z} = \frac{y''}{\cos Z}$$

$$y = y'' \, N \sin 1''$$

$$K = u'' \, N \sin 1''$$

in which—

L, L', M, M', represent the latitudes and longitudes of the two points;

u'', the distance between these points in seconds of arc;

K, the distance between these points in linear units;

x'', the number of seconds in the arc passing through the point of which L' is the latitude, and perpendicular to the meridian of the point of which L is the latitude;

y'', the seconds in the portion of this meridian between L and the foot of this perpendicular;

x, y, the same quantities in linear units;

Z, the azimuth of the second point, L', from the first, L; and

N, the normal at the middle latitude.

XL.—*To Compute the Length and Direction of a Line, &c.*—Con'd.

Particular attention must be paid to the sign (L — L′), for upon this depends the sign of $\frac{\beta}{2}$, and also to that of $(l - l')$ in the value of y'', so as to know whether the small quantity—

$$(-\tfrac{1}{2} \sin 1'' \, x''^{2} \tan l)$$

is to be added to or substracted from $(l - l')$.

The azimuth Z is counted from the south, round by the west, from 0° to 360°.

The azimuth Z′ (if required) is to be computed from Z, as in XXXIX, (2.)

This can be presented in a different form, thus:

$$\text{as, } (M' - M) \cos L' = u'' \sin Z$$

and,

$$u'' \cos Z = \frac{(L - L') - \frac{1}{2} u''^{2} \sin^{2} Z \cos^{2} L' \tan L \sin 1'' (1 + e^{2} \cos^{2} L)}{1 + e^{2} \cos^{2} L}$$

Substituting, in this last, the value of u'' sin Z, and dividing one by the other:

$$\tan Z = \frac{(M' - M) \cos L' (1 + e^{2} \cos^{2} L)}{(L - L') - \frac{1}{2} (M' - M)^{2} \cos^{2} L' \tan L \sin 1'' (1 + e^{2} \cos^{2} L)}$$

Then, knowing Z—

$$u'' = \frac{(M' - M) \cos L'}{\sin Z}$$

and—

$$K = u'' N \sin 1''$$

N being the normal for the mean latitude.

XLI.—*To Compute the Distance between Two Points, knowing their Latitudes and the Azimuth of one from the other.*

$$\frac{\beta}{2} = \frac{e^{2} (L - L') \cos^{2} \frac{1}{2} (L + L')}{2}$$

$$N = \frac{a}{[1 - e^{2} \sin^{2} \frac{1}{2} (L + L')]^{\frac{1}{2}}}$$

$$l = L - \frac{\beta}{2}; \qquad \tan \varphi = \frac{\tan l}{\cos Z}; \qquad l' = L' + \frac{\beta}{2}$$

$$\sin (\varphi - u'') = \frac{\sin l'}{\sin l} \sin \varphi; \quad K = u'' N \sin 1''$$

See the note to the preceding formulæ. The algebraic sign of the azimuth, Z, will determine the sign of φ, and consequently whether the quantity u'' is to be added to or subtracted from φ.

XLII.—*To Compute the Distance between Two Points, knowing the Latitude of one, the Azimuth from this to the other, and the Difference of their Longitudes.*

$$\tan \varphi = \sin \mathrm{L} \tan \mathrm{Z} \qquad \tan \mathrm{L}'' = \frac{\tan \mathrm{L} \sin (\varphi - m)}{\sin \varphi}$$

$$\beta = e^2 (\mathrm{L} - \mathrm{L}'') \cos^2 \tfrac{1}{2} (\mathrm{L} + \mathrm{L}'')$$

$$\mathrm{L}' = \mathrm{L}'' - \beta \qquad l = \mathrm{L} - \frac{\beta}{2} \qquad l' = \mathrm{L}' + \frac{\beta}{2}$$

$$u'' = \frac{m \cos l'}{\sin \mathrm{Z}} \qquad \mathrm{K} = u'' \mathrm{N} \sin 1''$$

m = the difference of longitude.

The azimuth, Z, is, as before, counted from the south round by the west; its algebraic sign will determine the sign of φ, and consequently whether it is to be increased or diminished by m.

Elements of the Figure of the Earth, Deduced by Captain A. R. Clarke, Royal Engineers, in Computing the Figures of the Meridians and of the Equator for Several Measured Arcs of Meridian, (Comparison of Standards of Length, &c., London, 1866.)

Semi-axes.	Length.		
	Feet.	*Toises.*	*Metres.*
Major semi-axis = a, of equator, (longitude 15° 34′ east)	20926350	3272537.3	6378294.0
Minor semi-axis = b, of equator, (longitude 105° 34′ east)	20919972	3271540.1	6376350.4
Polar semi-axis = c	20853429	3261133.8	6356068.1

$$\frac{a-c}{c} = \frac{1}{285.97} \qquad \frac{b-c}{c} = \frac{1}{313.38} \qquad \frac{a-b}{c} = \frac{1}{3269.5}$$

The length of the meridian quadrant passing through Paris is.......10001472.5 metres.
and the minimum quadrant, in longitude 105° 34′, is.......10000024.5 metres.

For a Spheroid of Revolution more nearly Corresponding to the Same Geodetic Measurements.

Semi-axes.	Length.		
	Feet.	*Toises.*	*Metres.*
Equatorial semi-axis = a	20926062	3272492.3	6378206.4
Polar semi-axis = b	20855121	3261398.4	6356583.8

$$\frac{b}{a} = \frac{293.98}{294.98} \qquad \frac{a-b}{a} = \frac{1}{294.98} = \text{ellipticity.}$$

FORMS FOR RECORD

Survey of ....

No. of triangle.	Position.	Names of stations.	No. of obs.	Observed angles.	Errors and their distribution.	Spherical angles.	Spherical excess.	Final plane angles.
				° ′ ″	″	″	″	° ′ ″
	Sought.	Cedar Point..	18	66 34 04.80	−0.36	04.44	1.58	66 34 02.86
XIII	Right..	Buck Hill ...	18	64 08 37.78	−0.36	37.43	1.58	64 08 35.84
	(Known side.)							
	Left ...	Fort Flats ...	18	49 17 23.24	−0.36	22.88	1.58	47 17 21.30
								180 0 0.00

Survey of ....

Names of stations.	Latitudes.	
	$L' = L - u''.(1 + e^2 \cos^2 L) \cos Z - \frac{1}{2} \sin 1'' \sin^2 Z\, u''^2 (1 + e^2 \cos^2 L) \tan L$	
Fort Flats...	Latitude L .... = 45° 39′ 13″.89	
	log K (yards) .... = 4.7295212	$\frac{1}{2}$ sin 1″ .... = 4.38454
	$\log \frac{1}{N \sin 1''}$ .... = 8.4701676	2 log sin Z ...... = 9.09522
	log u'' .... = 3.1996888	2 log u'' .... = 6.39936
	log $(1 + e^2 \cos^2 L)$ .... = 0.0014140	.... = 0.00141
	log cos Z .... (−) = 9.9711240	log tan L .... = 0.00991
	log 1st term .... = 3.1722268	log 2d term ..... = 9.89034
	1st term .... (+) = +1486″.71	2d term .... = 0″.77
	2d term .... (−) = − 0″.77	
	δ L .... = 0° 24′ 45″.94	
	L .... = 45° 39′ 13″.89	L + L′ ... = 91° 43′ 13″.72
Cedar Point.	Latitude L′ .... = 46° 03′ 59″.83	$\frac{L + L'}{2}$ ... = 45° 51′ 36″.86

AND COMPUTATION.

Calculation of Triangles.

Computing letter.	Logarithms of their sines.	Calculation of the sides.	Sides in yards.	Designation.
S	9.9626198	log R L ....= 4.7379524 comp log sin S ..= 0.0373802 log sin R ....= 9.9541886	= 54695.61	Buck Hill—Fort Flats.
R	9.9541886	log L S ....= 4.7295212	= 53644.00	Fort Flats—Cedar Point.
		log R L + comp log sin S = 4.7753326 log sin L ....= 9.8796760		
L	9.8796760	log R S ....= 4.6550086	= 45186.49	Buck Hill—Cedar Point.

Geodetic Determination of Positions. (Secondary.)

Longitudes.	Azimuths.	Remarks.
$M' = M + \dfrac{u'' \sin Z}{\cos L'}$	$Z' = 180° + Z - (\delta M) \sin \dfrac{L + L'}{2}$	
Longitude M = 84° 42′ 22″.19	Azimuth Z .....= 159° 20′ 13″.62	
log sin Z ....(+) = 9.5476117	180°	
log u'' ....= 3.1996888	180° + Z ......= 339° 20′ 13″.62	
	20° 39′ 46″.38	
2.7473005		
log cos L′ ....= 9.8412474	log sin $\frac{L + L'}{2}$ ......= 9.8559089	
log δ M ....= 2.9060529	...............(+) = 2.9060529	
	log δ Z ....= 2.7619618	
	− 578″.05	
δ M ....= 0° 13′ 25″.48	δ Z............= 0° 09′ 38″.05	
M ....= 84° 42′ 22″.19	180° + Z ......= 339° 20′ 13″.62	
Longitude M = 84° 55′ 47″.67	Azimuth Z′ ....= 339° 10′ 35″.57	

Normal or Radius of Curvature of the Perpendicular to the Meridian.

Ellipticity $= \frac{1}{300}$; equatorial radius $= 6974532$ yards; log $= 6.8435151$.
Log. to reduce yards to metres $= 9.9611283714$.

Latitude.		$N = \frac{a}{(1 - e^2 \sin^2 L)^{\frac{1}{2}}}$			$\log (1 + e^2 \cos^2 L)$	Difference for 10′.
		log N	Com. diff. for 10′.	$\log \frac{1}{N \sin 1''}$		
°	′					
20	0	6. 8436847	27. 3	8. 4707404	0. 0025521	55
	15	6888	27. 6	7363	5439	56
	30	6929	27. 8	7322	5356	55
	45	6971	28. 1	7280	5274	56
21	0	7013	28. 4	7238	5191	57
	15	7056	28. 7	7196	5106	57
	30	7099	29. 0	7153	5021	58
	45	7142	29. 3	7109	4934	58
22	0	7186	29. 5	7066	4847	58
	15	7230	29. 7	7021	4760	59
	30	7274	30. 0	6977	4671	59
	45	7319	30. 2	6932	4582	60
23	0	7365	30. 5	6887	4492	61
	15	7410	30. 7	6841	4401	62
	30	7457	31. 0	6795	4309	62
	45	7503	31. 2	6748	4217	62
24	0	7550	31. 5	6701	4124	63
	15	7597	31. 7	6654	4030	63
	30	7645	32. 0	6607	3935	63
	45	7693	32. 2	6559	3840	64
25	0	7741	32. 5	6510	3744	64
	15	7790	32. 7	6462	3648	65
	30	7839	32. 9	6413	3550	65
	45	7888	33. 2	6363	3452	66
26	0	7938	33. 4	6313	3353	66
	15	7988	33. 6	6263	3254	67
	30	8038	33. 8	6213	3154	67
	45	8089	34. 0	6162	3053	68
27	0	8140	34. 3	6111	2951	68
	15	8192	34. 5	6060	2849	69
	30	8243	34 7	6008	2746	69
	45	8295	34. 9	5956	2643	69
28	0	8348	35. 1	5904	2539	70
	15	8400	35. 3	5851	2434	70
	30	8453	35. 5	5798	2329	71
	45	6. 8438507	35. 7	8. 4705745	0. 0022223	71

Normal, &c.—Continued.

Latitude.		$N = \frac{a}{(1 - e^2 \sin^2 L)^{\frac{1}{2}}}$			$\log (1 + e^2 \cos^2 L)$	Difference for 10′.
		log N	Com. diff. for 10′.	$\log \frac{1}{N \sin 1''}$		
°	′					
29	0	6.8438560	35.9	8.4705691	0.0022117	71
	15	8614	36.1	5637	2010	72
	30	8668	36.3	5583	1902	72
	45	8723	36.5	5529	1794	72
30	0	8777	36.7	5474	1686	73
	15	8832	36.9	5419	1576	73
	30	8888	37.1	5364	1466	73
	45	8943	37.2	5308	1356	74
31	0	8999	37.4	5252	1245	74
	15	9055	37.6	5196	1134	75
	30	9111	37.7	5140	1022	75
	45	9168	37.9	5084	0910	75
32	0	9225	38.1	5027	0797	75
	15	9282	38.2	4970	0684	76
	30	9339	38.4	4912	0570	77
	45	9397	38.5	4855	0455	77
33	0	9454	38.7	4797	0340	77
	15	9512	38.9	4737	0225	77
	30	9571	39.0	4681	.0020109	77
	45	9629	39.1	4622	.0019993	77
34	0	9688	39.2	4564	9877	78
	15	9747	39.4	4505	9760	78
	30	9806	39.5	4446	9643	79
	45	9865	39.7	4387	9525	79
35	0	9924	39.8	4327	9407	80
	15	.8439984	40.0	4267	9288	80
	30	.8440044	40.0	4208	9169	80
	45	0104	40.1	4148	9050	80
36	0	0164	40.3	4087	8931	80
	15	0224	40.3	4027	8811	81
	30	0285	40.5	3966	8690	80
	45	0346	40.6	3906	8570	81
37	0	0406	40.7	3845	8449	81
	15	0467	40.7	3784	8328	81
	30	0529	40.9	3723	8206	81
	45	0590	41.0	3661	8084	81
38	0	0651	41.1	3600	7963	82
	15	0713	41.1	3538	7840	82
	30	0775	41.2	3477	7717	82
	45	0837	41.4	3415	7594	82
39	0	0898	41.4	3353	7471	82
	15	0961	41.4	3291	7348	83
	30	1023	41.5	3229	7224	82
	45	6.8441085	41.6	8.4703166	0.0017101	83

Normal, &c.—Continued.

Latitude.		$N = \frac{a}{(1 - e^2 \sin^2 L)^{\frac{1}{2}}}$			$\log (1 + e^2 \cos^2 L)$	Difference for 10′.
		log N	Com. diff. for 10′.	$\log \frac{1}{N \sin 1''}$		
°	′					
40	0	6.8441147	41.7	8.4703104	0.0016977	83
	15	1210	41.8	3041	6853	84
	30	1273	41.8	2979	6728	83
	45	1335	41.9	2916	6604	84
41	0	1398	41.9	2853	6479	84
	15	1461	41.9	2791	6354	84
	30	1524	42.0	2728	6229	84
	45	1587	42.1	2665	6104	84
42	0	1650	42.1	2602	5979	84
	15	1713	42.1	2539	5853	84
	30	1776	42.2	2475	5728	84
	45	1839	42.1	2412	5602	84
43	0	1903	42.2	2349	5477	84
	15	1967	42.2	2286	5351	84
	30	2029	42.3	2222	5225	84
	45	2093	42.3	2159	5099	84
44	0	2156	42.3	2095	4973	84
	15	2219	42.3	2032	4847	84
	30	2283	42.3	1969	4721	84
	45	2346	42.3	1905	4595	84
45	0	2410	42.3	1842	4469	84
	15	2473	42.3	1778	4343	84
	30	2537	42.3	1715	4217	84
	45	2600	42.3	1651	4091	84
46	0	2663	42.3	1588	3965	84
	15	2727	42.3	1525	3839	84
	30	2790	42.3	1461	3713	84
	45	2854	42.2	1398	3587	84
47	0	2917	42.2	1334	3461	84
	15	2980	42.1	1271	3336	84
	30	3043	42.1	1208	3210	84
	45	3107	42.1	1145	3084	84
48	0	3170	42.1	1082	2959	84
	15	3233	42.0	1018	2833	84
	30	3296	42.0	0955	2708	84
	45	3359	41.9	0892	2583	84
49	0	3422	41.9	0830	2458	84
	15	3485	41.9	0767	2333	83
	30	3547	41.8	0704	2209	84
	45	3610	41.8	0641	2084	83
50	0	6.8443673		8.4700579	0.0011960	

Radius of Curvature of the Meridian.

Ellipticity $= \frac{1}{300}$; equatorial radius $= 6974532$ yards.

$$R = \frac{a(1 - e^2)}{(1 - e^2 \sin^2 L)^{\frac{3}{2}}}$$

Latitude.		log R	Com. diff. for 10′.	$\log \frac{1}{R \sin 1''}$
°	′			
20	0	6.8411155	81.9	8.4733096
	15	1278	82.7	2973
	30	1402	83.5	2849
	45	1527	84.3	2724
21	0	1654	85.1	2598
	15	1781	86.0	2470
	30	1910	86.8	2341
	45	2040	87.5	2211
22	0	2172	88.4	2080
	15	2304	89.1	1947
	30	2438	90.0	1813
	45	2573	91.0	1679
23	0	2709	91.5	1543
	15	2846	92.3	1405
	30	2984	93.0	1267
	45	3124	93.6	1128
24	0	3264	94.6	0987
	15	3406	95.2	0845
	30	3549	96.0	0702
	45	3693	96.7	0559
25	0	3838	97.4	0414
	15	3984	98.1	0268
	30	4131	98.8	.4730120
	45	4279	99.4	.4729972
26	0	4428	100.1	9823
	15	4578	100.9	9673
	30	4730	101.5	9522
	45	4882	102.1	9370
27	0	5035	102.8	9216
	15	5189	103.4	9062
	30	5344	104.1	8907
	45	5500	104.7	8751
28	0	5657	105.3	8594
	15	5815	106.0	8436
	30	5974	106.5	8277
	45	6.8416134	107.1	8.4728117

Radius of Curvature of the Meridian—Continued.

$$R = \frac{a(1-e^2)}{(1-e^2 \sin^2 L)^{\frac{3}{2}}}$$

Latitude.		log R	Com. diff. for 10′.	$\log \frac{1}{R \sin 1''}$
°				
29	0	6.8416295	107.1	8.4727956
	15	6456	108.2	7795
	30	6619	108.6	7632
	45	6782	109.4	7469
30	0	6946	110.0	7305
	15	7111	110.5	7140
	30	7277	111.1	6974
	45	7444	111.6	6808
31	0	7611	112.2	6640
	15	7779	112.7	6472
	30	7948	113.1	6303
	45	8118	113.7	6133
32	0	8288	114.1	5963
	15	8460	114.6	5792
	30	8632	115.1	5620
	45	8804	115.6	5447
33	0	8978	116.0	5274
	15	9152	116.5	5100
	30	9326	116.9	4925
	45	9502	117.3	4750
34	0	9678	117.7	4574
	15	.8419854	118.1	4397
	30	.8420031	118.5	4220
	45	0209	119.1	4042
35	0	0387	119.3	3864
	15	0566	119.7	3685
	30	0746	120.0	3506
	45	0926	120.4	3325
36	0	1107	120.7	3145
	15	1288	121.1	2964
	30	1469	121.4	2782
	45	1651	121.7	2600
37	0	1834	122.0	2417
	15	2017	122.3	2234
	30	2200	122.7	2051
	45	2384	122.9	1867
38	0	2569	123.1	1683
	15	2753	123.5	1498
	30	2939	123.7	1313
	45	3124	124.0	1127
39	0	3310	124.1	0941
	15	3496	124.4	0755
	30	3683	124.6	0568
	45	6.8423870	124.8	8.4720382

Radius of Curvature of the Meridian—Continued.

$$R = \frac{a(1-e^2)}{(1-e^2\sin^2 L)^{\frac{3}{2}}}$$

Latitude.		log R	Com. diff. for 10′.	$\log \frac{1}{R \sin 1''}$
°	′			
40	0	6.8424057	125.0	8.4720194
	15	4244	125.1	.4720007
	30	4432	125.3	.4719819
	45	4620	125.5	9631
41	0	4808	125.7	9443
	15	4997	125.8	9254
	30	5186	126.0	9066
	45	5375	126.2	8877
42	0	5564	126.2	8687
	15	5753	126.3	8498
	30	5943	126.4	8309
	45	6132	126.6	8119
43	0	6322	126.6	7929
	15	6512	126.7	7739
	30	6702	126.7	7549
	45	6892	126.8	7359
44	0	7082	126.8	7169
	15	7273	126.9	6979
	30	7463	127.0	6788
	45	7653	126.9	6598
45	0	7844	127.0	6408
	15	8034	126.9	6217
	30	8224	126.9	6027
	45	8415	126.9	5837
46	0	8605	126.8	5647
	15	8795	126.8	5456
	30	8985	126.7	5266
	45	9175	126.7	5076
47	0	9365	126.6	4886
	15	9555	126.6	4696
	30	9745	126.4	4506
	45	.8429934	126.3	4317
48	0	.8430124	126.2	4127
	15	0313	126.1	3938
	30	0502	126.0	3749
	45	0691	125.8	3560
49	0	0880	125.7	3371
	15	1068	125.5	3183
	30	1257	125.3	2995
	45	1445	125.0	2807
50	0	6.8431632		8.4712619

XLIII.—*Projection of Maps.*

POLYCONIC PROJECTION.

In this development of the earth's surface each parallel of latitude is supposed to be represented on a plane by the development of a cone having the parallel for its base and its vertex in the point where a tangent to the parallel intersects the earth's axis. The map thus becomes the development of the surfaces of several successive cones, and the degrees of the parallel preserve their true length.

Normal .... $N = \dfrac{a}{(1 - e^2 \sin^2 L)^{\frac{1}{2}}}$

Radius of the meridian................ $R_m = N^3 \dfrac{(1 - e^2)}{a^2}$

Radius of the parallel................. $R_p = N \cos L$

Degree of the meridian.............. $D_m = \dfrac{\pi}{180} R_m$

$= 3600\ R_m \sin 1''$

Degree of the parallel................. $D_p = \dfrac{\pi}{180} R_p$

$= 3600\ R_p \sin 1''$

Radius of the developed parallel or side of tangent cone................. $r = N \cot L$

Designating by n any arc of the parallel, or difference of longitude, to be developed, and by θ the corresponding angle subtended by the developed parallel at the vertex of the cone; then the length of the given arc will be:

$$n\ R_p = n\ N \cos L$$

and also—

$$\theta r = \theta\ N \cot L$$

whence—

Angle of the developed parallel, $\theta = n \sin L$

and as the developed parallels are circular arcs, the co-ordinates of curvature are:

d_m, difference of meridians, $= x = r \sin \theta$

d_p, difference of parallels, $= y = r \text{ ver sin } \theta = x \tan \frac{1}{2} \theta$

For surfaces of small extent the arc of the parallel may be considered coincident with its chord; and as the angle between a

XLIII.—*Projection of Maps*—Continued.

tangent and a chord is half the angle at the center subtended by the chord,

δ_m, difference of meridians, $= x = D_p \cos \frac{1}{2} \theta$

δ_p, difference of parallels, $= y = D_p \sin \frac{1}{2} \theta$

The values of δ_m and δ_p and of D_m and D_p will be found in the following tables.

Example of their Use.—Let it be required to make a projection containing 40′ of longitude between the parallels of 41° 30′ and 42° 10′, to be subdivided to 5′.

Assume the center of the sheet to be the intersection of the middle parallel with the middle meridian of the proposed map, which point call A; in this case a point in the parallel of 41° 50′.

Through A draw the central meridian and a line at right angles to it.

Beginning at A, lay of above and below, on the central meridian, the values of D_m from 41° 50′ to 41° 55′; 41° 55′ to 42°; 42° to 42° 5′, etc.; and from 41° 50′ to 41° 45′; 41° 45′ to 41° 40′, etc.; these values to be taken from the table of *Meridional Arcs—Values of D_m in Yards*, by interpolation from the values there given for the middle latitudes of 41° and 42°.

Through each of the points ..., A^{ii}, A^{i}, A, A_{i}, A_{ii}, ... , thus found, lay off perpendiculars to the central meridian.

Now turn to the table of *Co-ordinates, δ_m and δ_p, in Yards*, and lay off, from each of the points ..., A^{ii}, A^{i}, A, A_{i}, A_{ii}, ..., to the right and left of the central meridian, the values of δ_m for successively 5′, 10′, 15′, and 20′, corresponding (by interpolation from the columns of 41° 30′ and 42°) to each parallel of latitude required; and, from the points thus found, the corresponding values of δ_p at right angles to the lines already drawn.

Lines passing through the extremities of δ_p will be the required meridians and parallels.

The projection being made, any point whose latitude and longitude are known will be projected on the map from elements taken from the tables of values of D_m and D_p, which are measured from the *meridians* and *parallels*, and not from the axes of co-ordinates used in making the projection.

Polyconic Projection—Co-ordinates, δ_m, δ_p, in Yards.

Longitude.	Latitude 22° 0′.		Latitude 22° 30′.		Latitude 23° 0′.	
	δ_m	δ_p	δ_m	δ_p	δ_m	δ_p
° ′						
1	1882.0	0.1	1875.3	0.1	1868.5	0.1
2	3763.9	0.4	3750.6	0.4	3737.0	0.4
3	5645.9	0.9	5625.9	1.0	5605.4	1.0
4	7527.8	1.6	7501.2	1.7	7473.9	1.7
5	9409.8	2.6	9376.4	2.6	9342.4	2.7
6	11291.8	3.7	11251.7	3.7	11210.9	3.8
7	13173.7	5.0	13127.0	5.1	13079.4	5.2
8	15055.7	6.6	15002.3	6.7	14947.8	6.8
9	16937.6	8.3	16877.6	8.5	16816.3	8.6
10	18819.6	10.3	18752.9	10.5	18684.8	10.6
11	20701.6	12.4	20628.2	12.6	20553.3	12.8
12	22583.5	14.8	22503.5	15.0	22421.8	15.3
13	24465.5	17.3	24378.8	17.6	24290.2	17.9
14	26347.4	20.1	26254.1	20.5	26158.7	20.8
15	28229.4	23.1	28129.3	23.5	28027.2	23.9
16	30111.4	26.2	30004.6	26.7	29895.7	27.2
17	31993.3	29.6	31879.9	30.2	31764.2	30.7
18	33875.3	33.2	33755.2	33.3	33632.6	34.4
19	35757.2	37.0	35630.5	37.7	35501.1	38.3
20	37639.2	41.0	37505.8	41.7	37369.6	42.5
25	47049.0	64.1	46675.8	65.2	46712.0	66.4
30	56458.7	92.3	56258.7	93.9	56054.3	95.6
40	75278.2	164.1	75011.5	167.0	74739.0	169.9
50	94097.7	256.3	93764.2	260.9	93423.7	265.5
1 00	112917.0	369.1	112516.9	375.8	112108.2	382.3
1 20	150555.4	656.2	150021.9	668.0	149476.9	679.6
1 30	169374.4	830.5	168774.2	845.4	168161.1	860.1
1 40	188193.3	1025.4	187526.3	1043.8	186845.1	1061.8
2 00	225830.5	1476.5	225030.0	1503.0	224212.5	1529.1
2 30	282284.7	2307.1	281284.0	2348.5	280261.9	2389.2
3 00	338736.6	3322.2	337535.6	3381.8	336309.0	3440.4
3 30	395186.0	4521.9	393784.5	4603.0	392353.1	4682.8
4 00	451632.0	5906.2	450029.9	6012.1	448393.7	6116.2

Polyconic Projection—Co ordinates, δ_m, δ_p, in Yards.

Longitude.	Latitude 23° 30′.		Latitude 24° 0′.		Latitude 24° 30′.	
	δ_m	δ_p	δ_m	δ_p	δ_m	δ_p
° ′						
1	1861.5	0.1	1854.4	0.1	1847.2	0.1
2	3723.1	0.4	3708.9	0.4	3694.4	0.4
3	5584.6	1.0	5563.3	1.0	5541.6	1.0
4	7446.1	1.8	7417.7	1.8	7388.8	1.8
5	9307.6	2.7	9272.2	2.7	9236.0	2.8
6	11169.2	3.9	11126.6	3.9	11083.2	4.0
7	13030.7	5.3	12981.0	5.4	12930.4	5.4
8	14892.2	6.9	14835.5	7.0	14777.6	7.1
9	16753.7	8.7	16689.9	8.9	16624.8	9.0
10	18615.3	10.8	18544.3	11.0	18472.0	11.1
11	20476.8	13.1	20398.8	13.3	20319.2	13.5
12	22338.3	15.6	22253.2	15.8	22166.4	16.0
13	24199.9	18.2	24107.6	18.5	24013.6	18.8
14	26061.4	21.2	25962.1	21.5	25860.8	21.8
15	27922.9	24.3	27816.5	24.7	27708.0	25.1
16	29784.4	27.7	29670.9	28.1	29555.2	28.5
17	31646.0	31.2	31525.4	31.7	31402.4	32.2
18	33507.5	35.0	33379.8	35.5	33249.6	36.0
19	35369.0	39.0	35234.2	39.6	35096.8	40.2
20	37230.5	43.2	37088.7	43.9	36944.0	44.6
25	46538.1	67.5	46360.8	68.6	46179.9	69.6
30	55845.8	97.2	55632.9	98.7	55415.9	100.3
40	74460.9	172.7	74177.1	175.5	73887.7	178.3
50	93076.0	269.9	92721.3	274.3	92359.5	278.5
1 00	111691.0	388.7	111265.3	394.9	110831.2	401.1
1 20	148920.6	690.9	148353.0	702.1	147774.1	713.0
1 30	167535.2	874.5	166896.6	888.6	166245.4	902.4
1 40	186149.7	1079.6	185440.1	1097.0	184716.5	1114.1
2 00	223377.9	1554.6	222526.4	1579.7	221658.0	1604.3
2 30	279218.6	2429.1	278154.1	2468.3	277068.4	2506.8
3 00	335056.8	3497.9	333779.1	3554.4	332476.1	3609.8
3 30	390892.0	4761.1	389401.1	4837.9	387880.6	4913.3
4 00	446723.4	6218.5	445019.2	6318.9	443281.1	6417.4

Polyconic Projection—Co-ordinates, δ_m, δ_p, in Yards.

Longitude.	Latitude 25° 0′.		Latitude 25° 30′.		Latitude 26° 0′.	
	δ_m	δ_p	δ_m	δ_p	δ_m	δ_p
° ′						
1	1839.8	0.1	1832.3	0.1	1824.7	0.1
2	3679.6	0.5	3664.6	0.5	3649.3	0.5
3	5519.5	1.0	5496.9	1.0	5474.0	1.0
4	7359.3	1.8	7329.2	1.8	7298.6	1.9
5	9199.1	2.8	9161.5	2.9	9123.3	2.9
6	11038.9	4.1	10993.8	4.1	10947.9	4.2
7	12878.8	5.5	12826.1	5.6	12772.6	5.7
8	14718.6	7.2	14658.5	7.3	14597.2	7.4
9	16558.4	9.2	16490.8	9.3	16421.9	9.4
10	18398.2	11.3	18323.1	11.5	18246.5	11.6
11	20238.0	13.7	20155.4	13.9	20071.2	14.1
12	22077.9	16.3	21987.7	16.5	21895.8	16.8
13	23917.7	19.1	23820.0	19.4	23720.5	19.7
14	25757.5	22.2	25652.3	22.5	25545.1	22.8
15	27597.3	25.4	27484.6	25.8	27369.8	26.2
16	29437.1	29.0	29316.9	29.4	29194.4	29.8
17	31277.0	32.7	31149.2	33.2	31019.1	33.6
18	33116.8	36.6	32981.5	37.2	32843.7	37.7
19	34956.6	40.8	34813.8	41.4	34668.4	42.0
20	36796.4	45.2	36646.1	45.9	36493.0	46.5
25	45995.5	70.7	45807.6	71.7	45616.2	72.7
30	55194.6	101.8	54969.1	103.2	54739.5	104.7
40	73592.7	180.9	73292.0	183.6	72985.8	186.1
50	91990.7	282.7	91614.9	286.8	91232.1	290.8
1 00	110388.6	407.1	109937.6	413.0	109478.3	418.8
1 20	147184.0	723.8	146582.7	734.3	145970.3	744.6
1 30	165581.5	916.0	164905.0	929.3	164216.0	942.3
1 40	183978.8	1130.9	183227.1	1147.3	182461.5	1163.4
2 00	220772.7	1628.5	219870.6	1652.1	218951.9	1675.2
2 30	275961.6	2544.5	274833.9	2581.4	273685.3	2617.6
3 00	331147.8	3664.1	329794.3	3717.3	328415.8	3769.3
3 30	386330.6	4987.2	384751.3	5059.6	383142.7	5130.5
4 00	441509.4	6513.9	439704.0	6608.5	437865.3	6701.0

Polyconic Projection—Co-ordinates, δ_m, δ_p, in Yards.

Longitude.	Latitude 26° 30′.		Latitude 27° 0′.		Latitude 27° 30′.	
	δ_m	δ_p	δ_m	δ_p	δ_m	δ_p
° ′						
1	1816.9	0.1	1808.9	0.1	1800.9	0.1
2	3633.7	0.5	3617.9	0.5	3601.7	0.5
3	5450.6	1.1	5426.8	1.1	5402.6	1.1
4	7267.4	1.9	7235.7	1.9	7203.4	1.9
5	9084.3	2.9	9044.6	3.0	9004.3	3.0
6	10901.2	4.2	10853.6	4.3	10805.1	4.4
7	12718.0	5.8	12662.5	5.9	12606.0	5.9
8	14534.9	7.5	14471.4	7.6	14406.9	7.7
9	16351.7	9.5	16280.3	9.7	16207.7	9.8
10	18168.6	11.8	18089.3	11.9	18008 6	12.1
11	19985.5	14.3	19898.2	14.5	19809.4	14.6
12	21802.3	17.0	21707.1	17.2	21610.3	17.4
13	23619.2	19.9	23516.0	20.2	23411.1	20.4
14	25436.0	23.1	25325.0	23.4	25212.0	23.7
15	27252.9	26.5	27133.9	26.9	27012.8	27.2
16	29069.8	30.2	28942.8	30.6	28813.7	30.9
17	30886.6	34.1	30751.7	34.5	30614.6	34.9
18	32703.5	38.2	32560.7	38.7	32415.4	39.2
19	34520.3	42.6	34369.6	43.1	34216.3	43.6
20	36337.2	47.2	36178.5	47.8	36017.1	48.4
25	45421.4	73.7	45223.1	74.7	45021.4	75.6
30	54505.6	106.1	54267.7	107.5	54025.6	108.8
40	72674.1	188.6	72356.8	191.1	72034.0	193.5
50	90842.4	294.8	90445.8	298.6	90042.4	302.4
1 00	109010.7	424.5	108534.8	430.0	108050.6	435.4
1 20	145346.7	754.6	144712.1	764.4	144066.5	774.0
1 30	163514.5	955.1	162800.5	967.5	162074.2	979.6
1 40	181682.0	1179.1	180888.7	1194.4	180081.7	1209.4
2 00	218016.4	1697.9	217064.4	1720.0	216095.9	1741.6
2 30	272515.9	2652.9	271325.7	2687.5	270114.9	2721.2
3 00	327012.2	3820.2	325583.8	3870.0	324130.7	3918.6
3 30	381505.0	5199.8	379838.2	5267.5	378142.5	5333.6
4 00	435993.2	6791.5	434088.0	6880.0	432149.7	6966.3

Polyconic Projection—Co-ordinates, δ_m, δ_p, in Yards.

Longitude.	Latitude 28° 0′.		Latitude 28° 30′.		Latitude 29°.	
	δ_m	δ_p	δ_m	δ_p	δ_m	δ_p
° ′						
1	1792.7	0.1	1784.3	0.1	1775.8	0.1
2	3585.3	0.5	3568.6	0.5	3551.7	0.5
3	5378.0	1.1	5352.9	1.1	5327.5	1.1
4	7170.6	2.0	7137.2	2.0	7103.3	2.0
5	8963.3	3.1	8921.5	3.1	8879.1	3.1
6	10755.9	4.4	10705.8	4.5	10655.0	4.5
7	12548.6	6.0	12490.2	6.1	12430.8	6.1
8	14341.2	7.8	14274.5	7.9	14206.6	8.0
9	16133.9	9.9	16058.8	10.0	15982.5	10.1
10	17926.5	12.2	17843.1	12.4	17758.3	12.5
11	19719.2	14.8	19627.4	15.0	19534.1	15.2
12	21511.8	17.6	21411.7	17.8	21309.9	18.0
13	23304.5	20.7	23196.0	20.9	23085.8	21.2
14	25097.1	24.0	24980.3	24.3	24861.6	24.5
15	26889.8	27.5	26764.6	27.9	26637.4	28.2
16	28682.4	31.3	28548.9	31.7	28413.2	32.1
17	30475.1	35.4	30333.2	35.8	30189.1	36.2
18	32267.7	39.7	32117.5	40.1	31964.9	40.6
19	34060.3	44.2	33901.8	44.7	33740.7	45.2
20	35853.0	49.0	35686.1	49.5	35516.6	50.1
25	44816.2	76.5	44607.6	78.4	44395.7	78.3
30	53779.4	110.1	53529.1	111.4	53274.8	112.7
40	71705.8	195.8	71372.1	198.1	71032.9	200.3
50	89632.0	306.0	89214.9	309.6	88790.9	313.0
1 00	107558.2	440.7	107057.6	445.8	106548.8	450.8
1 20	143410.0	783.4	142742.5	792.5	142064.1	801.4
1 30	161335.6	991.5	160584.6	1003.0	159821.5	1014.3
1 40	179260.9	1224.1	178426.5	1238.3	177578.6	1252.2
2 00	215110.9	1762.6	214109.6	1783.2	213092.0	1803.1
2 30	268883.6	2754.1	267631.8	2786.2	266359.6	2817.4
3 00	322652.8	3965.9	321150.5	4012.1	319623.7	4057.1
3 30	376418.1	5398.1	374665.0	5460.9	372883.4	5522.1
4 00	430178.5	7050.6	428174.6	7132 7	426138.2	7212.6

Polyconic Projection—Co-ordinates, δ_m, δ_p, in Yards.

Longitude.	Latitude 29° 30′.		Latitude 30° 0′.		Latitude 30° 30′.	
	δ_m	δ_p	δ_m	δ_p	δ_m	δ_p
° ′						
1	1767.2	0.1	1758.5	0.1	1749.6	0.1
2	3534.4	0.5	3516.9	0.5	3499.2	0.5
3	5301.6	1.1	5275.4	1.2	5248.8	1.2
4	7068.9	2.0	7033.9	2.0	6998.3	2.1
5	8836.1	3.2	8792.3	3.2	8747.9	3.2
6	10603.3	4.6	10550.8	4.6	10497.5	4.6
7	12370.5	6.2	12309.3	6.3	12247.1	6.3
8	14137.7	8.1	14067.8	8.2	13996.7	8.2
9	15904.9	10.3	15826.2	10.4	15746.3	10.5
10	17672.2	12.7	17584.7	12.8	17495.9	12.9
11	19439.4	15.3	19343.1	15.5	19245.4	15.6
12	21206.6	18.2	21101.6	18.4	20995.0	18.6
13	22973.8	21.4	22860.1	21.6	22744.6	21.8
14	24741.0	24.8	24618.5	25.1	23494.2	25.3
15	26508.2	28.5	26377.0	28.8	25243.8	29.1
16	28275.4	32.4	28135.5	32.7	26993.3	33.1
17	30042.6	36.6	29893.9	37.0	28742.9	37.4
18	31809.9	41.0	31652.4	41.4	30492.5	41.8
19	33577.1	45.7	33410.9	46.2	32242.1	46.6
20	35344.3	51.6	35169.3	51.2	33991.7	51.7
25	44180.3	79.1	43961.6	79.9	43239.6	80.7
30	53016.4	113.9	52753.9	115.1	52487.4	116.3
40	70688.4	202.5	70338.4	204.6	69983.1	206.6
50	88360.2	316.4	87922.8	319.7	87478.7	322.9
1 00	106032.0	455.6	105507.1	460.4	104974.1	464.9
1 20	141375.0	810.0	140675.1	818.4	139964.4	826.6
1 30	159046.2	1025.2	158258.7	1035.8	157459.2	1046.1
1 40	176717.1	1265.7	175842.2	1278.8	174953.8	1291.5
2 00	212058.1	1822.6	211008.1	1841.5	209942.1	1859.8
2 30	265067.1	2847.8	263754.5	2877.3	262421.8	2905.9
3 00	318072.5	4100.8	316497.1	4143.3	314897.6	4184.5
3 30	371073.4	5581.7	369235.2	5639.5	367368.9	5695.6
4 00	424069.3	7290.3	421968.0	7365.9	419834.7	7439.1

Polyconic Projection—Co-ordinates, δ_m, δ_p, in Yards.

Longitude.	Latitude 31° 0′.		Latitude 31° 30′.		Latitude 32° 0′.	
	δ_m	δ_p	δ_m	δ_p	δ_m	δ_p
° ′						
1	1740.6	0.1	1731.4	0.1	1722.1	0.1
2	3481.1	0.5	3462.8	0.5	3444.3	0.5
3	5221.7	1.2	5194.3	1.2	5166.4	1.2
4	6962.3	2.1	6925.7	2.1	6888.6	2.1
5	8702.9	3.3	8657.1	3.3	8610.7	3.3
6	10443.4	4.7	10388.5	4.7	10332.8	4.8
7	12184.0	6.4	12119.9	6.4	12055.0	6.5
8	13924.6	8.3	13851.4	8.4	13777.1	8.5
9	15665.1	10.6	15582.8	10.7	15499.3	10.8
10	17405.7	13.0	17314.2	13.2	17221.4	13.3
11	19146.3	15.8	19045.6	15.9	18943.6	16.1
12	20886.8	18.8	20777.1	18.9	20665.7	19.1
13	22627.4	22.0	22508.5	22.2	22387.8	22.4
14	24368.0	25.6	24239.9	25.8	24110.0	26.0
15	26108.5	29.3	25971.3	29.6	25832.1	29.9
16	27849.1	33.4	27702.7	33.7	27554.3	34.0
17	29589.7	37.7	29434.2	38.0	29276.4	38.4
18	31330.3	42.2	31165.6	42.6	30998.5	43.0
19	33070.8	47.1	32897.0	47.5	32720.7	47.9
20	34811.4	52.2	34628.4	52.6	34442.8	53.1
25	43514.2	81.5	43286.0	82.3	43053.5	83.0
30	52217.0	117.3	51942.5	118.4	51664.1	119.5
40	69622.5	208.6	69256.6	210.5	68885.4	212.4
50	87027.9	326.0	86570.5	328.9	86106.5	331.8
1 00	104433.2	469.4	103884.3	473.7	103327.4	477.8
1 20	139243.1	834.5	138511.2	842.1	137768.8	849.5
1 30	156647.8	1056.1	155824.4	1065.8	154989.1	1075.1
1 40	174052.2	1303.8	173137.2	1315.8	172209.1	1327.3
2 00	208860.0	1877.5	207762.0	1894.7	206648.2	1911.3
2 30	261069.1	2933.7	259696.5	2960.5	258304.1	2986.5
3 00	313274.2	4224.5	311626.9	4263.1	309955.8	4300.5
3 30	365474.6	5750.0	363552.4	5802.6	361602.5	5853.5
4 00	417669.4	7510.2	415472.3	7578.9	413243.4	7645.3

Polyconic Projection—Co-ordinates, δ_m, δ_p, in Yards.

Longitude.		Latitude 32° 30′.		Latitude 33° 0′.		Latitude 33° 30′.	
		δ_m	δ_p	δ_m	δ_p	δ_m	δ_p
°	′						
	1	1712.7	0.1	1703.2	0.1	1693.5	0.1
	2	3425.5	0.5	3406.4	0.5	3387.0	0.5
	3	5138.2	1.2	5109.6	1.2	5080.6	1.2
	4	6850.9	2.1	6812.8	2.2	6774.1	2.2
	5	8563.7	3.3	8515.9	3.4	8467.6	3.4
	6	10276.4	4.8	10219.1	4.9	10161.1	4.9
	7	11989.1	6.6	11922.3	6.6	18154.6	6.7
	8	13701.8	8.6	13625.5	8.6	13548.1	8.7
	9	15414.6	10.8	15328.7	10.9	15241.7	11.0
	10	17127.3	13.4	17031.9	13.5	16935.2	13.6
	11	18840.0	16.2	18735.1	16.3	18628.7	16.4
	12	20552.8	19.3	20438.3	19.4	20322.2	19.6
	13	22265.5	22.6	22141.4	22.8	22015.7	23.0
	14	23978.2	26.2	23844.6	26.4	23709.2	26.6
	15	25690.9	30.1	25547.8	30.4	25402.7	30.6
	16	27403.7	34.3	27251.0	34.5	27096.3	34.9
	17	29116.4	38.7	28954.2	39.0	28789.8	39.3
	18	30829.1	43.4	30657.4	43.7	30483.3	44.0
	19	32541.9	48.5	32360.6	49.0	32176.8	49.1
	20	34254.6	53.5	34063.8	54.0	33870.3	54.4
	25	42818.2	83.6	42579.6	84.3	42337.9	85.0
	30	51381.8	120.5	51095.5	121.4	50805.4	122.3
	40	68508.9	214.1	68127.2	215.9	67740.4	217.5
	50	85635.9	334.6	85158.8	337.3	84675.2	339.9
1	00	102762.7	481.8	102190.2	485.7	101609.9	489.4
1	20	137015.8	856.6	136252.4	863.5	135478.6	870.1
1	30	154142.0	1084.1	153283.1	1092.8	152412.6	1101.2
1	40	171267.9	1338.4	170313.6	1349.2	169346.3	1359.5
2	00	205518.7	1927.4	204373.5	1942.8	203212.7	1957.7
2	30	256892.0	3011.5	255460.4	3035.6	254009.2	3058.8
3	00	308261.1	4336.6	306542.9	4371.3	304801.4	4404.7
3	30	359625.1	5902.6	357620.3	5949.9	355588.2	5995.3
4	00	410983.2	7709.5	408691.6	7771.2	406368.8	7830.6

Polyconic Projection—Co-ordinates, δ_m, δ_p, in Yards.

Longitude.	Latitude 34° 0′.		Latitude 34° 30′.		Latitude 35° 0′.	
	δ_m	δ_p	δ_m	δ_p	δ_m	δ_p
° ′						
1	1683.7	0.1	1673.8	0.1	1663.7	0.1
2	3367.4	0.5	3347.6	0.6	3327.5	0.6
3	5051.1	1.2	5021.4	1.2	4991.2	1.2
4	6734.9	2.2	6695.1	2.2	6654.9	2.2
5	8418.6	3.4	8368.9	3.4	8318.7	3.5
6	10102.3	4.9	10042.7	5.0	9982.4	5.0
7	11786.0	6.7	11716.5	6.8	11646.1	6.8
8	13469.7	8.8	13390.3	8.8	13309.8	8.9
9	15153.4	11.1	15064.1	11.2	14973.6	11.2
10	16837.2	13.7	16737.9	13.8	16637.3	13.9
11	18520.9	16.6	18411.6	16.7	18301.0	16.8
12	20204.6	19.7	20085.4	19.9	19964.8	20.0
13	21888.3	23.1	21759.2	23.3	21628.5	23.5
14	23572.0	26.8	23433.0	27.0	23292.2	27.2
15	25255.7	30.8	25106.8	31.0	24955.9	31.2
16	26939.4	35.1	26780.6	35.3	26619.7	35.5
17	28623.1	39.6	28454.4	39.8	28283.4	40.1
18	30306.9	44.4	30128.1	44.7	29947.1	45.0
19	31990.6	49.4	31801.9	49.8	31610.9	50.1
20	33674.3	54.8	33475.7	55.2	33274.6	55.5
25	42092.8	85.6	41844.6	86.2	41593.2	86.7
30	50511.4	123.2	50213.5	125.1	49911.8	124.9
40	67348.3	219.1	66951.1	220.6	66548.9	222.1
50	84185.1	342.3	83688.7	344.7	83185.8	347.0
1 00	101021.8	493.0	100426.0	496.4	99822.6	499.7
1 20	134694.5	876.4	133900.1	882.5	133095.5	888.3
1 30	151530.5	1109.2	150636.8	1116.9	149731.5	1124.2
1 40	168366.1	1369.4	167373.1	1378.9	166367.3	1387.9
2 00	202036.4	1971.9	200844.7	1985.6	199637.7	1998.6
2 30	252538.7	3081.1	251049.0	3102.5	249540.1	3122.8
3 00	303036.6	4436.8	301248.7	4467.5	299437.8	4496.9
3 30	353529.0	6039.0	351442.8	6080.8	349329.8	6120.8
4 00	404015.1	7887.7	401630.5	7942.3	399215.4	7994.5

Polyconic Projection—Co-ordinates, δ_m, δ_p, in Yards.

Longitude.	Latitude 35° 30′.		Latitude 36° 0′.		Latitude 36° 30′.	
	δ_m	δ_p	δ_m	δ_p	δ_m	δ_p
° ′						
1	1653.5	0.1	1643.2	0.1	1632.8	0.1
2	3307.1	0.6	3286.5	0.6	3265.6	0.6
3	4960.6	1.3	4929.7	1.3	4898.4	1.3
4	6614.2	2.2	6572.9	2.2	6531.2	2.3
5	8267.7	3.5	8216.2	3.5	8164.0	3.5
6	9921.3	5.0	9859.4	5.1	9796.8	5.1
7	11574.8	6.8	11502.7	6.9	11429.6	6.9
8	13228.4	8.9	13145.9	9.0	13062.4	9.0
9	14881.9	11.3	14789.1	11.4	14695.2	11.4
10	16535.5	14.0	16432.4	14.0	16328.0	14.1
11	18189.0	16.9	18075.6	17.0	17960.8	17.0
12	19842.5	20.1	19718.8	20.2	19593.6	20.3
13	21496.1	23.6	21362.0	23.7	21226.4	23.9
14	23149.6	27.4	23005.3	27.5	22859.2	27.7
15	24803.2	31.4	24648.5	31.6	24492.0	31.8
16	26456.7	35.7	26291.8	36.0	26124.8	36.2
17	28110.3	40.4	27935.0	40.6	27757.6	40.8
18	29763.8	45.2	29578.2	45.5	29390.4	45.8
19	31417.3	50.4	31221.5	50.7	31023.2	51.0
20	33070.9	55.9	32864.7	56.2	32656.0	56.5
25	41338.6	87.2	41080.8	87.8	40819.9	88.3
30	49606.2	125.7	49296.9	126.4	48983.9	127.1
40	66141.5	223.4	65729.1	224.8	65311.7	226.0
50	82676.6	349.1	82161.1	351.2	81639.3	353.1
1 00	99211.5	502.7	98592.9	505.7	97966.7	508.5
1 20	132280.7	893.8	131455.9	899.1	130621.0	904.1
1 30	148814.9	1131.2	147886.9	1137.9	146947.7	1144.2
1 40	165348.8	1396.6	164317.7	1404.8	163274.0	1412.6
2 00	198415.4	2011.1	197178.0	2022.9	195925.6	2034.1
2 30	248012.1	3142.3	246465.3	3160.8	244899.6	3178.3
3 00	297604.0	4524.9	295747.6	4551.6	293868.6	4576.8
3 30	347190.2	6158.9	345024.1	6195.2	342831.7	6229.5
4 00	396769.7	8044.3	394293.8	8091.6	391787.8	8136.5

Polyconic Projection—Co-ordinates, δ_m, δ_p, in Yards.

Longitude.	Latitude 37° 0′.		Latitude 37° 30′.		Latitude 38° 0′.	
	δ_m	δ_p	δ_m	δ_p	δ_m	δ_p
° ′						
1	1622.2	0.1	1611.6	0.1	1600.7	0.1
2	3244.5	0.6	3223.1	0.6	3201.5	0.6
3	4866.7	1.3	4834.7	1.3	4802.2	1.3
4	6489.0	2.3	6446.2	2.3	6403.0	2.3
5	8111.2	3.5	8057.8	3.6	8003.7	3.6
6	9733.4	5.1	9669.3	5.1	9604.5	5.2
7	11355.7	7.0	11280.9	7.0	11205.2	7.0
8	12977.9	9.1	12892.4	9.1	12806.0	9.2
9	14600.2	11.5	14504.0	11.6	14406.7	11.6
10	16222.4	14.2	16115.6	14.3	16007.5	14.3
11	17844.6	17.2	17727.1	17.3	17608.2	17.3
12	19466.9	20.4	19338.6	20.5	19209.0	20.6
13	21089.1	24.0	20950.2	24.1	20809.7	24.2
14	22711.4	27.8	22561.8	28.0	22410.5	28.1
15	24333.6	31.9	24173.3	32.1	24011.2	32.3
16	25955.8	36.4	25784.9	36.5	25612.0	36.7
17	27578.1	41.0	27396.4	41.2	27212.7	41.4
18	29200.3	46.0	29008.0	46.2	28813.4	46.4
19	30822.6	51.3	30619.5	51.5	30414.2	51.7
20	32444.8	56.8	32231.1	57.1	32015.0	57.3
25	40555.9	88.7	40288.8	89.1	40018.6	89.6
30	48667.1	127.8	48346.5	128.4	48022.3	129.0
40	64889.2	227.2	64461.9	228.3	64029.6	229.3
50	81111.3	355.0	80577.1	356.7	80036.7	358.3
1 00	97333.1	511.2	96692.0	513.7	96043.6	516.0
1 20	129776.1	908.8	128921.3	913.2	128056.7	917.4
1 30	145997.2	1150.2	145035.5	1155.8	144062.8	1161.0
1 40	162217.9	1419.9	161149.4	1426.9	160068.5	1433.4
2 00	194658.2	2044.7	193375.9	2054.7	192078.9	2064.1
2 30	243315.2	3194.9	241712.2	3210.5	240090.8	3225.1
3 00	291967.1	4600.7	290043.4	4623.1	288097.5	4644.1
3 30	340613.1	6262.0	338368.4	6292.6	336098.0	6321.2
4 00	389251.9	8178.9	386686.3	8218.8	384091.2	8256.3

Polyconic Projection—Co-ordinates, δ_m, δ_p, in Yards.

Longitude.	Latitude 38° 30′.		Latitude 39° 0′.		Latitude 39° 30′.	
	δ_m	δ_p	δ_m	δ_p	δ_m	δ_p
° ′						
1	1589.8	0.1	1578.8	0.1	1567.6	0.1
2	3179.6	0.6	3157.5	0.6	3135.2	0.6
3	4769.5	1.3	4736.3	1.3	4702.8	1.3
4	6359.3	2.3	6315.1	2.3	6270.4	2.3
5	7949.1	3.6	7893.8	3.6	7838.0	3.6
6	9538.9	5.2	9472.6	5.2	9405.6	5.2
7	11128.7	7.1	11051.4	7.1	10973.2	7.1
8	12718.6	9.2	12630.1	9.2	12540.8	9.3
9	14308.4	11.7	14208.9	11.7	14108.4	11.7
10	15898.2	14.4	15787.7	14.5	15676.0	14.5
11	17488.0	17.4	17366.4	17.5	17243.6	17.5
12	19077.8	20.7	18945.2	20.8	18811.1	20.9
13	20667.6	24.3	20524.0	24.4	20378.7	24.5
14	22257.4	28.2	22102.7	28.3	21946.3	28.4
15	23847.3	32.4	23681.5	32.5	23513.9	32.6
16	25437.1	36.8	25260.3	37.0	25081.5	37.1
17	27026.9	41.6	26839.0	41.8	26649.1	41.9
18	28616.7	46.6	28417.8	46.8	28216.7	47.0
19	30206.5	52.0	29996.6	52.2	29784.3	52.3
20	31796.3	57.6	31575.3	57.8	31351.9	58.0
25	39745.4	90.0	39469.1	90.3	39189.8	90.6
30	47694.4	129.5	47362.9	130.1	47027.7	130.5
40	63592.4	230.3	63150.3	231.2	62703.4	232.0
50	79490.2	359.9	78937.6	361.3	78379.0	362.6
1 00	95387.8	518.2	94724.7	520.2	94054.4	522.1
1 20	127182.3	921.2	126298.1	924.8	125404.3	928.2
1 30	143079.0	1165.9	142084.4	1170.5	141078.8	1174.7
1 40	158975.5	1439.4	157870.3	1445.1	156753.0	1450.2
2 00	190767.1	2072.8	189440.8	2080.9	188100.1	2088.3
2 30	238451.0	3238.7	236793.0	3251.4	235116.9	3263.0
3 00	286129.6	4663.8	284139.8	4682.0	282128.3	4698.8
3 30	333801.8	6347.9	331480.2	6372.7	329133.2	6395.6
4 00	381466.7	8291.2	378813.1	8323.6	376130.5	8353.4

Polyconic Projection—Co-ordinates, δ_m, δ_p, in Yards.

Longitude.	Latitude 40° 0′.		Latitude 40° 30′.		Latitude 41° 0′.	
	δ_m	δ_p	δ_m	δ_p	δ_m	δ_p
° ′						
1	1556.3	0.1	1544.9	0.1	1533.4	0.1
2	3112.6	0.6	3089.8	0.6	3066.7	0.6
3	4668.9	1.3	4634.7	1.3	4600.1	1.3
4	6225.2	2.3	6179.6	2.3	6133.5	2.3
5	7781.5	3.6	7724.5	3.6	7666.8	3.7
6	9337.8	5.2	9269.4	5.3	9200.2	5.3
7	10894.1	7.1	10814.3	7.2	10733.6	7.2
8	12450.4	9.3	12359.2	9.3	12266.9	9.4
9	14006.7	11.8	13904.0	11.8	13800.3	11.9
10	15563.0	14.6	15448.9	14.6	15333.7	14.6
11	17119.3	17.6	16993.8	17.7	16867.0	17.7
12	18675.6	21.0	18538.7	21.0	18400.4	21.1
13	20231.9	24.6	20083.6	24.7	19933.7	24.7
14	21788.2	28.5	21628.5	28.6	21467.1	28.7
15	23344.5	32.7	23173.4	32.8	23000.4	32.9
16	24900.8	37.2	24718.3	37.4	24533.8	37.5
17	26457.1	42.0	26263.2	42.2	26067.2	42.3
18	28013.4	47.1	27808.1	47.3	27600.5	47.4
19	29569.8	52.5	29352.9	52.7	29133.9	52.8
20	31126.1	58.2	30897.8	58.4	30667.3	58.5
25	38907.5	90.9	38622.2	91.2	38334.0	91.4
30	46689.0	130.9	46346.7	131.3	46000.8	131.7
40	62251.8	232.8	61795.3	233.5	61334.2	234.1
50	77814.4	363.7	77243.9	364.8	76667.4	365.8
1 00	93376.9	523.8	92692.2	525.3	92000.4	526.7
1 20	124500.9	931.2	123588.0	933.9	122665.7	936.4
1 30	140062.5	1178.5	139035.5	1182.0	137997.8	1185.1
1 40	155623.7	1455.0	154482.6	1459.3	153329.6	1463.1
2 00	186744.9	2095.2	185375.4	2101.4	183991.8	2106.9
2 30	233422.9	3273.7	231710.9	3283.4	229998.3	3292.0
3 00	280095.3	4714.1	278040.9	4728.1	275965.1	4740.6
3 30	326761.1	6416.5	324364.1	6435.4	321942.2	6452.4
4 00	373419.3	8380.7	370679.4	8405.5	367911.3	8427.7

Polyconic Projection—Co-ordinates, δ_m, δ_p, in Yards.

Longitude.	Latitude 41° 30′.		Latitude 42° 0′.		Latitude 42° 30′.	
	δ_m	δ_p	δ_m	δ_p	δ_m	δ_p
° ′						
1	1521.7	0.1	1510.0	0.1	1498.1	0.1
2	3043.4	0.6	3019.9	0.6	2996.2	0.6
3	4565.2	1.3	4529.9	1.3	4494.2	1.3
4	6086.9	2.3	6039.8	2.4	5992.3	2.4
5	7608.6	3.7	7549.8	3.7	7490.4	3.7
6	9130.3	5.3	9059.7	5.3	8988.5	5.3
7	10652.0	7.2	10569.7	7.2	10486.5	7.2
8	12173.8	9.4	12079.6	9.4	11984.6	9.4
9	13695.5	11.9	13589.6	11.9	13482.7	11.9
10	15217.2	14.7	15099.6	14.7	14980.8	14.7
11	16738.9	17.7	16609.5	17.8	16478.8	17.8
12	18260.6	21.1	18119.5	21.2	17976.9	21.2
13	19782.3	24.8	19629.4	24.8	19475.0	24.9
14	21304.0	28.7	21139.4	28.8	20973.1	28.8
15	22825.8	33.0	22649.3	33.1	22471.1	33.1
16	24347.5	37.5	24159.3	37.6	23969.2	37.6
17	25869.2	42.4	25669.2	42.5	25467.3	42.5
18	27390.9	47.5	27179.2	47.6	26965.4	47.7
19	28912.6	52.9	28689.1	53.0	28463.4	53.1
20	30434.3	58.6	30199.1	58.8	29961.5	58.9
25	38042.9	91.6	37748.8	91.8	37451.3	92.0
30	45651.4	132.0	45298.5	132.3	44942.2	132.5
40	60868.3	234.6	60397.8	235.1	59922.7	235.5
50	76085.1	366.6	75496.9	367.4	74903.0	368.0
1 00	91301.6	527.9	90595.9	529.0	89883.2	529.9
1 20	121733.9	938.6	120792.9	940.5	119842.6	942.1
1 30	136949.6	1187.9	135890.9	1190.3	134821.8	1192.3
1 40	152164.9	1466.5	150988.5	1469.5	149800.6	1472.0
2 00	182594.1	2111.8	181182.5	2116.1	179756.9	2119.7
2 30	228234.1	3299.7	226469.4	3306.4	224687.4	3312.0
3 00	273868.3	4751.6	271750.5	4761.2	269612.0	4769.3
3 30	319495.6	6467.5	317024.7	6480.5	314529.5	6491.6
4 00	365115.0	8447.3	362290.8	8464.4	359438.9	8478.8

Polyconic Projection—Co-ordinates, δ_m, δ_p, in Yards.

Longitude.		Latitude 43° 0′.		Latitude 43° 30′.		Latitude 44° 0′.	
		δ_m	δ_p	δ_m	δ_p	δ_m	δ_p
°	′						
	1	1486.1	0.1	1474.0	0.1	1461.8	0.1
	2	2972.2	0.6	2948.0	0.6	2923.5	0.6
	3	4458.3	1.3	4421.9	1.3	4385.3	1.3
	4	5944.3	2.4	5895.9	2.4	5847.0	2.4
	5	7430.4	3.7	7369.9	3.7	7308.8	3.7
	6	8916.5	5.3	8843.9	5.3	8770.5	5.3
	7	10402.6	7.2	10317.8	7.2	10232.3	7.2
	8	11888.7	9.4	11791.8	9.5	11694.1	9.5
	9	13374.8	11.9	13265.8	12.0	13155.8	12.0
	10	14860.8	14.7	14739.8	14.8	14617.6	14.8
	11	16346.9	17.8	16213.7	17.8	16079.3	17.9
	12	17833.0	21.2	17687.7	21.2	17541.1	21.3
	13	19319.1	24.9	19161.7	24.9	19002.8	25.0
	14	20805.2	28.9	20635.7	28.9	20464.6	28.9
	15	22291.2	33.2	22109.6	33.2	21926.3	33.2
	16	23777.3	37.7	23583.6	37.8	23388.1	37.8
	17	25263.4	42.6	25057.6	42.6	24849.9	42.7
	18	26749.5	47.8	26531.6	47.8	26311.6	47.9
	19	28235.6	53.2	28005.5	53.3	27773.4	53.3
	20	29721.7	59.0	29479.5	59.0	29235.1	59.1
	25	37102.0	92.1	36849.3	92.2	36543.8	92.3
	30	44582.4	132.7	44219.2	132.8	43852.6	132.9
	40	59443.0	235.8	58958.7	236.1	58469.9	236.3
	50	74303.4	368.5	73698.0	368.9	73087.0	369.2
1	00	89163.6	530.7	88437.1	531.2	87703.9	531.7
1	20	118883.1	943.4	117914.5	944.4	116936.9	945.2
1	30	133742.4	1194.0	132652.7	1195.3	131552.9	1196.3
1	40	148601.2	1474.1	147390.5	1475.7	146168.4	1476.9
2	00	178317.6	2122.7	176864.7	2125.0	175398.2	2126.7
2	30	222888.2	3316.7	221071.9	3320.3	219238.7	3323.0
3	00	267452.8	4776.0	265273.1	4781.3	263073.1	4785.1
3	30	312010.3	6500.7	309467.1	6507.9	306900.3	6513.0
4	00	356559.5	8490.7	353652.8	8500.1	350719.0	8506.8

Polyconic Projection—Co ordinates, δ_m, δ_p, in Yards.

Longitude.	Latitude 44° 30′.		Latitude 45° 0′.		Latitude 45° 30′.	
	δ_m	δ_p	δ_m	δ_p	δ_m	δ_p
° ′						
1	1449.4	0.1	1437.0	0.1	1424.4	0.1
2	2898.9	0.6	2874.0	0.6	2848.9	0.6
3	4348.3	1.3	4311.0	1.3	4273.3	1.3
4	5797.7	2.4	5747.9	2.4	5697.7	2.4
5	7247.1	3.7	7184.9	3.7	7122.2	3.7
6	8696.6	5.3	8621.9	5.3	8546.6	5.3
7	10146.0	7.2	10058.9	7.2	9971.0	7.2
8	11595.4	9.5	11495.9	9.5	11395.4	9.5
9	13044.8	12.0	12932.9	12.0	12819.9	12.0
10	14494.3	14.8	14369.8	14.8	14244.3	14.8
11	15943.7	17.9	15806.8	17.9	15668.7	17.9
12	17393.1	21.3	17243.8	21.3	17093.1	21.3
13	18842.5	25.0	18680.8	25.0	18517.6	25.0
14	20292.0	29.0	20117.7	29.0	19942.0	29.0
15	21741.4	33.2	21554.7	33.3	21366.4	33.2
16	23190.8	37.8	22991.7	37.8	22790.8	37.8
17	24640.2	42.7	24428.7	42.7	24215.3	42.7
18	26089.7	47.9	25865.7	47.9	25639.7	47.9
19	27539.1	53.3	27302.7	53.4	27064.1	53.3
20	28988.5	59.1	28739.6	59.1	28488.6	59.1
25	36235.6	92.3	35924.5	92.4	35610.6	92.3
30	43482.6	133.0	43109.3	133.0	42732.7	133.0
40	57976.6	236.4	57478.9	236.5	56976.8	236.4
50	72470.4	369.4	71848.3	369.5	71220.6	369.4
1 00	86964.0	531.9	86217.4	532.0	85464.2	532.0
1 20	115950.3	945.7	114954.9	945.8	113950.6	945.7
1 30	130442.9	1196.8	129323.0	1197.1	128193.2	1196.9
1 40	144935.2	1477.6	143690.8	1477.9	142435.5	1477.7
2 00	173918.3	2127.7	172425.0	2128.1	170918.5	2127.8
2 30	217388.7	3324.6	215522.0	3325.2	213638.8	3324.8
3 00	260853.0	4787.4	258612.0	4788.3	256352.9	4787.7
3 30	304310.0	6516.2	301696.3	6517.3	299059.6	6516.5
4 00	347758.4	8510.9	344771.2	8512.5	341757.6	8511.4

Polyconic Projection—Co-ordinates, δ_m, δ_p, in Yards.

Longitude.	Latitude 46° 0′.		Latitude 46° 30′.		Latitude 47° 0′.	
	δ_m	δ_p	δ_m	δ_p	δ_m	δ_p
° ′						
1	1411.8	0.1	1399.0	0.1	1386.1	0.1
2	2823.5	0.6	2798.0	0.6	2772.2	0.6
3	4235.3	1.3	4197.0	1.3	4158.4	1.3
4	5647.1	2.4	5596.0	2.4	5544.5	2.4
5	7058.8	3.7	6995.0	3.7	6930.6	3.7
6	8470.6	5.3	8394.0	5.3	8316.7	5.3
7	9882.4	7.2	9793.0	7.2	9702.8	7.2
8	11294.1	9.5	11192.0	9.4	11089.0	9.4
9	12705.9	12.0	12591.0	12.0	12475.1	11.9
10	14117.7	14.8	13990.0	14.8	13861.2	14.7
11	15529.4	17.9	15389.0	17.9	15247.3	17.8
12	16941.2	21.3	16788.0	21.3	16633.4	21.2
13	18353.0	25.0	18186.9	24.9	18019.5	24.9
14	19764.7	28.9	19585.9	28.9	19405.7	28.9
15	21176.5	33.2	20984.9	33.2	20791.8	33.2
16	22588.3	37.8	22383.9	37.8	22177.9	37.7
17	24000.0	42.7	23782.9	42.7	23564.0	42.6
18	25411.8	47.9	25181.9	47.8	24950.1	47.8
19	26823.6	53.3	26580.9	53.3	26336.2	53.2
20	28235.3	59.1	27979.9	59.0	27722.4	59.0
25	35294.1	92.3	24974.8	92.2	34652.9	92.2
30	42352.9	132.9	41969.8	132.8	41583.4	132.7
40	56470.3	236.3	55959.5	236.2	55444.3	235.9
50	70587.5	369.3	69949.0	369.0	69305.1	368.6
1 00	84704.5	531.7	83938.2	531.3	83165.6	530.8
1 20	112937.6	945.3	111915.9	944.6	110885.7	943.6
1 30	127053.6	1196.4	125904.2	1195.5	124747.2	1194.3
1 40	141169.2	1477.0	139892.1	1476.0	138604.3	1474.4
2 00	169399.0	2126.9	167866.4	2125.4	166321.0	2123.2
2 30	211739.3	3323.3	209823.5	3320.9	207891.7	3317.5
3 00	254073.4	4785.6	251774.4	4782.1	249456.0	4777.2
3 30	296400.0	6513.8	293717.6	6509.0	291012.8	6502.7
4 00	338717.8	8507.8	335652.1	8501.5	332560.6	8492.7

Polyconic Projection—Co-ordinates, δ_m, δ_p, in Yards.

Longitude.	Latitude 47° 30′.		Latitude 48° 0′.		Latitude 48° 30′.	
	δ_m	δ_p	δ_m	δ_p	δ_m	δ_p
° ′						
1	1373.1	0.1	1360.0	0.1	1346.9	0.1
2	2746.3	0.6	2720.1	0.6	2693.7	0.6
3	4119.4	1.3	4080.1	1.3	4040.6	1.3
4	5492.5	2.4	5440.2	2.4	5387.4	2.4
5	6865.7	3.7	6800.2	3.7	6734.3	3.7
6	8238.8	5.3	8160.3	5.3	8081.1	5.3
7	9612.0	7.2	9520.3	7.2	9428.0	7.2
8	10985.1	9.4	10880.4	9.4	10774.8	9.4
9	12358 2	11.9	12240.4	11.9	12121.7	11.9
10	13731.4	14.7	13600.5	14.7	13468.5	14.7
11	15104.5	17.8	14960.5	17.8	14815.4	17.8
12	16477.6	21.2	16320.5	21.2	16162.2	21.1
13	17850.8	24.9	17680.6	24.8	17509.1	24.8
14	19223.9	28.9	19040.6	28.8	18855.9	28.8
15	20597.0	33.1	20400.7	33.1	20202.8	33.0
16	21970.1	37.7	21760.7	37.6	21549.6	37.6
17	23343.3	42.6	23120.7	42.5	22896.5	42.4
18	24716.4	47.7	24480.8	47.6	24243.3	47.5
19	26089.5	53.1	25840.8	53.1	25590.2	53.0
20	27462.7	58.9	27200.9	58.8	26937.0	58.7
25	34328.3	92.0	34001.0	91.9	33671.2	91.7
30	41193.9	132.5	40801.2	132.3	40405.4	132.0
40	54925.0	235.6	54401.4	235.2	53873.6	234.7
50	68655.8	368.1	68001.4	367.5	67341.7	366.8
1 00	82386.5	530.1	81601.1	529.2	80809.5	528.2
1 20	109846.9	942.4	108799.7	940.8	107744.2	939.0
1 30	123576.6	1192.7	122398.5	1190.7	121211.0	1188.4
1 40	137305.8	1472.4	135996.8	1470.0	134677.3	1467.1
2 00	164762.8	2120.3	163191.9	2116.8	161608.6	2112.7
2 30	205943.9	3313.0	203980.3	3307.5	202001.0	3301.1
3 00	247118.6	4770.7	244762.2	4762.9	242387.0	4753.5
3 30	288285.6	6493.5	285536.3	6482.8	282765.2	6470.1
4 00	329443.7	8481.3	326301.5	8467.3	323134.3	8450.7

Polyconic Projection—Co-ordinates, δ_m, δ_p, in Yards.

Longitude.		Latitude 49° 0'.		Latitude 49° 30'.		Latitude 50° 0'.	
		δ_m	δ_p	δ_m	δ_p	δ_m	δ_p
°	'						
	1	1333.6	0.1	1320.2	0.1	1306.7	0.1
	2	2667.1	0.6	2640.3	0.6	2213.3	0.6
	3	4000.7	1.3	2960.5	1.3	3920.0	1.3
	4	5334.2	2.3	5280.6	2.3	5226.6	2.3
	5	6667.8	3.7	6600.8	3.7	6533.3	3.6
	6	8001.3	5.3	7920.9	5.3	7839.9	5.2
	7	9334.9	7.2	9241.1	7.2	9146.6	7.1
	8	10668.4	9.4	10561.2	9.3	10453.2	9.3
	9	12002.0	11.9	11881.4	11.8	11759.9	11.8
	10	13335.5	14.6	13201.6	14.6	13066.5	14.6
	11	14669.1	17.7	14521.7	17.7	14373.2	17.6
	12	16002.7	21.1	15841.8	21.0	15679.8	21.0
	13	17336.2	24.7	17162.0	24.7	16986.5	24.6
	14	18669.7	28.7	18482.1	28.6	18293.1	28.5
	15	20003.3	32.9	19802.3	32.8	19599.8	32.8
	16	21336.8	37.5	21122.4	37.4	20906.4	37.3
	17	22670.4	42.3	22442.6	42.2	22213.1	42.1
	18	24004.0	47.4	23762.8	47.3	23519.7	47.2
	19	25337.5	52.8	25082.9	52.7	24826.4	52.6
	20	26671.1	58.6	26403.1	58.4	26133.0	58.2
	25	33338.8	91.5	33003.8	91.2	32666.2	91.0
	30	40006.5	131.7	39604.5	131.4	39199.4	131.0
	40	53341.7	234.2	52805.7	233.6	52265.7	232.9
	50	66676.8	365.9	66006.8	365.0	65331.7	364.0
1	00	80011.6	527.0	79207.6	525.6	78397.5	524.1
1	20	106680.4	936.8	105608.3	934.4	104528.2	931.7
1	30	120074.2	1185.7	118808.2	1182.6	117593.0	1179.2
1	40	133347.5	1463.8	132007.5	1460.0	130657.4	1455.8
2	00	160012.8	2107.9	158404.8	2102.5	156784.7	2096.4
2	30	200006.3	3293.6	197996.2	3285.1	195970.8	3275.6
3	00	239993.2	4742.8	237581.0	4730.5	235150.6	4716.9
3	30	279972.3	6455.4	277158.0	6438.8	274322.4	6420.2
4	00	319942.3	8431.6	316725.8	8409.9	313485.0	8385.6

Arcs of Parallel—Values of D_p *in Yards.*

′ ″	L. 20° 30′.	L. 21° 0′.	L. 21° 30′.	L. 22° 0′.	L. 22° 30′.	L. 23° 0′.
7	221.8	221.1	220.3	219.6	218.8	218.0
8	253.5	252.6	251.8	250.9	250.0	249.1
9	285.2	284.2	283.3	282.3	281.3	280.3
10	316.9	315.8	314.7	313.7	312.5	311.4
20	633.7	631.6	629.5	627.3	625.1	622.8
30	950.6	947.4	944.2	941.0	937.6	934.2
40	1267.4	1263.2	1259.0	1254.6	1250.2	1245.7
50	1584.3	1579.1	1573.7	1568.3	1562.7	1557.1
60	1901.1	1894.9	1888.5	1881.9	1875.3	1868.5
7 00	13307.8	13264.1	13219.4	13173.7	13127.0	13079.4
8 00	15208.9	15159.0	15107.9	15055.7	15002.3	14947.9
9 00	17110.0	17053.8	16996.4	16937.6	16877.6	16816.3
10 00	19011.1	18948.7	18884.9	18819.6	18752.9	18684.8
20 00	38022.1	37897.4	37769.7	37639.2	37505.8	37369.6
30 00	57033.2	56846.1	56654.6	56458.8	36258.8	56054.4
40 00	76044.3	75794.8	75539.5	75278.4	75011.7	74739.3
50 00	95055.4	94743.4	94424.3	94098.0	93764.6	93424.1
60 00	114066.4	113692.1	113309.2	112917.7	112517.6	112108.9

Meridional Arcs—Values of D_m *in Yards.*

′ ″		L. 21° 0′.		L. 22° 0′.		L. 23° 0′.
7		235.4		235.4		235.5
8		269.0		269.1		269.1
9		302.7		302.7		302.8
10		336.3		336.4		336.4
20		672.6		672.7		672.8
30		1008.9		1009.1		1009.2
40		1345.2		1345.4		1345.6
50		1681.6		1681.8		1682.0
60		2017.9		2018.1		2018.4
7 00		14125.0		14126.7		14128.5
8 00		16142.9		16144.8		16146.8
9 00		18160.8		18162.9		18165.2
10 00		20178.6		20181.0		20183.5
20 00		40357.2		40362.1		40367.1
30 00		60535.9		60543.1		60550.6
40 00		80714.5		80724.1		80734.1
50 00		100893.1		100905.2		100917.6
60 00		121071.7		121086.2		121101.2

Intermediate minutes and seconds will be found by moving the decimal point.

Arcs of Parallel—Values of D_p *in Yards.*

′ ″	L. 23° 30′.	L. 24° 0′.	L. 24° 30′.	L. 25° 0′.	L. 25° 30′.	L. 26° 0′.
7	217.2	216.4	215.4	214.6	213.8	212.9
8	248.2	247.3	246.3	245.3	244.3	243.3
9	279.2	278.2	277.1	276.0	274.8	273.7
10	310.3	309.1	307.9	306.6	305.4	304.1
20	620.5	618.1	615.7	613.3	610.8	608.2
30	930.8	927.6	923.6	919.9	916.2	912.3
40	1241.0	1236.3	1231.5	1226.5	1221.5	1216.4
50	1551.3	1545.4	1539.3	1533.2	1526.9	1520.5
60	1861.5	1854.4	1847.2	1839.8	1832.3	1824.7
7 00	13030.7	12981.0	12930.4	12878.8	12826.1	12772.6
8 00	14892.2	14835.5	14777.6	14718.6	14658.5	14597.2
9 00	16753.8	16689.9	16624.8	16558.4	16490.8	16421 9
10 00	18615.3	18544.3	18472.0	18398.2	18323.1	18246.5
20 00	37230.6	37088.7	36944.0	36796.5	36646.1	36493.0
30 00	55845.8	55633.0	55416.0	55194.7	54969.2	54739.6
40 00	74461.1	74177.4	73887.9	73592.9	73292.3	72986.1
50 00	93076.4	92721.7	92359.9	91991.1	91615.3	91232.6
60 00	111691.7	111266.0	110831.9	110389.4	109938.4	109479.1

Meridional Arcs—Values of D_m *in Yards.*

′ ″	L. 24° 0′.	L. 25° 0′.	L. 26° 0′.
7	235.5	235.5	235.6
8	269.1	269.2	269.2
9	302.8	302.8	302.9
10	336.4	336.5	336.5
20	672.9	673.0	673.1
30	1009.3	1009.4	1009.6
40	1345.7	1345.9	1346.1
50	1682.2	1682.4	1682.6
60	2018.6	2018.9	2019.2
7 00	14130.3	14132.1	14134.1
8 00	16148.9	16151.0	16153.2
9 00	18167.5	18169.9	18172.4
10 00	20186.1	20188.8	20191.5
20 00	40372.2	40377.5	40383.0
30 00	60558.3	60566.3	60574.6
40 00	80744.4	80755.1	80766.1
50 00	100930.5	100943.9	100957.6
60 00	121116.6	121132.6	121149.1

Intermediate minutes and seconds will be found by moving the decimal-point.

Arcs of Parallel—Values of D_p *in Yards.*

′ ″	L. 26° 30′.	L. 27° 0′.	L. 27° 30′.	L. 28° 0′.	L. 28° 30′.	L. 29° 0′.
7	212.0	211.0	210.1	209.1	208.2	207.2
8	242.2	241.2	240.1	239.0	237.9	236.8
9	272.5	271.3	270.1	268.9	267.6	266.4
10	302.8	301.5	300.1	298.8	297.4	296.0
20	605.6	603.0	600.3	597.6	594.8	591.9
30	908.4	904.5	900.4	896.3	892.2	887.9
40	1211.2	1206.0	1200.6	1195.1	1189.5	1183.9
50	1514.0	1507.4	1500.7	1493.9	1486.9	1479.9
60	1816.9	1808.9	1800.9	1792.7	1784.3	1775.8
7 00	12718.0	12662.5	12606.0	12548.6	12490.1	12430.8
8 00	14534.9	14471.4	14406.9	14341.2	14274.5	14206.6
9 00	16351.7	16280.3	16207.7	16133.9	16058.8	15982.5
10 00	18168.6	18089.3	18008.6	17926.5	17843.1	17758.3
20 00	36337.2	36178.5	36017.1	35853.0	35686.2	35516.6
30 00	54505.8	54267.8	54025.7	53779.5	53529.3	53274.9
40 00	72674.4	72357.1	72034.3	71706.1	71372.4	71033.2
50 00	90843.0	90446.3	90042.9	89632.6	89215.4	88791.5
60 00	109011.5	108535.6	108051.4	107559.1	107058.5	106549.8

Meridional Arcs—Values of D_m *in Yards.*

′ ″		L. 27° 0′.		L. 28° 0′.		L. 29° 0′.
7		235.6		235.6		235.7
8		269.3		269.3		269.3
9		302.9		303.0		303.0
10		336.6		336.6		336.7
20		673.1		673.2		673.3
30		1009.7		1009.9		1010.0
40		1346.3		1346.5		1346.7
50		1682.9		1683.1		1683.3
60		2019.4		2019.7		2020.0
7 00		14136.0		14138.1		14140.1
8 00		16155.5		16157.8		16160.2
9 00		18174.9		18177.5		18180.2
10 00		20194.3		20197.2		20200.2
20 00		40388.7		40394.5		40400.4
30 00		60583.0		60591.7		60600.6
40 00		80777.4		80788.9		80800.8
50 00		100971.7		100986.2		101001.0
60 00		121166.0		121183.4		121201.2

Intermediate minutes and seconds will be found by moving the decimal-point.

Arcs of Parallel—Values of D_p in Yards.

′ ″	L. 29° 30′.	L. 30° 0′.	L. 30° 30′.	L. 31° 0′.	L. 31° 30′.	L. 32° 0′.
7	206.2	205.2	204.1	203.1	202.0	200.9
8	235.6	234.5	233.3	232.1	230.9	229.6
9	265.1	263.8	262.4	261.1	259.7	258.3
10	294.5	293.1	291.6	290.1	288.6	287.0
20	589.1	586.2	583.2	580.2	577.1	574.0
30	883.6	879.2	874.8	870.3	865.7	861.1
40	1178.1	1172.3	1166.4	1160.4	1154.3	1148.1
50	1472.7	1465.4	1458.0	1450.5	1442.9	1435.1
60	1767.2	1758.5	1749.6	1740.6	1731.4	1722.1
7 00	12370.5	12309.3	12247.1	12184.0	12120.0	12055.0
8 00	14137.7	14067.7	13996.7	13924.6	13851.4	13777.1
9 00	15904.9	15826.2	15746.3	15665.1	15582.8	15499.3
10 00	17672.2	17584.7	17495.9	17405.7	17314.2	17221.4
20 00	35344.3	35169.4	34991.7	34811.4	34628.4	34442.9
30 00	53016.5	52754.0	52487.6	52217.1	51942.7	51664.3
40 00	70688.7	70338.7	69983.4	69622.8	69256.9	68885.7
50 00	88360.8	87923.4	87479.3	87028.5	86571.1	86107.1
60 00	106033.0	105508.1	104975.2	104434.2	103885.3	103328.6

Meridional Arcs—Values of D_m in Yards.

′ ″	L. 30° 0′.	L. 31° 0′.	L. 32° 0′.
7	235.7	235.7	235.8
8	269.4	269.4	269.5
9	303.0	303.1	303.1
10	336.7	336.8	336.8
20	673.4	673.5	673.6
30	1010.2	1010.3	1010.5
40	1346.9	1347.1	1347.3
50	1683.6	1683.9	1684.1
60	2020.3	2020.6	2020.9
7 00	14142.3	14144.4	14146.6
8 00	16162.6	16165.1	16167.6
9 00	18182.9	18185.7	18188.5
10 00	20203.2	20206.3	20209.5
20 00	40406.5	40412.6	40418.9
30 00	60609.7	60619.0	60628.4
40 00	80812.9	80825.3	80837.9
50 00	101016.1	101031.6	101047.4
60 00	121219.4	121237.9	121256.8

Intermediate minutes and seconds will be found by moving the decimal-point.

Arcs of Parallel—Values of D_p *in Yards.*

′ ″	L. 32° 30′.	L. 33° 0′.	L. 33° 30′.	L. 34° 0′.	L. 34° 30′.	L. 35° 0′.
7	199.8	198.7	197.6	196.4	195.3	194.1
8	228.4	227.1	225.8	224.5	223.2	221.8
9	256.9	254.5	254.0	252.6	251.1	249.6
10	285.5	283.9	282.3	280.6	279.0	277.3
20	570.9	567.7	564.5	561.2	557.9	554.6
30	856.4	851.6	846.8	841.9	836.9	831.9
40	1141.8	1135.5	1129.0	1122.5	1115.9	1109.2
50	1427.3	1419.3	1411.3	1403.1	1394.8	1386.4
60	1712.7	1703.2	1693.5	1683.7	1673.8	1663.7
7 00	11989.1	11922.3	11854.6	11786.0	11716.5	11646.1
8 00	13701.9	13625.5	13548.1	13469.7	13390.3	13309.8
9 00	15414.6	15328.7	15241.7	15153.5	15064.1	14973.6
10 00	17127.3	17031.9	16935.2	16837.2	16737.9	16637.3
20 00	34254.6	34063.8	33870.4	33674.3	33475.8	33274.6
30 00	51381.9	51095.7	50805.5	50511.5	50213.6	49911.9
40 00	68509.3	68127.6	67740.7	67348.7	66951.5	66549.2
50 00	85636.6	85159.5	84675.9	84185.8	83689.4	83186.5
60 00	102763.9	102191.4	101611.1	101023.0	100427.3	99823.8

Meridional Arcs—Values of D_m *in Yards.*

′ ″		L. 33° 0′.		L. 34° 0′.		L. 35° 0′.
7		235.8		235.9		235.9
8		269.5		269.5		269.6
9		303.2		303.2		303.3
10		336.9		336.9		337.0
20		673.8		673.9		674.0
30		1010.6		1010.8		1011.0
40		1347.5		1347.7		1347.9
50		1684.4		1684.7		1684.9
60		2021.3		2021.6		2021.9
7 00		14148.9		14151.2		14153.5
8 00		16170.1		16172.7		16175.4
9 00		18191.4		18194.3		18197.3
10 00		20212.7		20215.9		20219.2
20 00		40425.4		40431.9		40438.5
30 00		60638.0		60647.8		60657.7
40 00		80850.7		80863.7		80877.0
50 00		101063.4		101079.7		101096.2
60 00		121276.1		121295.6		121315.4

Intermediate minutes and seconds will be found by moving the decimal-point.

Arcs of Parallel—Values of D_p *in Yards.*

′ ″	L. 35° 30′.	L. 36° 0′.	L. 36° 30′.	L. 37° 0′.	L. 37° 30.	L. 38° 0′.
7	192.9	191.7	190.5	189.3	188.0	186.8
8	220.5	219.1	217.7	216.3	214.9	213.4
9	248.0	246.5	244.9	243.3	241.7	240.1
10	275.6	273.9	272.1	270.4	268.6	266.8
20	551.2	547.7	544.3	540.7	537.2	533.6
30	826.8	821.6	816.4	811.1	805.8	800.4
40	1102.4	1095.5	1088.5	1081.5	1074.4	1067.2
50	1378.0	1369.4	1360.7	1351.9	1343.0	1334.0
60	1653.5	1643.2	1632.8	1622.2	1611.6	1600.7
7 00	11574.8	11502.7	11429.6	11355.7	11280.9	11205.2
8 00	13228.4	13145.9	13062.4	12977.9	12892.5	12806.0
9 00	14881.9	14789.1	14695.2	14600.2	14504.0	14406.7
10 00	16535.5	16432.4	16328.0	16222.4	16115.6	16007.5
20 00	33070.9	32864.7	32656.0	32444.8	32231.1	32015.0
30 00	49606.4	49297.1	48984.0	48667.2	48346.7	48022.5
40 00	66141.9	65729.5	65312.1	64889.6	64462.3	64030.0
50 00	82677.3	82161.8	81640.1	81112.0	80577.8	80037.5
60 00	99212.8	98594.2	97968.1	97334.4	96693.4	96045.0

Meridional Arcs—Values of D_m *in Yards.*

′ ″	L. 36° 0′.	L. 37° 0′.	L. 38° 0′.
7	235.9	236.0	236.0
8	269.6	269.7	269.7
9	303.3	303.4	303.4
10	337.0	337.1	337.2
20	674.1	674.2	674.3
30	1011.1	1011.3	1011.5
40	1348.2	1348.4	1348.6
50	1685.2	1685.5	1685.8
60	2022.3	2022.6	2022.9
7 00	14155.8	14158.2	14160.6
8 00	16178.1	16180.8	16183.5
9 00	18200.3	18203.4	18206.5
10 00	20222.6	20226.0	20229.4
20 00	40445.2	40451.9	40458.8
30 00	60667.5	60677.9	60688.2
40 00	80890.3	80903.9	80917.6
50 00	101112.9	101129.9	101147.0
60 00	121335.5	121355.8	121376.4

Intermediate minutes and seconds will be found by moving the decimal-point.

Arcs of Parallel—Values of D_p in Yards.

′ ″	L. 38° 30′.	L. 39° 0′.	L. 39° 30′.	L. 40° 0′.	L. 40° 30′.	L. 41° 0′.
7	185.5	184.2	182.9	181.6	180.2	178.9
8	212.0	210.5	209.0	207.5	206.0	204.4
9	238.5	236.8	235.1	233.4	231.7	230.0
10	265.0	263.1	261.3	259.4	257.5	255.6
20	529.9	526.3	522.5	518.8	515.0	511.1
30	794.9	789.4	783.8	778.2	772.4	766.7
40	1059.9	1052.5	1045.1	1037.5	1029.9	1022.2
50	1324.9	1315.6	1306.3	1296.9	1287.4	1277.8
60	1589.8	1578.8	1567.6	1556.3	1544.9	1533.4
7 00	11128.7	11051.4	10973.2	10894.1	10814.3	10733.6
8 00	12718.6	12630.2	12540.8	12450.4	12359.2	12266.9
9 00	14308.4	14208.9	14108.4	14006.8	13904.1	13800.3
10 00	15898.2	15787.7	15676.0	15563.1	15448.9	15333.4
20 00	31796.4	31575.4	31351.9	31126.1	30897.9	30667.3
30 00	47694.6	47363.1	47027.9	46689.2	46346.8	46001.0
40 00	63592.8	63150.8	62703.9	62252.2	61795.8	61334.6
50 00	79491.0	78938.4	78379.9	77815.3	77244.7	76668.3
60 00	95389.2	94726.1	94055.8	93378.3	92693.7	92001.9

Meridional Arcs—Values of D_m in Yards.

′ ″		L. 39° 0′.		L. 40° 0′.		L. 41° 0′.
7		236.0		236.1		236.1
8		269.8		269.8		269.9
9		303.5		303.5		303.6
10		337.2		337.3		337.3
20		674.4		674.5		674.7
30		1011.6		1011.8		1012.0
40		1348.9		1349.1		1349.3
50		1686.1		1686.4		1686.7
60		2023.3		2023.6		2024.0
7 00		14163.0		14165.4		14167.9
8 00		16186.3		16189.1		16191.9
9 00		18209.6		18212.7		18215.8
10 00		20232.8		20236.3		20239.8
20 00		40465.7		40472.7		40479.7
30 00		60698.5		60709.0		60719.5
40 00		80931.4		80945.3		80959.3
50 00		101164.2		101181.6		101199.2
60 00		121397.1		121418.0		121439.0

Intermediate minutes and seconds will be found by moving the decimal-point.

Arcs of Parallel—Values of D_p *in Yards.*

′ ″	L. 41° 30′.	L. 42° 0′.	L. 42° 30′.	L. 43° 0′.	L. 43° 30′.	L. 44° 0′.
7	177.5	176.2	174.8	173.4	172.0	170.5
8	202.9	201.3	199.7	198.1	196.5	194.9
9	228.3	226.5	224.7	222.9	221.1	219.3
10	253.6	251.7	249.7	247.7	245.7	243.6
20	507.2	503.3	449.4	495.4	491.3	487.3
30	760.9	755.0	749.0	743.0	737.0	730.9
40	1014.5	1006.6	998.7	990.7	982.7	974.5
50	1268.1	1258.3	1248.4	1238.4	1228.3	1218.1
60	1521.7	1510.0	1498.1	1486.1	1474.0	1461.8
7 00	10652.0	10569.7	10486.6	10402.6	10317.9	10232.3
8 00	12173.8	12079.7	11984.6	11888.7	11791.8	11694.1
9 00	13695.5	13589.6	13482.7	13374.8	13265.8	13155.8
10 00	15217.2	15099.6	14980.8	14860.9	14739.8	14617.6
20 00	30434.4	30199.1	29961.6	29721.7	29479.6	29235.2
30 00	45651.6	45298.7	44942.4	44582.6	44219.4	43852.8
40 00	60868.8	60398.3	59923.2	59443.4	58959.2	58470.4
50 00	76086.0	75497.9	74903.9	74304.3	73698.9	73088.0
60 00	91303.2	90597.4	89884.7	89165.1	88438.7	87705.6

Meridional Arcs—Values of D_m *in Yards.*

′ ″		L. 42° 0′.		L. 43° 0′.		L. 44° 0′.
7		236.2		236.2		236.3
8		269.9		270.0		270.0
9		303.7		303.7		303.8
10		337.4		337.4		337.5
20		674.8		674.9		675.0
30		1012.2		1012.3		1012.5
40		1349.6		1349.8		1350.0
50		1686.9		1687.2		1687.5
60		2024.3		2024.7		2025.0
7 00		14170.3		14172.8		14175.3
8 00		16194.7		16197.5		16200.3
9 00		18219.0		18222.2		18225.4
10 00		20243.4		20246.9		20250.4
20 00		40486.7		40493.8		40500.9
30 00		60730.1		60740.7		60751.3
40 00		80973.4		80987.5		81001.7
50 00		101216.8		101234.4		101252.2
60 00		121460.1		121481.3		121502.6

Intermediate minutes and seconds will be found by moving the decimal-point.

Arcs of Parallel—Values of D_p *in Yards.*

′ ″	L. 44° 30′.	L. 45° 0′.	L. 45° 30′.	L. 46° 0′.	L. 46° 30′.	L. 47° 0′.
7	169.1	167.6	166.2	164.7	163.2	161.7
8	193.3	191.6	189.9	188.2	186.5	184.8
9	217.4	215.5	213.7	211.8	209.8	207.9
10	241.6	239.5	237.4	235.3	233.2	331.0
20	483.1	479.0	474.8	470.6	466.3	462.0
30	724.7	718.5	712.2	705.9	699.5	693.1
40	966.3	958.0	949.6	941.2	932.7	924.1
50	1207.9	1197.5	1187.0	1176.5	1165.8	1155.1
60	1449.4	1437.0	1424.4	1411.8	1399.0	1386.1
7 00	10146.0	10058.9	9971.0	9882.4	9793.0	9702.8
8 00	11595.4	11495.9	11395.5	11294.2	11192.0	11089.0
9 00	13044.8	12932.9	12819.9	12705.9	12591.0	12475.1
10 00	14494.3	14369.8	14244.3	14117.7	13990.0	13861.2
20 00	28988.6	28739.7	28488.6	28235.4	27980.0	27722.4
30 00	43482.8	43109.5	42733.0	42353.1	41970.0	41583.6
40 00	57977.1	57479.4	56977.3	56470.8	55960.0	55444.8
50 00	72471.4	71849.2	71221.6	70588.5	69950.0	69306.0
60 00	86965.7	86219.1	85465.9	84706.2	83940.0	83167.3

Meridional Arcs—Values of D_m *in Yards.*

′ ″		L. 45° 0′.		L. 46° 0′.		L. 47° 0′.
7		236.3		236.3		236.4
8		270.1		270.1		270.1
9		303.8		303.9		303.9
10		337.6		337.6		337.7
20		675.1		675.3		675.4
30		1012.7		1012.9		1013.1
40		1350.3		1350.5		1350.7
50		1687.8		1688.1		1688.4
60		2025.4		2025.8		2026.1
7 00		14177.8		14180.3		14182.8
8 00		16203.2		16206.0		16208.9
9 00		18228.6		18231.8		18235.0
10 00		20254.0		20257.5		20261.1
20 00		40508.0		40515.1		40522.2
30 00		60761.9		60772.6		60783.2
40 00		81015.9		81030.1		81044.3
50 00		101269.9		101287.7		101305.4
60 00		121523.9		121545.2		121566.5

Intermediate minutes and seconds will be found by moving the decimal point.

Arcs of Parallel—Values of D_p *in Yards.*

′ ″	L. 47° 30′.	L. 48° 0′.	L. 48° 30′.	L. 49° 0′.	L. 49° 30′.	L. 50° 0′.
7	160.2	158.7	157.1	155.6	154.0	152.4
8	183.1	181.3	179.6	177.8	176.0	174.2
9	206.0	204.0	202.0	200.0	198.0	196.0
10	228.9	226.7	224.5	222.3	220.0	217.8
20	457.7	453.3	449.0	444.5	440.1	435.6
30	686.6	680.0	673.4	666.8	660.1	653.3
40	915.4	906.7	897.9	889.0	880.1	871.1
50	1144.3	1133.4	1122.4	1111.3	1100.1	1088.9
60	1373.1	1360.0	1346.9	1333.6	1320.2	1306.7
7 00	9612.0	9520.3	9428.0	9334.9	9241.1	9146.6
8 00	10985.1	10880.4	10774.8	10668.5	10561.2	10453.2
9 00	12358.2	12240.4	12121.7	12002.0	11881.4	11759.9
10 00	13731.4	13600.5	13468.5	13335.6	13201.6	13066.5
20 00	27462.7	27200.9	26937.1	26671.1	26403.1	26133.1
30 00	41194.1	40801.4	40405.6	40006.7	39604.7	39199.6
40 00	54925.5	54401.9	53874.1	53342.3	52806.2	52266.2
50 00	68656.8	68002.4	67342.7	66677.8	66007.8	65332.7
60 00	82388.2	81602.8	80811.2	80013.4	79209.4	78399.3

Meridional Arcs—Values of D_m *in Yards.*

′ ″		L. 48° 0′.		L. 49° 0′.		L. 50° 0′.
7		236.4		236.5		236.5
8		270.2		270.2		270.3
9		304.0		304.0		304.1
10		337.7		337.8		337.9
20		675.5		675.6		675.7
30		1013.2		1013.4		1013.6
40		1351.0		1351.2		1351.4
50		1688.7		1689.0		1689.3
60		2026.5		2026.8		2027.2
7 00		14185.2		14187.7		14190.2
8 00		16211.7		16214.5		16217.3
9 00		18238.2		18241.3		18244.5
10 00		20264.6		20268.1		20271.7
20 00		40529.2		40536.3		40543.3
30 00		60793.9		60804.4		60815.0
40 00		81058.5		81072.6		81086.6
50 00		101323.1		101340.7		101358.9
60 00		121587.7		121608.9		121629 9

Intermediate minutes and seconds will be found by moving the decimal-point.

Lengths in Nautical Miles and Statute Miles of Degrees of Latitude and Longitude in Different Latitudes.

DEGREE OF THE PARALLEL.			DEGREE OE THE MERIDIAN.		
Latitude of Parallel.	Nautical miles.	Statute miles.	Latitude of middle point.	Nautical miles.	Statute miles.
°			°		
20	56.404	65.018	20	59.664	68.777
21	56.039	64.598			
22	55.657	64.158			
23	55.258	63.698			
24	54.843	63.219			
25	54.411	62.721	25	59.706	68.825
26	53.962	62.204			
27	53.497	61.668			
28	53.016	61.113			
29	52.518	60.540			
30	52.005	59.948	30	59.749	68.875
31	51.476	59.338			
32	50.931	58.709			
33	50.370	58.063			
34	49.794	57.399			
35	49.203	56.718	35	59.796	68.929
36	48.597	56.019			
37	47.976	55.304			
38	47.341	54.571			
39	46.690	53.822			
40	46.026	53.056	40	59.847	68.987
41	45.348	52.274			
42	44.654	51.476			
43	43.949	50.662			
44	43.230	49.833			
45	42.497	48.988	45	59.899	69.048
46	41.752	48.128			
47	40.993	47.254			
48	40.222	46.365			
49	39.439	45.462			
50	38.643	44.545	50	59.951	69.108

A degree of longitude at the equator = 69.163 statute miles.
A second of time at the equator = 1521.6 feet.

Co-ordinates of Curvature in Statute Miles for Maps of Large Extent.

Longitude.	Latitude 20°.		Latitude 22°.		Latitude 24°.		Latitude 26°.	
	d_m	d_p	d_m	d_p	d_m	d_p	d_m	d_p
°								
2	130.0	0.8	128.3	0.8	126.4	0.9	124.4	0.9
4	260.0	3.1	256.6	3.3	252.8	3.6	248.8	3.8
6	390.0	6.9	384.8	7.5	379.2	8.1	373.1	8.6
8	520.0	12.4	513.0	13.4	505.5	14.4	497.3	15.2
10	649.8	19.4	641.1	21.0	631.7	22.4	621.4	23.8
12	779.7	27.8	769.1	30.2	757.9	32.2	745.4	34.2
14	909.2	38.0	896.9	41.0	883.6	43.9	869.2	46.6
16	1039.2	49.6	1024.5	53.6	1009.9	57.4	992.8	60.8
18	1168.1	62.8	1152.2	67.9	1134.8	72.6	1116.1	77.0
20	1298.0	77.6	1279.5	83.8	1261.2	89.7	1239.2	95.0
r	10892		9813		8905		8130	

Longitude.	Latitude 28°.		Latitude 30°.		Latitude 32°.		Latitude 34°.	
	d_m	d_p	d_m	d_p	d_m	d_p	d_m	d_p
°								
2	122.2	1.0	119.8	1.0	117.4	1.1	114.8	1.1
4	244.4	4.0	239.7	4.2	234.8	4.3	229.5	4.5
6	366.5	9.0	359.5	9.4	352.0	9.8	344.2	10.1
8	488.6	16.0	479.2	16.7	469.3	17.3	458.7	17.9
10	610.4	25.0	598.7	26.1	586.3	27.1	573.1	28.0
12	732.4	36.0	718.0	37.6	703.5	39.1	687.2	40.3
14	853.7	49.0	837.1	51.2	819.6	53.1	801.1	54.8
16	975.7	64.1	956.0	66.9	936.8	69.5	914.7	71.6
18	1096.0	80.9	1074.6	84.6	1051.9	87.8	1027.9	90.5
20	1218.8	100.1	1192.9	104.3	1169.2	108.6	1140.7	111.7
r	7458		6869		6348		5881	

Co-ordinates of Curvature in Statute Miles for Maps of Large Extent.

Longitude.	Latitude 36°.		Latitude 38°.		Latitude 40°.		Latitude 42°.	
	d_m	d_p	d_m	d_p	d_m	d_p	d_m	d_p
°								
2	112.0	1.2	109.1	1.2	106.1	1.2	102.9	1.2
4	224.0	4.6	218.2	4.7	212.2	4.8	205.8	4.8
6	335.9	10.3	327.2	10.6	318.1	10.7	308.6	10.8
8	447.7	18.4	436.0	18.8	423.9	18.9	411.2	19.2
10	559.2	28.7	544.7	29.3	529.4	29.7	513.6	30.0
12	670.5	41.3	653.0	42.2	634.7	42.8	615.7	43.2
14	781.6	56.2	761.1	57.4	739.7	58.2	717.5	58.8
16	892.3	73.4	868.8	74.9	844.3	76.0	818.8	76.7
18	1002.6	92.8	976.2	94.7	948.5	96.1	919.8	97.0
20	1112.5	114.5	1083.0	116.8	1052.3	118.5	1020.2	119.7
r	5461		5079		4729		4408	

Longitude.	Latitude 44°.		Latitude 46°.		Latitude 48°.		Latitude 50°.	
	d_m	d_p	d_m	d_p	d_m	d_p	d_m	d_p
°								
2	99.7	1.2	96.2	1.2	92.7	1.2	89.1	1.2
4	198.9	4.8	192.4	4.8	185.4	4.8	178.1	4.8
6	298.7	10.9	288.5	10.9	277.9	10.8	267.0	10.7
8	398.0	19.3	384.4	19.3	370.3	19.2	355.7	19.0
10	497.1	30.2	480.0	30.2	462.3	30.0	444.1	29.7
12	595.9	43.4	575.4	43.4	554.1	43.2	532.3	42.8
14	694.3	59.1	670.3	59.1	645.6	58.8	620.0	58.2
16	792.3	77.1	764.9	77.1	736.5	76.7	707.3	75.9
18	889.9	97.5	859.0	97.5	827.0	97.0	794.1	96.0
20	986.9	120.2	952.5	120.2	916.9	119.6	880.3	118.4
r	4110		3833		3575		3332	

XLIV.—*Trigonometrical Leveling.*

Reciprocal zenith-distances measured at two stations at the same time give the best results. When reciprocal, but not simultaneous, the observations should be made on different days, in order to obtain as far as possible an average value of the refraction as well as the mean value of the difference between the respective angles; the same with zenith-distances measured at one station only.

The refraction being greater and more variable at sunrise and sunset, and comparatively stationary between the hours of 10 a. m. and 4 p. m., the best time for observation is between those hours.

The condition of the atmosphere and the relative refraction may be so different at stations more than twenty miles apart, that, as a general rule, the difference of level, determined even by reciprocal observations, cannot be relied upon for much accuracy at greater distances unless a very large number of measurements have been made under the most favorable circumstances. The higher the elevations the more reliable the results.

1.—*To obtain the Difference of Level of Two Points from Reciprocal Zenith-Distances, simultaneous or not.*

Let—

Z, Z' = the measured zenith-distances of the telescopes at the two stations, of which Z is the smaller;

K = distance in yards between the two stations;

R_z = radius of curvature of the arc joining the two stations;

C = angle at the earth's center subtended by the arc; and

dh = difference of level of the two stations;

then—

$$C = \frac{K}{R_z \sin 1''}; \qquad dh = \frac{K \sin \frac{1}{2}(Z' - Z)}{\cos \frac{1}{2}(Z' - Z + C)}$$

XLIV.—*Trigonometrical Leveling*—Continued.

If the telescope is not observed upon, but some other object near, the measured zenith-distance can be reduced to the telescope by the formula:

$$\text{Reduction to telescope, in seconds,} = \pm \frac{r}{K \sin 1''}$$

in which r represents the distance of the object (in yards and decimals) above or below the telescope.

Logarithmic values of R_z, in yards, which depend on the inclination to the meridian of the arc joining the two stations and their mean latitude, are given in the following table:

Angle of inclination.	Lat. 25°.	Lat. 30°.	Lat. 35°.	Lat. 40°.	Lat. 45°.	Lat. 50°.
	log R_z	log R_z	log R_z	log R_z	log R_z	log R_z
°						
0	6.841384	6.841695	6.842039	6.842406	6.842785	6.843164
10	1456	1760	2098	2458	2829	3200
20	1663	1950	2267	2606	2955	3304
30	1981	2240	2527	2833	3148	3464
40	2370	2596	2845	3111	3386	3631
50	2785	2975	3184	3408	3639	3870
60	3176	3331	3504	3687	3877	4066
70	3494	3622	3764	3915	4069	4227
80	3702	3812	3934	4063	4197	4331
90	6.843774	6.843878	6.843993	6.844115	6.844241	6.844368

$$\log \sin 1'' = 4.685575$$

2.—*By the Zenith-Distance measured at one Station.*

Let—

$Z =$ the measured zenith-distance of the signal or object;

$K =$ the distance between the two stations in yards;

$m =$ the co-efficient of refraction $= 0.071$; and

$dh =$ difference in height between the two stations;

then—

$$C = \frac{K}{R_z \sin 1''}; \qquad dh = \frac{K \cos (Z + m\,C - \frac{1}{2}\,C)}{\sin (Z + m\,C - C)}$$

XLIV.—*Trigonometrical Leveling*—Continued.

3.—*By the Observed Zenith-Distance of the Sea-Horizon.*

Let—

$Z =$ the measured zenith-distance;

$R_z =$ the radius of curvature of the arc; and

$m =$ the co-efficient of refraction $= 0.078$;

then—

$$h = \frac{R_z}{2(1-m)^2} \tan^2 (Z - 90°)$$

4.—*By Observed Angles of Elevation or Depression.*

Let—

$A =$ the observed angle expressed in seconds; and

$K =$ the distance in yards between the two stations.

$$dh = 0.00000485 \text{ K A} \pm 0.000000667 \text{ K}^2$$

$$[\log 4.68574] \qquad [\log 2.82413]$$

This gives the difference in heights between stations not more than ten or fifteen miles apart, with a probable error less than the uncertainty in the co-efficient of refraction.

Co-efficient of Refraction.

The co-efficient of refraction, or proportion of the intercepted arc, is determined from the observed zenith-distances of two stations, the relative altitudes of which have been determined by the spirit-level; or from reciprocal zenith-distances, simultaneous or not, under the assumption that the mean of a number of observations taken under favorable conditions will eliminate the difference of refraction which is found to exist, even at the same moment at two stations a few miles apart.

The longer the distance the greater is the error caused by any uncertainty in the co-efficient, or in the actual refraction; consequently there is a limit to the distance for which any assumed mean value of the refraction can be depended on for accurate results.

XLIV.—*Trigonometrical Leveling*—Continued.

The average value of the co-efficient from the Coast Survey observations in the New England States—

Between primary stations = 0.071
Of small elevations...... = 0.075
Of the sea-horizon...... = 0.078

In the trigonometrical survey of Massachusetts Mr. Borden used 0.0784 as a mean co-efficient for the sea-coast, and 0.0697 for the interior of the State.

1.—*To determine the Co-efficient of Refraction from Reciprocal Zenith-Distances.*

Let—

C = angle at earth's center subtended by arc;
F = angle of refraction; and
m = co-efficient of refraction;

then—

$$C = \frac{K}{R_z \sin 1''}; \quad F = \frac{C}{2} - \tfrac{1}{2}(Z' + Z - 180°); \quad m = \frac{F}{C}$$

2.—*To determine the Co-efficient from the Zenith-Distance observed at one Station, when the Altitude of the Two Stations above Tide, or their Difference in Height, have been determined by the Leveling-Instrument.*

Compute the true zenith-distances, Z_0' and Z_0, of the two given points, and the difference between the true and the observed zenith-distances will be the angle of refraction F.

$$\tfrac{1}{2}(Z_0' + Z_0) = 90° + \frac{C}{2}$$

$$\tfrac{1}{2}(Z_0' - Z_0) = \tan^{-1}\left\{\frac{h' - h}{K}\left(1 - \frac{h' + h}{2R} - \frac{K^2}{12R^2}\right)\right\}$$

$$Z_0 - Z = F; \qquad m = \frac{F}{C}$$

h and h' having been determined by the leveling-instrument.

Difference between the Apparent and the True Level.

Correction for curvature............ $= \frac{D^2}{2R}$

Correction for refraction............ $= \frac{D^2}{R} \cdot m$

Correction for curvature and refraction $= (1 - 2m)\frac{D^2}{2R}$

where—

D = the distance;

R = mean radius of the earth; and

m = co-efficient of refraction;

or—

m being = 0.075, and log R in feet = 7.31991307,

Correction for curvature, in feet, = log D^2 — const log [7.6209430]

Correction for refraction, in feet, = log D^2 + const log [1.5551483]

Corrections for Curvature and Refraction, showing the Differences of the Apparent and True Level, in Feet and Decimals of a Foot, for Distances in Miles.

Distance, miles.	Difference in feet for—			Distance, miles.	Difference in feet for—		
	Curvature.	Refraction.	Curvature and refraction.		Curvature.	Refraction.	Curvature and refraction.
1	0.7	0.1	0.6	13	112.8	16.9	95.9
2	2.7	0.4	2.3	14	130.8	19.6	111.2
3	6.0	0.9	5.1	15	150.2	22.5	127.7
4	10.6	1.6	9.0	16	170.8	25.6	145.2
5	16.7	2.5	14.2	17	192.9	28.9	164.1
6	24.0	3.6	20.4	18	216.2	32.4	183.8
7	32.7	4.9	27.8	19	240.9	36.1	204.8
8	42.7	6.4	36.3	20	266.9	40.0	226.9
9	54.1	8.1	44.0	21	294.3	44.1	250.2
10	66.7	10.0	56.7	22	323.0	48.4	274.6
11	80.7	12.1	68.6	23	353.0	52.9	300.1
12	96.1	14.4	81.7	24	384.4	57.7	326.7

Reduction, in Feet and Decimals, upon 100 *Feet, for the following Vertical Angles.*

Angle.	Reduc.	Angle.	Reduc.	Angle.	Reduc.	Angle.	Reduc.
° ′		° ′		° ′		° ′	
3 0	.137	7 30	.856	12 0	2.185	16 30	4.118
3 15	.161	7 45	.913	12 15	2.277	16 45	4.243
3 30	.187	8 0	.973	12 30	2.370	17 0	4.370
3 45	.214	8 15	1.035	12 45	2.466	17 15	4.498
4 0	.244	8 30	1.098	13 0	2.553	17 30	4.628
4 15	.275	8 45	1.164	13 15	2.662	17 45	4.760
4 30	.308	9 0	1.231	13 30	2.763	18 0	4.894
4 45	.343	9 15	1.300	13 45	2.866	18 15	5.030
5 0	.381	9 30	1.371	14 0	2.970	18 30	5.168
5 15	.420	9 45	1.444	14 15	2.077	18 45	5.307
5 30	.460	10 0	1.519	14 30	3.185	19 0	5.448
5 45	.503	10 15	1.596	14 45	3.295	19 15	5.591
6 0	.548	10 30	1.675	15 0	3.407	19 30	5.736
6 15	.594	10 45	1.755	15 15	3.521	19 45	5.882
6 30	.643	11 0	1.837	15 30	3.637	20 0	6.031
6 45	.693	11 15	1.921	15 45	3.754		
7 0	.745	11 30	2.008	16 0	3.874		
7 15	.800	11 45	2.095	16 15	3.995		

Ratio of Slopes for the following Vertical Angles.

Angle.	To one perpendicular.	Angle.	To one perpendicular.	Angle.	To one perpendicular.	Angle.	To one perpendicular.
° ′		° ′		° ′		° ′	
0 15	229	3 35	16	8 8	7	18 26	3
0 30	115	3 49	15	8 45	6¼	19 59	2¾
0 45	76	4 6	14	9 27	6	21 48	2½
1 0	57	4 24	13	9 52	5¾	23 58	2¼
1 15	46	4 45	12	10 18	5½	26 34	2
1 30	39	5 0	11½	10 47	5¼	29 44	1¾
1 45	33	5 12	11	11 19	5	33 42	1½
2 0	28	5 27	10½	11 53	4¾	38 40	1¼
2 15	25	5 42	10	12 32	4½	45 0	1
2 30	23	6 0	9½	13 15	4¼	53 8	¾
2 45	21	6 21	9	14 2	4	63 28	½
3 0	19	6 43	8½	14 55	3¾	75 58	¼
3 15	18	7 7	8	15 56	3½	78 41	⅕
3 28	17	7 36	7½	17 6	3¼		

XLV.—*Barometrical Measurement of Heights.*

TO OBTAIN THE DIFFERENCE IN THE HEIGHT OF TWO PLACES BY MEANS OF THE BAROMETER.

The following tables have been condensed from those in the appendix of Lieutenant-Colonel Williamson's Treatise on the Use of the Barometer, etc., Professional Papers, Corps of Engineers, No. 15, and are those of Plantamour (Guyot's tables D, 72–79) re-arranged and adapted to English measures.

They are based upon Bessel's formula, which differs from that of La Place principally in containing a factor depending upon the humidity of the air. The modifications introduced by Plantamour consist in some slight changes in the values of the barometric constants in accordance with the supposed more accurate results obtained from the experiments of Regnault.

La Place's formula reduced to English measures, as given by Guyot, (page D, 35,) is:

$$Z = \log\frac{h}{H} \times 60158.6 \text{ English feet} \begin{cases} \left(1 + \dfrac{t + t' - 64}{900}\right) \\ (1 + 0.00260 \cos 2\,L) \\ \left(1 + \dfrac{Z + 52252}{20886860} + \dfrac{h}{10443430}\right) \end{cases} \quad (1)$$

Williamson adopts the same convenient form in his reduction of Plantamour's formula to English measures; thus:

$$Z = \log\frac{h}{H} \times 60384.3 \text{ English feet} \begin{cases} \left(1 + \dfrac{t + t' - 64}{982.2647}\right) \\ (1 + 0.0026257 \cos 2\,L) \\ \left(1 + \dfrac{Z + 52252}{20886860} + \dfrac{s}{10443430}\right) \\ (1 + m) \end{cases} \quad (2)$$

where h and H are the heights of the barometer reduced to 32° Fahrenheit; t, t', the temperatures of the air at the two stations; and m, a factor depending upon the humidity of the stratum of air between them; L the latitude of the place.

XLV.—*Barometrical Measurement, &c.*—Continued.

APPLICATION.—1. Reduce the readings of the barometer to 32° Fahrenheit by Table I.

2. Representing by A the constant 60384.3, Table II gives the value of A log h or A log H, and consequently their difference, $A \frac{\log h}{\log H}$. The numbers have been diminished by a constant quantity, which does not affect their differences.

3. Table III, column B, gives values of the factor $\frac{t+t'-64}{982.2647}$ of the temperature-term, to be used only in connection with the tables that give the corrections for humidity.

Column C gives values of the factor $\frac{t+t'-64}{900}$, and is used where no observations of atmospheric humidity are made.

4. Table IV gives $A \log \frac{h}{H} \times 0.0026257 \cos 2L$, the correction due to gravity at the sea-level in the mean latitude between the two stations. It is positive from 45° to the equator, and negative from 45° to the poles.

5. Table V shows the correction $A \log \frac{h}{H} \times \frac{Z+52252}{20886860}$, to be added to the approximate difference of altitude, on account of the decrease of gravity on a vertical acting on the density of the mercurial column.

6. Table VI furnishes the small correction $A \log \frac{h}{H} \times \frac{s}{10443430}$ for the decrease of gravity on a vertical acting on the density of the air; s representing the approximate difference of altitude. It is always additive

7. Table VII gives the relative atmospheric humidity in fractions of unity. This table is different from any given in Guyot's collection; for, though based upon Regnault's table of maximum force of vapor, and so far the same as Guyot's, it has been calculated with factors determined by Glaisher.

8. Tables VIII and IX are intended to give the correction $A \log \frac{h}{H} \times m$, due to the relative humidity of the stratum of air between the two stations.

XLV.—*Barometrical Measurement, &c.*—Continued.

These hypsometrical tables represent the full formula of Plantamour. If, as is often the case, the observations for the relative humidity are not given with those of the barometer and dry thermometer, then the tables III, (column B,) VIII, and IX should not be used, but the temperature-correction must be calculated from the formula A log $\frac{h}{H} \times \frac{t+t'-64}{900}$ with the aid of column C of Table III.

With the temperature-term so calculated, the results will differ from those by Guyot's table, on account of the different value given to the barometric constant of the pressure term.

Abnormal and Horary Oscillations of the Weight of the Atmosphere.—The first is a gradual change generally extending over a period of two to seven days, causing the barometer to rise or fall gradually during that time, although sometimes more or less sudden, and occupying perhaps but a few hours; the second, a regular horary oscillation occurring at about the same hours every day, and having a magnitude entirely independent of this gradual change.

The abnormal change usually extends over large tracts of country, and in settled weather the barometer rises and falls so gradually that the forces that produce the motion can be separated with more or less accuracy from the horary changes by assuming the portion of this great wave corresponding to 24 hours to be a straight line; generally inclined, however, since the observations at any time of a barometric day differs from that of the next one at the same hour.

To eliminate this movement, subtract the barometric reading (reduced to 32°) at the beginning of one day from that at the same hour on the next succeeding one, and divide the difference by 24. The result is the correction for one hour, to be applied with its proper sign to the hour succeeding the initial hour. The correction for two hours is twice the correction for one hour, etc.

EXAMPLE.—Barometer at 7 a. m., August 7, = 29.743
Barometer at 7 a. m., August 8, = 29.487

Difference for 24 hours..... = +0.256
For 1 hour..... = +0.0106

and correction at 8 a. m., = + 0.011; at 9, = + 0.021; at 10, = + 0.032, etc.

This correction Williamson names "reduction to level."

XLV.—*Barometrical Measurement, &c.*—Continued.

In the horary oscillation there are two maximum and two minimum points during the 24 hours. Near the sea-level the barometer attains its maximum about 9 or 10 a. m. In the afternoon there is a minimum about 3, 4, or 5 p. m. It then rises until from 10 to midnight, when it falls again until about 4 a. m., and again rises to attain its morning maximum, the day-fluctuations being the larger. The oscillation is greatest nearer the equator and diminishes toward the poles. Its amount within the limits of the United States varies from 40 to 120 thousandths of an inch of the barometric column. It is not equal at all places of the same latitude.

In a series of barometric observations at any place, the mean barometric reading is better obtained from daily horary curves, by plotting the separate readings of each day reduced to 32°, and corrected for the abnormal change by reduction to level. These would present an approximate horary curve for every day of the series, from which erroneous or erratic observations could be detected and rejected if necessary.

EXAMPLE OF THE USE OF THESE TABLES.

Geneva and the Grand St. Bernard.

h = 28.600 in. t = 48°.2 F. Relative humidity, a = 0.77
H = 22.191 in. t' = 28°.6 F. Relative humidity, a' = 0.80

Lat = 46° $t + t'$ = 76°.8 F. $a + a'$ = 1.57

Table II, with argument h, gives	27557.3
Table II, with argument H, gives	20903.7
Difference = first approximate difference of altitude...	6653.6
Table III, col. B, with argument 76°.8, gives + 0.0130	
0.0130 × 6653.6 =	+ 86.5
Second approximate difference of altitude.......... =	6740.1
Table IV, arguments 46° lat. and 6700 feet, gives.... —	0.6
Table V, argument 6740, gives	19.0
Table VI, argument 6700 feet and 28.6 inches......	0.8
Third approximate difference of altitude........... =	6759.3

XLV.—*Barometrical Measurement, &c.*—Continued.

Table VIII, arguments 22.2 in. and 28.6 in., gives 79
Table IX, arguments 79 and 76°.8........... 11.9

11.9×1.57 = vapor correction =	18.7
Difference of altitude............................	6778.0

The altitude by level is stated to be 6791.5 feet.

The same readings of the barometer and the same air-temperature being used, but calculating the value of the temperature-term from column C, table III,

$$\text{temperature-term} = 6653.6 \times 0.0142 = 94.4$$

The value of the temperature-term, as calculated in the above example, increased by the vapor-correction, is 105.2, a larger result than by the method of La Place, because the sum of the observed relative humidities of the stations was greater than that assumed by him.

In a dry climate the reverse would have been the case.

Calculation of the same Observations by Guyot's Tables.

First table of Guyot gives, for h.................		27454.4
First table of Guyot gives, for H................		20825.6
First approximate difference of altitude...........	=	6628.8
$6628.8 \times \dfrac{t + t' - 64}{900}$	=	94.3
Second approximate difference of altitude..........	=	6733.1
Table III of Guyot gives	—	0.6
Table IV of Guyot gives		19.0
Table V of Guyot gives		0.8
Difference of altitude...........................		6752.3

This result is 25.6 less than by the following tables, and 39.2 less than by the spirit-level.

Aneroid Barometer.

The best aneroids are, as nearly as possible, compensated by the maker for differences of temperature, so that the index shall remain at the same reading on the dial when it is heated and cooled, and are intended to be adjusted to read uniformly inches of mercury at a temperature of 32° Fahrenheit at the level of the sea in 45° latitude. But before using any aneroid for

XLV.—*Barometrical Measurement, &c.*—Continued.

accurate observations it should be tested under an air-pump, together with a mercurial column, at a known temperature, and its scale-errors carefully noted. In many of them there is an additional scale of altitudes in feet outside of the scale, corresponding to the inches of mercury, generally in the best English instruments divided and marked according to a table prepared for the purpose by Professor Airy. There are some, however, very erroneously marked.

As the aneroid is not affected in its reading by the variation in the force of gravity, it needs no correction for the latitude, nor for the decrease of gravity in altitude acting on the mercurial column. The correction for the decrease of gravity in altitude acting on the density of the air, and the correction for humidity, remain; but the first being small in amount, it can be omitted, and the second combined with the correction for temperature, as is done in the formula of La Place.

The formula for the aneroid would then be:

$$Z = \log \frac{h}{H} \times 60384.3 \text{ Eng. feet} \left(1 + \frac{t + t' - 64}{900}\right)$$

and the tabular quantities may be taken from these tables.

Professor Airy's table, made for the purpose of graduating aneroids to a scale of feet, gives the height of the corrected mercurial column in inches for each fifty feet of altitude at 50° Fahrenheit. The formula is:

$$Z = \log \frac{h}{H} \times 62759 \text{ Eng. feet} \left(1 + \frac{t + t' - 100}{1000}\right)$$

As it is sometimes convenient to have an approximate formula that can be used without any tables whatever, the following may be found useful:

$$Z = 55032 \frac{H - h}{H + h} \text{ Eng. feet, at } 55° \text{ Fahrenheit,}$$

with a correction of $\pm \frac{1}{435}$ for each degree of mean temperature above 55°; or, nearly—

$$Z = 55000 \frac{H - h}{H + h} \pm \frac{1}{500}$$

a formula easily remembered, but only useful for altitudes not exceeding 3000 feet.

TABLE I.—*Reduction of the English Barometer to the Freezing-Point.*

Degrees of Fahrenheit.	English inches.							Degrees of Fahrenheit.
	17.5	18	18.5	19	19.5	20	20.5	
0	+.045	+.046	+.047	+.049	+.050	+.051	+.053	0
1	.043	.045	.046	.047	.048	.049	.051	1
2	.042	.043	.044	.045	.046	.048	.049	2
3	.040	.041	.042	.044	.045	.046	.047	2
4	.039	.040	.041	.042	.043	.044	.045	4
5	.037	.038	.039	.040	.041	.042	.043	5
6	.035	.036	.037	.038	.039	.040	.041	6
7	.034	.035	.036	.037	.038	.039	.040	7
8	.032	.033	.034	.035	.036	.037	.038	8
9	.031	.032	.032	.033	.034	.035	.036	9
10	.029	.030	.031	.032	.032	.033	.034	10
11	.028	.028	.029	.030	.031	.031	.032	11
12	.026	.027	.027	.028	.029	.030	.030	12
13	.024	.025	.026	.026	.027	.028	.029	13
14	.023	.023	.024	.025	.025	.026	.027	14
15	.021	.022	.022	.023	.024	.024	.025	15
16	.020	.020	.021	.021	.022	.022	.023	16
17	.018	.019	.019	.020	.020	.021	.021	17
18	.017	.017	.017	.018	.018	.019	.019	18
19	.015	.015	.016	.016	.017	.017	.018	19
20	.013	.014	.014	.015	.015	.015	.016	20
21	.012	.012	.012	.013	.013	.013	.014	21
22	.010	.011	.011	.011	.011	.012	.012	22
23	.009	.009	.009	.009	.010	.010	.010	23
24	.007	.007	.007	.008	.008	.008	.008	24
25	.006	.006	.006	.006	.006	.006	.006	25
26	.004	.004	.004	.004	.004	.005	.005	26
27	.002	.002	.003	.003	.003	.003	.003	27
28	+.001	+.001	+.001	+.001	+.001	+.001	+.001	28
29	−.001	−.001	−.001	−.001	−.001	−.001	−.001	29
30	.002	.002	.002	.003	.003	.003	.003	30
31	.004	.004	.004	.004	.004	.004	.005	31
32	.005	.006	.006	.006	.006	.006	.006	32
33	.007	.007	.007	.008	.008	.008	.008	33
34	.009	.009	.009	.009	.010	.010	.010	34
35	.010	.010	.011	.011	.011	.012	.012	35
36	.012	.012	.012	.013	.013	.013	.014	36
37	.013	.014	.014	.014	.015	.015	.016	37
38	.015	.015	.016	.016	.017	.017	.017	38
39	.016	.017	.017	.018	.018	.019	.019	39
40	.018	.019	.019	.020	.020	.021	.021	40
41	.020	.020	.021	.021	.022	.022	.023	41
42	.021	.022	.022	.023	.024	.024	.025	42
43	.023	.023	.024	.025	.025	.026	.027	43
44	.024	.025	.026	.026	.027	.028	.028	44
45	.026	.027	.027	.028	.029	.030	.030	45
46	.027	.028	.029	.030	.031	.031	.032	46
47	.029	.030	.031	.031	.032	.033	.034	47
48	.031	.031	.032	.033	.034	.035	.036	48
49	.032	.033	.034	.035	.036	.037	.038	49
50	−.034	−.035	−.036	−.037	−.038	−.038	−.039	50

TABLE I.—*Reduction of the English Barometer to the Freezing-Point*—Continued.

Degrees of Fahrenheit.	English inches.							Degrees of Fahrenheit.
	17.5	18	18.5	19	19.5	20	20.5	
51	−.035	−.036	−.037	−.038	−.039	−.040	−.041	51
52	.037	.038	.039	.040	.041	.042	.043	52
53	.038	.039	.041	.042	.043	.044	.045	53
54	.040	.041	.042	.043	.044	.046	.047	54
55	.041	.043	.044	.045	.046	.047	.049	55
56	.043	.044	.045	.047	.048	.049	.050	56
57	.045	.046	.047	.048	.050	.051	.052	57
58	.046	.047	.049	.050	.051	.053	.054	58
59	.048	.049	.050	.052	.053	.055	.056	59
60	.049	.051	.052	.054	.055	.056	.058	60
61	.051	.052	.054	.055	.057	.058	.060	61
62	.052	.054	.055	.057	.058	.060	.061	62
63	.054	.055	.057	.059	.060	.062	.063	63
64	.056	.057	.059	.060	.062	.063	.065	64
65	.057	.059	.060	.062	.064	.065	.067	65
66	.059	.060	.062	.064	.065	.067	.069	66
67	.060	.062	.064	.065	.067	.069	.071	67
68	.062	.064	.065	.067	.069	.071	.072	68
69	.063	.065	.067	.069	.071	.072	.074	69
70	.065	.067	.069	.070	.072	.074	.076	70
71	.066	.068	.070	.072	.074	.076	.078	71
72	.068	.070	.072	.074	.076	.078	.080	72
73	.070	.072	.074	.075	.077	.079	.081	73
74	.071	.073	.075	.077	.079	.081	.083	74
75	.073	.075	.077	.079	.081	.083	.085	75
76	.074	.076	.078	.081	.083	.085	.087	76
77	.076	.078	.080	.082	.084	.087	.089	77
78	.077	.080	.082	.084	.086	.088	.091	78
79	.079	.081	.083	.086	.088	.090	.092	79
80	.080	.083	.085	.087	.090	.092	.094	80
81	.082	.084	.087	.089	.091	.094	.096	81
82	.084	.086	.088	.091	.093	.095	.098	82
83	.085	.088	.090	.092	.095	.097	.100	83
84	.087	.089	.092	.094	.097	.099	.101	84
85	.088	.091	.093	.096	.098	.101	.103	85
86	.090	.092	.095	.097	.100	.103	.105	86
87	.091	.094	.096	.099	.102	.104	.107	87
88	.093	.095	.098	.101	.103	.106	.109	88
89	.094	.097	.100	.102	.105	.108	.111	89
90	.096	.099	.101	.104	.107	.110	.112	90
91	.097	.100	.103	.106	.109	.111	.114	91
92	.099	.102	.105	.108	.110	.113	.116	92
93	.101	.103	.106	.109	.112	.115	.118	93
94	−.102	−.105	−.108	.111	.114	.117	.120	94
95				.113	.116	.118	.121	95
96				.114	.117	.120	.123	96
97				.116	.119	.122	.125	97
98				.118	.121	.124	.127	98
99				−.119	−.122	−.126	−.129	99

TABLE I.—*Reduction of the English Barometer to the Freezing-Point*—Continued.

Degrees of Fahrenheit.	English inches.							Degrees of Fahrenheit.
	21	21.5	22	22.5	23	23.5	24	
0	+.054	+.055	+.055	+.058	+.059	+.060	+.062	0
1	.052	.053	.053	.056	.057	.058	.059	1
2	.050	.051	.051	.054	.055	.056	.057	2
3	.048	.049	.049	.052	.053	.054	.055	3
4	.046	.047	.047	.050	.051	.052	.053	4
5	.044	.045	.045	.048	.049	.050	.051	5
6	.042	.044	.044	.046	.047	.048	.049	6
7	.041	.042	.042	.044	.044	.045	.046	7
8	.039	.040	.040	.041	.042	.043	.044	8
9	.037	.038	.038	.039	.040	.041	.042	9
10	.035	.036	.036	.037	.038	.039	.040	10
11	.033	.034	.034	.035	.036	.037	.038	11
12	.031	.032	.032	.033	.034	.035	.036	12
13	.029	.030	.030	.031	.032	.033	.033	13
14	.027	.028	.028	.029	.030	.031	.031	14
15	.025	.026	.026	.027	.028	.029	.029	15
16	.024	.024	.024	.025	.026	.026	.027	16
17	.022	.022	.022	.023	.024	.024	.025	17
18	.020	.020	.020	.021	.022	.022	.023	18
19	.018	.018	.018	.019	.020	.020	.020	19
20	.016	.016	.016	.017	.018	.018	.018	20
21	.014	.014	.014	.015	.016	.016	.016	21
22	.012	.013	.013	.013	.013	.014	.014	22
23	.010	.011	.011	.011	.011	.012	.012	23
24	.008	.009	.009	.009	.009	.010	.010	24
25	.007	.007	.007	.007	.007	.007	.008	25
26	.005	.005	.005	.005	.005	.005	.005	26
27	.003	.003	.003	.003	.003	.003	.003	27
28	+.001	+.001	+.001	+.001	+.001	+.001	+.001	28
29	−.001	−.001	−.001	−.001	−.001	−.001	−.001	29
30	.003	.003	.003	.003	.003	.003	.003	30
31	.005	.005	.005	.005	.005	.005	.005	31
32	.007	.007	.007	.007	.007	.007	.008	32
33	.008	.009	.009	.009	.009	.009	.010	33
34	.010	.011	.011	.011	.011	.012	.012	34
35	.012	.013	.013	.013	.013	.014	.014	35
36	.014	.014	.015	.015	.015	.016	.016	36
37	.016	.016	.017	.017	.018	.018	.018	37
38	.018	.018	.019	.019	.020	.020	.020	38
39	.020	.020	.021	.021	.022	.022	.023	39
40	.022	.022	.023	.023	.024	.024	.025	40
41	.023	.024	.025	.025	.026	.026	.027	41
42	.025	.026	.027	.027	.028	.028	.029	42
43	.027	.028	.029	.029	.030	.031	.031	43
44	.029	.030	.031	.031	.032	.033	.033	44
45	.031	.032	.032	.033	.034	.035	.035	45
46	.033	.034	.034	.035	.036	.037	.038	46
47	.035	.036	.036	.037	.038	.039	.040	47
48	.037	.038	.038	.039	.040	.041	.042	48
49	.039	.039	.040	.041	.042	.043	.044	49
50	−.040	−.041	−.042	−.043	−.044	−.045	−.046	50

TABLE I.—*Reduction of the English Barometer to the Freezing-Point*—Continued.

Degrees of Fahrenheit.	English inches.							Degrees of Fahrenheit.
	21	21.5	22	22.5	23	23.5	24	
51	−.042	−.043	−.044	−.045	−.046	−.047	−.048	51
52	.044	.045	.046	.047	.048	.049	.050	52
53	.046	.047	.048	.049	.050	.051	.053	53
54	.048	.049	.050	.051	.052	.054	.055	54
55	.050	.051	.052	.053	.055	.056	.057	55
56	.052	.053	.054	.055	.057	.058	.059	56
57	.054	.055	.056	.057	.059	.060	.061	57
58	.055	.057	.058	.059	.061	.062	.063	58
59	.057	.059	.060	.061	.063	.064	.065	59
60	.059	.061	.062	.063	.065	.066	.068	60
61	.061	.062	.064	.065	.067	.068	.070	61
62	.063	.064	.066	.067	.069	.070	.072	62
63	.065	.066	.068	.069	.071	.072	.074	63
64	.067	.068	.070	.071	.073	.075	.076	64
65	.068	.070	.072	.073	.075	.077	.078	65
66	.070	.072	.074	.075	.077	.079	.080	66
67	.072	.074	.076	.077	.079	.081	.083	67
68	.074	.076	.078	.079	.081	.083	.085	68
69	.076	.078	.080	.081	.083	.085	.087	69
70	.078	.080	.082	.083	.085	.087	.089	70
71	.080	.082	.083	.085	.087	.089	.091	71
72	.082	.084	.085	.087	.089	.091	.093	72
73	.083	.085	.087	.089	.091	.093	.095	73
74	.085	.087	.089	.091	.093	.095	.097	74
75	.087	.089	.091	.093	.095	.098	.100	75
76	.089	.091	.093	.095	.098	.100	.102	76
77	.091	.093	.095	.097	.100	.102	.104	77
78	.093	.095	.097	.099	.102	.104	.106	78
79	.095	.097	.099	.101	.104	.106	.108	79
80	.096	.099	.101	.103	.106	.108	.110	80
81	.098	.101	.103	.105	.108	.110	.112	81
82	.100	.103	.105	.107	.110	.112	.115	82
83	.102	.105	.107	.109	.112	.114	.117	83
84	.104	.106	.109	.111	.114	.116	.119	84
85	.106	.108	.111	.113	.116	.118	.121	85
86	.108	.110	.113	.115	.118	.120	.123	86
87	.110	.112	.115	.117	.120	.123	.125	87
88	.111	.114	.117	.119	.122	.125	.127	88
89	.113	.116	.119	.121	.124	.127	.129	89
90	.115	.118	.121	.123	.126	.129	.132	90
91	.117	.120	.123	.125	.128	.131	.134	91
92	.119	.122	.124	.127	.130	.133	.136	92
93	.121	.124	.126	.129	.132	.135	.138	93
94	.123	.125	.128	.131	.134	.137	.140	94
95	.124	.127	.130	.133	.136	.139	.142	95
96	.126	.129	.132	.135	.138	.141	.144	96
97	.128	.131	.134	.137	.140	.143	.146	97
98	.130	.133	.136	.139	.142	.145	.149	98
99	−.132	−.135	−.138	−.141	−.144	−.148	−.151	99

TABLE I.—*Reduction of the English Barometer to the Freezing-Point*—Continued.

Degrees of Fahrenheit.	English inches.							Degrees of Fahrenheit.
	24.5	25	25.5	26	26.5	27	27.5	
0	+.063	+.064	+.065	+.067	+.068	+.069	+.071	0
1	.061	.062	.063	.064	.066	.067	.068	1
2	.058	.060	.061	.062	.063	.064	.066	2
3	.056	.057	.058	.060	.061	.062	.063	3
4	.054	.055	.056	.057	.058	.060	.061	4
5	.052	.053	.054	.055	.056	.057	.058	5
6	.050	.051	.052	.053	.054	.055	.056	6
7	.047	.048	.049	.050	.051	.052	.053	7
8	.045	.046	.047	.048	.049	.050	.051	8
9	.043	.044	.045	.046	.046	.047	.048	9
10	.041	.042	.042	.043	.044	.045	.046	10
11	.039	.039	.040	.041	.042	.042	.043	11
12	.036	.037	.038	.039	.039	.040	.041	12
13	.034	.035	.036	.036	.037	.038	.038	13
14	.032	.033	.033	.034	.035	.035	.036	14
15	.030	.030	.031	.032	.032	.033	.033	15
16	.028	.028	.029	.029	.030	.030	.031	16
17	.025	.026	.026	.027	.027	.028	.028	17
18	.023	.024	.024	.025	.025	.025	.026	18
19	.021	.021	.022	.022	.023	.023	.023	19
20	.019	.019	.019	.020	.020	.021	.021	20
21	.017	.017	.017	.018	.018	.018	.019	21
22	.014	.015	.015	.015	.015	.016	.016	22
23	.012	.012	.013	.013	.013	.013	.014	23
24	.010	.010	.010	.011	.011	.011	.011	24
25	.008	.008	.008	.008	.008	.009	.009	25
26	.006	.006	.006	.006	.006	.006	.006	26
27	.003	.003	.003	.004	.004	.004	.004	27
28	+.001	+.001	+.001	+.001	+.001	+.001	+.001	28
29	−.001	−.001	−.001	−.001	−.001	−.001	−.001	29
30	.003	.003	.003	.003	.004	.004	.004	30
31	.005	.006	.006	.006	.006	.006	.006	31
32	.008	.008	.008	.008	.008	.008	.009	32
33	.010	.010	.010	.010	.011	.011	.011	33
34	.012	.012	.013	.013	.013	.013	.014	34
35	.014	.015	.015	.015	.015	.016	.016	35
36	.016	.017	.017	.017	.018	.018	.018	36
37	.019	.019	.019	.020	.020	.021	.021	37
38	.021	.021	.022	.022	.023	.023	.023	38
39	.023	.024	.024	.024	.025	.025	.026	39
40	.025	.026	.026	.027	.027	.028	.028	40
41	.027	.028	.029	.029	.030	.030	.031	41
42	.030	.030	.031	.031	.032	.033	.033	42
43	.032	.032	.033	.034	.034	.035	.036	43
44	.034	.035	.035	.036	.037	.037	.038	44
45	.036	.037	.038	.038	.039	.040	.041	45
46	.038	.039	.040	.041	.042	.042	.043	46
47	.041	.041	.042	.043	.044	.045	.046	47
48	.043	.044	.044	.045	.046	.047	.048	48
49	.045	.046	.047	.048	.049	.050	.050	49
50	−.047	−.048	−.049	−.050	−.051	−.052	−.053	50

TABLE I.—*Reduction of the English Barometer to the Freezing-Point*—Continued.

Degrees of Fahrenheit.	English inches.							Degrees of Fahrenheit.
	24.5	25	25.5	26	26.5	27	27.5	
51	−.049	−.050	−.051	−.052	−.053	−.054	−.055	51
52	.052	.053	.054	.055	.056	.057	.058	52
53	.054	.055	.056	.057	.058	.059	.060	53
54	.056	.057	.058	.059	.060	.062	.063	54
55	.058	.059	.060	.062	.063	.064	.065	55
56	.060	.061	.063	.064	.065	.066	.068	56
57	.062	.064	.065	.066	.068	.069	.070	57
58	.065	.066	.067	.069	.070	.071	.073	58
59	.067	.068	.070	.071	.072	.074	.075	59
60	.069	.070	.072	.073	.075	.076	.077	60
61	.071	.073	.074	.076	.077	.078	.080	61
62	.073	.075	.076	.078	.079	.081	.082	62
63	.076	.077	.079	.080	.082	.083	.085	63
64	.078	.079	.081	.082	.084	.086	.087	64
65	.080	.082	.083	.085	.086	.088	.090	65
66	.082	.084	.085	.087	.089	.090	.092	66
67	.084	.086	.088	.089	.091	.093	.095	67
68	.086	.088	.090	.092	.093	.095	.097	68
69	.089	.090	.092	.094	.096	.098	.099	69
70	.091	.093	.095	.096	.098	.100	.102	70
71	.093	.095	.097	.099	.101	.102	.104	71
72	.095	.097	.099	.101	.103	.105	.107	72
73	.097	.099	.101	.103	.105	.107	.109	73
74	.100	.102	.104	.106	.108	.110	.112	74
75	.102	.104	.106	.108	.110	.112	.114	75
76	.104	.106	.108	.110	.112	.114	.117	76
77	.106	.108	.110	.113	.115	.117	.119	77
78	.108	.110	.113	.115	.117	.119	.121	78
79	.110	.113	.115	.117	.119	.122	.124	79
80	.113	.115	.117	.119	.122	.124	.126	80
81	.115	.117	.119	.122	.124	.126	.129	81
82	.117	.119	.122	.124	.126	.129	.131	82
83	.119	.122	.124	.126	.129	.131	.134	83
84	.121	.124	.126	.129	.131	.134	.136	84
85	.123	.126	.128	.131	.134	.136	.139	85
86	.126	.128	.131	.133	.136	.138	.141	86
87	.128	.130	.133	.136	.138	.141	.143	87
88	.130	.133	.135	.138	.141	.143	.146	88
89	.132	.135	.138	.140	.143	.146	.148	89
90	.134	.137	.140	.143	.145	.148	.151	90
91	.136	.139	.142	.145	.148	.150	.153	91
92	.139	.141	.144	.147	.150	.153	.156	92
93	.141	.144	.147	.149	.152	.155	.158	93
94	.143	.146	.149	.152	.155	.158	.160	94
95	.145	.148	.151	.154	.157	.160	.163	95
96	.147	.150	.153	.156	.159	.162	.165	96
97	.149	.153	.156	.159	.162	.165	.168	97
98	.152	.155	.158	.161	.164	.167	.170	98
99	−.154	−.157	−.160	−.163	−.166	−.170	−.173	99

TABLE I.—*Reduction of the English Barometer to the Freezing-Point*—Continued.

Degrees of Fahrenheit.	English inches.							Degrees of Fahrenheit.
	28	28.5	29	29.5	30	30.5	31	
0	+.072	+.073	+.074	+.076	+.077	+.078	+.080	0
1	.069	.071	.072	.073	.074	.075	.077	1
2	.067	.068	.069	.070	.072	.073	.074	2
3	.064	.065	.067	.068	.069	.070	.071	3
4	.062	.063	.064	.065	.066	.067	.068	4
5	.059	.060	.061	.062	.063	.064	.066	5
6	.057	.058	.059	.060	.061	.062	.063	6
7	.054	.055	.056	.057	.058	.059	.060	7
8	.052	.053	.053	.054	.055	.056	.057	8
9	.049	.050	.051	.052	.053	.053	.054	9
10	.047	.047	.048	.049	.050	.051	.052	10
11	.044	.045	.046	.046	.047	.048	.049	11
12	.042	.042	.043	.044	.044	.045	.046	12
13	.039	.040	.040	.041	.042	.042	.043	13
14	.036	.037	.038	.038	.039	.040	.040	14
15	.034	.035	.035	.036	.036	.037	.038	15
16	.031	.032	.033	.033	.034	.034	.035	16
17	.029	.029	.030	.030	.031	.032	.032	17
18	.026	.027	.027	.028	.028	.029	.029	18
29	.024	.024	.025	.025	.026	.026	.027	19
20	.021	.022	.022	.023	.023	.023	.024	20
21	.019	.019	.020	.020	.020	.021	.021	21
22	.016	.017	.017	.017	.018	.018	.018	22
23	.014	.014	.014	.015	.015	.015	.015	23
24	.011	.012	.012	.012	.012	.012	.013	24
25	.009	.009	.009	.009	.009	.010	.010	25
26	.006	.006	.007	.007	.007	.007	.007	26
27	.004	.004	.004	.004	.004	.004	.004	27
28	+.001	+.001	+.001	+.001	+.001	+.001	+.001	28
29	−.001	−.001	−.001	−.001	−.001	−.001	−.001	29
30	.004	.004	.004	.004	.004	.004	.004	30
31	.006	.006	.006	.007	.007	.007	.007	31
32	.009	.009	.009	.009	.009	.010	.010	32
33	.011	.011	.012	.012	.012	.012	.012	33
34	.014	.014	.014	.015	.015	.015	.015	34
35	.016	.017	.017	.017	.017	.018	.018	35
36	.019	.019	.019	.020	.020	.020	.021	36
37	.021	.022	.022	.022	.023	.023	.024	37
38	.024	.024	.025	.025	.026	.026	.026	38
39	.026	.027	.027	.028	.028	.029	.029	39
40	.029	.029	.030	.030	.031	.031	.032	40
41	.031	.032	.032	.033	.034	.034	.035	41
42	.034	.034	.035	.036	.036	.037	.037	42
43	.036	.037	.038	.038	.039	.040	.040	43
44	.039	.040	.040	.041	.042	.042	.043	44
45	.041	.042	.043	.044	.044	.045	.046	45
46	.044	.045	.045	.046	.047	.048	.049	46
47	.046	.047	.048	.049	.050	.050	.051	47
48	.049	.050	.051	.051	.052	.053	.054	48
49	.051	.052	.053	.054	.055	.056	.057	49
50	−.054	−.055	−.056	−.057	−.058	−.059	−.060	50

TABLE I.—*Reduction of the English Barometer to the Freezing-Point*—Continued.

Degrees of Fahrenheit.	English inches.							Degrees of Fahrenheit.
	28	28.5	29	29.5	30	30.5	31	
51	−.056	−.057	−.058	−.059	−.060	−.061	−.062	51
52	.059	.060	.061	.062	.063	.064	.065	52
53	.061	.062	.064	.065	.066	.067	.068	53
54	.064	.065	.066	.067	.068	.070	.071	54
55	.066	.068	.069	.070	.071	.072	.073	55
56	.069	.070	.071	.073	.074	.075	.076	56
57	.071	.073	.074	.075	.076	.078	.079	57
58	.074	.075	.076	.078	.079	.080	.082	58
59	.076	.078	.079	.080	.082	.083	.085	59
60	.079	.080	.082	.083	.084	.086	.087	60
61	.081	083	.084	.086	.087	.089	.090	61
62	.084	.085	.087	.088	.090	.091	.093	62
63	.086	.088	.089	.091	.092	.094	.096	63
64	.089	.090	.092	.094	.095	.097	.098	64
65	.091	.093	.095	.096	.098	.099	.101	65
66	.094	.095	.097	.099	.101	.102	.104	66
67	.096	.098	.100	.101	.103	.105	.107	67
68	.099	.101	.102	.104	.106	.108	.109	68
69	.101	.103	.105	.107	.109	.110	.112	69
70	.104	.106	.107	.109	.111	.113	.115	70
71	.106	.108	.110	.112	.114	.116	.118	71
72	.109	.111	.113	.115	.117	.118	.120	72
73	.111	.113	.115	.117	.119	.121	.123	73
74	.114	.116	.118	.120	.122	.124	.126	74
75	.116	.118	.120	.122	.125	.127	.129	75
76	.119	.121	.123	.125	.127	.129	.131	76
77	.121	.123	.126	.128	.130	.132	.134	77
78	.124	.126	.128	.130	.133	.135	.137	78
79	.126	.128	.131	.133	.135	.137	.140	79
80	.129	.131	.133	.136	.138	.140	.143	80
81	.131	.133	.136	.138	.141	.143	.145	81
82	.134	.136	.138	.141	.143	.146	.148	82
83	.136	.139	.141	.143	.146	.148	.151	83
84	.139	.141	.144	.146	.148	.151	.154	84
85	.141	.144	.146	.149	.151	.154	.156	85
86	.144	.146	.149	.151	.154	.156	.159	86
87	.146	.149	.151	.154	.156	.159	.162	87
88	.149	.151	.154	.156	.159	.162	.165	88
89	.151	.154	.156	.159	.162	.164	.167	89
90	.153	.156	.159	.162	.164	.167	.170	90
91	.156	.159	.162	.164	.167	.170	.173	91
92	.158	.161	.164	.167	.170	.173	.175	92
93	.161	.164	.167	.170	.172	.175	.178	93
94	.163	.166	.169	.172	.175	.178	.180	94
95	.166	.169	.172	.175	.178	.181	.183	95
96	.168	.171	.174	.177	.180	.183	.186	96
97	.171	.174	.177	.180	.183	.186	.189	97
98	.173	.176	.180	.183	.186	.189	.191	98
99	−.176	−.179	−.182	−.185	−.188	−.191	−.194	−99

Table II.

$$D_1 = 60384.3 \times \log H \text{ or } h.$$

Barometer in English inches.	Hundredths of an inch.										Thousandths of an inch.	
	0.00	0.01	0.02	0.03	0.04	0.05	0.06	0.07	0.08	0.09	15.0	
	Peet.	*Peet.*	*Peet.*	*Peet.*	*Peet.*	*Peet.*	*Peet.*	*Peet.*	*Peet.*	*Peet.*		
17.0	13915.5	13930.9	13946.3	13961.7	13977.1	13992.5	14007.9	14023.3	14038.6	14054.0	1	1.5
.1	14069.3	14084.6	14100.0	14115.3	14130.6	14145.9	14161.2	14176.4	14191.7	14207.0	2	3.0
.2	14222.2	14237.5	14252.7	14267.9	14283.1	14298.3	14313.5	14328.7	14343.9	14359.1	3	4.5
.3	14374.2	14389.4	14404.5	14419.7	14434.8	14449.9	14465.0	14480.1	14495.2	14510.3	4	6.0
.4	14525.4	14540.5	14555.5	14570.6	14585.6	14600.6	14615.7	14630.7	14645.7	14660.7	5	7.5
.5	14675.7	14690.7	14705.6	14720.6	14735.6	14750.5	14765.4	14780.4	14795.3	14810.2	6	9.0
.6	14825.1	14840.0	14854.9	14869.8	14884.6	14899.5	14914.4	14929.2	14944.0	14958.9	7	10.5
.7	14973.7	14988.5	15003.3	15018.1	15032.9	15047.7	15062.4	15077.2	15092.0	15106.7	8	12.0
.8	15121.4	15136.2	15150.9	15165.6	15180.3	15195.0	15209.7	15224.4	15239.0	15253.7	9	13.5
.9	15268.4	15283.0	15297.6	15312.3	15326.9	15341.5	15335.1	15370.7	15385.3	15399.9		
18.0	15414.5	15429.0	15443.6	15458.1	15472.7	15487.2	15501.7	15516.2	15530.7	15545.2	14.2	
.1	15559.7	15574.2	15588.7	15603.2	15617.6	15632.1	15646.5	15661.0	15675.4	15689.8	1	1.4
.2	15704.2	15718.6	15733.0	15747.4	15761.8	15776.2	15790.5	15804.9	15819.2	15833.6	2	2.8
.3	15847.9	15862.3	15876.6	15890.9	15905.2	15919.5	15933.8	15948.0	15962.3	15976.6	3	4.3
.4	15990.8	16005.1	16019.3	16033.6	16047.8	16062.0	16076.2	16090.4	16104.6	16118.8	4	5.7
.5	16133.0	16147.1	16161.3	16175.5	16189.6	16203.8	16217.9	16232.0	16246.1	16260.2	5	7.1
.6	16274.4	16288.4	16302.5	16316.6	16330.7	16344.8	16358.8	16372.9	16386.9	16400.9	6	8.5
.7	16415.0	16429.0	16443.0	16457.0	16471.0	16485.0	16499.0	16512.9	16526.9	16540.9	7	9.9
.8	16554.8	16568.8	16582.7	16596.6	16610.6	16624.5	16638.4	16652.3	16666.2	16680.1	8	11.4
.9	16694.0	16707.8	16721.7	16735.5	16749.4	16763.2	16777.1	16790.9	16804.7	16818.5	9	12.8

TABLE II—Continued.

$$D_1 = 60384.3 \times \log H \text{ or } h.$$

Barometer in English inches.	Hundredths of an inch. 0.00	0.01	0.02	0.03	0.04	0.05	0.06	0.07	0.08	0.09	Thousandths of an inch.	
	Feet.	*Feet.*	*Feet.*	*Feet.*	*Feet.*	*Feet.*	*Feet.*	*Feet.*	*Feet.*	*Feet.*	13.5	
19.0	16832.3	16846.1	16859.9	16873.7	16887.5	16901.3	16915.0	16928.8	16942.5	16956.3	1	1.3
.1	16970.0	16983.7	16997.4	17011.2	17024.9	17038.6	17052.3	17065.9	17079.6	17093.3	2	2.7
.2	17106.9	17120.6	17134.2	17147.9	17161.5	17175.2	17188.8	17202.4	17216.0	17229.6	3	4.0
.3	17243.2	17256.8	17270.3	17283.9	17297.5	17311.0	17324.6	17338.1	17351.7	17365.2	4	5.4
.4	17378.7	17392.2	17405.7	17419.2	17432.7	17446.2	17459.7	17473.2	17486.6	17500.1	5	6.7
.5	17513.5	17527.0	17540.4	17553.9	17567.3	17580.7	17594.1	17607.5	17620.9	17634.3	6	8.1
.6	17647.7	17661.1	17674.4	17687.8	17701.1	17714.5	17727.8	17741.2	17754.5	17767.8	7	9.4
.7	17781.1	17794.4	17807.7	17821.0	17834.3	17847.6	17860.9	17874.2	17887.4	17900.7	8	10.8
.8	17913.9	17927.2	17940.4	17953.6	17966.8	17980.1	17993.3	18006.5	18019.7	18032.9	9	12.1
.9	18046.0	18059.2	18072.4	18085.5	18098.7	18111.8	18125.0	18138.1	18151.2	18164.4		
20.0	18177.5	18190.6	18203.7	18216.8	18229.9	18243.0	18256.0	18269.1	18282.2	18295.2	12.8	
.1	18308.3	18321.3	18334.4	18347.4	18360.4	18373.4	18386.4	18399.5	18412.5	18425.4	1	1.3
.2	18438.4	18451.4	18464.4	18477.3	18490.3	18503.3	18516.2	18529.1	18542.1	18555.0	2	2.6
.3	18567.9	18580.8	18593.8	18606.7	18619.6	18632.4	18645.3	18658.2	18671.1	18683.9	3	3.8
.4	18696.8	18709.7	18722.5	18735.3	18748.2	18761.0	18773.8	18786.6	18799.4	18812.2	4	5.1
.5	18825.0	18837.8	18850.6	18863.4	18876.2	18888.9	18901.7	18914.4	18927.2	18939.9	5	6.4
.6	18952.7	18965.4	18978.1	18990.8	19003.5	19016.2	19028.9	19041.6	19054.3	19067.0	6	7.7
.7	19079.6	19092.3	19105.0	19117.6	19130.3	19142.9	19155.6	19168.2	19180.8	19193.4	7	9.0
.8	19206.0	19218.6	19231.2	19243.8	19256.4	19269.0	19281.6	19294.1	19306.7	19319.3	8	10.2
.9	19331.8	19344.4	19356.9	19369.4	19382.0	19394.5	19407.0	19419.5	19432.0	19444.5	9	11.5

Table II—Continued.

$$D_1 = 60384.3 \times \log H \text{ or } h.$$

Barometer in English inches.	0.00	0.01	0.02	0.03	0.04	0.05	0.06	0.07	0.08	0.09
	Hundredths of an inch.									
	Feet.	*Feet.*	*Feet.*	*Feet.*	*Feet.*	*Feet.*	*Feet.*	*Feet.*	*Feet.*	*Feet.*
21.0	19457.0	19469.5	19482.0	19494.4	19506.9	19519.4	19531.8	19544.3	19556.7	19569.1
.1	19581.6	19594.0	19606.4	19618.8	19631.2	19643.6	19656.0	19668.4	19680.8	19693.2
.2	19705.6	19717.9	19730.3	19742.6	19755.0	19767.3	19779.7	19792.0	19804.3	19816.7
.3	19829.0	19841.3	19853.6	19865.9	19878.2	19890.5	19902.7	19915.0	19927.3	19939.5
.4	19951.8	19964.1	19976.3	19988.5	20000.8	20013.0	20025.2	20037.4	20049.7	20061.9
.5	20074.1	20086.3	20098.4	20110.6	20122.8	20135.0	20147.1	20159.3	20171.5	20183.6
.6	20195.8	20207.9	20220.1	20232.2	20244.3	20256.4	20268.5	20280.6	20292.7	20304.8
.7	20316.9	20329.0	20341.0	20353.1	20365.2	20377.2	20389.3	20401.3	20413.4	20425.4
.8	20437.5	20449.5	20461.5	20473.5	20485.5	20497.5	20509.5	20521.5	20533.5	20545.5
.9	20557.5	20569.5	20581.4	20593.4	20605.3	20617.3	20629.2	20641.2	20653.1	20665.0
22.0	20677.0	20688.9	20700.8	20712.7	20724.6	20736.5	20748.4	20760.3	20772.1	20784.0
.1	20795.9	20807.7	20819.6	20831.5	20843.3	20855.2	20867.0	20878.8	20890.6	20902.5
.2	20914.3	20926.1	20937.9	20949.7	20961.5	20973.3	20985.1	20996.8	21008.6	21020.4
.3	21032.1	21043.9	21055.7	21067.4	21079.1	21090.9	21102.6	21114.3	21126.1	21137.8
.4	21149.5	21161.2	21172.9	21184.6	21196.3	21208.0	21219.6	21231.3	21243.0	21254.6
.5	21266.3	21277.9	21289.6	21301.2	21312.9	21324.5	21336.1	21347.8	21359.4	21371.0
.6	21382.6	21394.2	21405.8	21417.4	21429.0	21440.5	21452.1	21463.7	21475.3	21486.8
.7	21498.4	21509.9	21521.5	21533.0	21544.5	21556.1	21567.6	21579.1	21590.6	21602.1
.8	21613.6	21625.1	21636.6	21648.1	21659.6	21671.1	21682.6	21694.0	21705.5	21717.0
.9	21728.4	21739.9	21751.3	21762.7	21774.2	21785.6	21797.0	21808.5	21819.9	21831.3

Thousandths of an inch.

12.2	
1	1.2
2	2.4
3	3.7
4	4.9
5	6.1
6	7.3
7	8.5
8	9.8
9	11.0

11.7	
1	1.2
2	2.3
3	3.5
4	4.7
5	5.8
6	7.0
7	8.2
8	9.4
9	10.5

TABLE II—Continued.

$$D_1 = 60384.3 \times \log H \text{ or } h.$$

Barometer in English inches.	Hundredths of an inch.									
	0.00	0.01	0.02	0.03	0.04	0.05	0.06	0.07	0.08	0.09
	Feet.	*Feet.*	*Feet.*	*Feet.*	*Feet.*	*Feet.*	*Feet.*	*Feet.*	*Feet.*	*Feet.*
23.0	21842.7	21854.1	21865.5	21876.9	21888.3	21899.6	21911.0	21922.4	21933.7	21945.1
.1	21956.5	21967.8	21979.2	21990.5	22001.8	22013.2	22024.5	22035.8	22047.1	22058.4
.2	22069.7	22081.0	22092.3	22103.6	22114.9	22126.2	22137.5	22148.7	22160.0	22171.3
.3	22182.5	22193.8	22205.0	22216.3	22227.5	22238.7	22250.0	22261.2	22272.4	22283.6
.4	22294.8	22306.0	22317.2	22328.4	22339.6	22350.8	22362.0	22373.2	22384.3	22395.5
.5	22406.7	22417.8	22429.0	22440.1	22451.3	22462.4	22473.5	22484.7	22495.8	22506.9
.6	22518.0	22529.1	22540.2	22551.3	22562.4	22573.5	22584.6	22595.7	22606.8	22617.8
.7	22628.9	22640.0	22651.0	22662.1	22673.1	22684.2	22695.2	22706.3	22717.3	22728.3
.8	22739.3	22750.4	22761.4	22772.4	22783.4	22794.4	22805.4	22816.4	22827.3	22838.3
.9	22849.3	22860.3	22871.2	22882.2	22893.1	22904.1	22915.0	22926.0	22936.9	22947.9
24.0	22958.8	22969.7	22980.6	22991.6	23002.5	23013.4	23024.3	23035.2	23046.1	23056.9
.1	23067.8	23078.7	23089.6	23100.5	23111.3	23122.2	23133.0	23143.9	23154.7	23165.6
.2	23176.4	23187.3	23198.1	23208.9	23219.7	23230.5	23241.4	23252.2	23263.0	23273.8
.3	23284.6	23295.4	23306.1	23316.9	23327.7	23338.5	23349.2	23360.0	23370.8	23381.5
.4	23392.3	23403.0	23413.7	23424.5	23435.2	23445.9	23456.7	23467.4	23478.1	24488.8
.5	23499.5	23510.2	23520.9	23531.6	23542.3	23553.0	23563.7	23574.3	23585.0	23595.7
.6	23606.3	23617.0	23627.7	23638.3	23648.9	23659.6	23670.2	23680.9	23691.5	23702.1
.7	23712.7	23723.3	23734.0	23744.6	23755.2	23765.8	23776.4	23786.9	23797.5	23808.1
.8	23818.7	23829.3	23839.8	23850.4	23861.0	23871.5	23882.1	23892.6	23903.1	23913.7
.9	23924.2	23934.7	23945.3	23955.8	23966.3	23976.8	23987.3	23997.8	24008.3	24018.8

Thousandths of an inch.	
11.2	
1	1.1
2	2.2
3	3.4
4	4.5
5	5.6
6	6.7
7	7.8
8	9.0
9	10.1
10.7	
1	1.1
2	2.1
3	3.2
4	4.3
5	5.4
6	6.4
7	7.5
8	8.6
9	9.6

TABLE II—Continued.

$$D_1 = 60384.3 \times \log H \text{ or } h.$$

Barometer in English inches.	Hundredths of an inch. 0.00	0.01	0.02	0.03	0.04	0.05	0.06	0.07	0.08	0.09	Thousandths of an inch.	
	Feet.	*Feet.*	*Feet.*	*Feet.*	*Feet.*	*Feet.*	*Feet.*	*Feet.*	*Feet.*	*Feet.*	10.3	
25.0	24029.3	24039.8	24050.3	24060.8	24071.3	24081.7	24092.2	24102.7	24113.1	24123.6	1	1.0
.1	24134.0	24144.5	24154.9	24165.3	24175.8	24186.2	24196.6	24207.1	42217.5	24227.9	2	2.1
.2	24238.3	24248.7	24259.1	24269.5	24279.9	24290.3	24300.7	24311.0	42321.4	24331.8	3	3.1
.3	24342.2	24352.5	24362.9	24373.2	24383.6	24393.9	24404.3	24414.6	24424.9	24435.3	4	4.1
.4	24445.6	24455.9	24466.2	24476.6	24486.9	24497.2	24507.5	24517.8	24528.1	24538.4	5	5.1
											6	6.2
.5	24548.6	24558.9	24569.2	24579.5	24589.7	24600.0	24610.3	24620.5	24630.8	24641.0	7	7.2
.6	24651.3	24661.5	24671.8	24682.0	24692.2	24702.5	24712.7	24722.9	24733.1	24743.3	8	8.2
.7	24753.5	24763.7	24773.9	24784.1	24794.3	24804.5	24814.7	24824.9	24835.0	24845.2	9	9.3
.8	24855.4	24865.5	24875.7	24885.8	24896.0	24906.1	24916.3	24926.4	24936.6	24946.7		
.9	24956.8	24966.9	24977.1	24987.2	24997.3	25007.4	25017.5	25027.6	25037.7	25047.8		
26.0	25057.9	25068.0	25078.0	25088.1	25098.2	25108.3	25118.3	25128.4	25138.4	25148.5	9.9	
.1	25158.5	25168.6	25178.6	25188.7	25198.7	25208.7	25218.8	25228.8	25238.8	25248.8		
.2	25258.8	25268.8	25278.8	25288.8	25298.8	25308.8	25318.8	25328.8	25338.8	25348.8	1	1.0
.3	25358.7	25368.7	25378.7	25388.6	25398.6	25408.5	25418.5	25428.4	25438.4	25448.3	2	2.0
.4	25458.3	25468.2	25478.1	25488.0	25498.0	25507.9	25517.8	25527.7	25537.6	25547.5	3	3.0
											4	4.0
.5	25557.4	25567.3	25577.2	25587.1	25597.0	25606.8	25616.7	25626.6	25636.5	25646.3	5	4.9
.6	25656.2	25666.0	25675.9	25685.7	25695.6	25705.4	25715.3	25725.1	25734.9	25744.8	6	5.9
.7	25754.6	25764.4	25774.2	25784.0	25793.8	25803.6	25813.4	25823.2	25833.0	25842.8	7	6.9
.8	25852.6	25862.4	25872.2	25882.0	25891.7	25901.5	25911.3	25921.0	25930.8	25940.5	8	7.9
.9	25950.3	25960.0	25969.8	25979.5	25989.3	25999.0	26008.7	26018.4	26028.2	26037.9	9	8.9

TABLE II—Continued.

$$D_1 = 60384.3 \times \log H \text{ or } h.$$

Barometer in English inches.	Hundredths of an inch.										Thousandths of an inch.	
	0.00	0.01	0.02	0.03	0.04	0.05	0.06	0.07	0.08	0.09		
	Feet.	*Feet.*	*Feet.*	*Feet.*	*Feet.*	*Feet.*	*Feet.*	*Feet.*	*Feet.*	*Feet.*	9.5	
27.0	26047.6	26057.3	26067.0	26076.7	26086.4	26096.1	26105.8	26115.5	26125.2	26134.9	1	1.0
.1	26144.5	26154.2	26163.9	26173.6	26183.2	26192.9	26202.5	26212.2	26221.8	26231.5	2	1.9
.2	26241.1	26250.8	26260.4	26270.0	26279.7	26289.3	26298.9	26308.5	26318.2	26327.8	3	2.9
.3	26337.4	26347.0	26356.6	26366.2	26375.8	26385.4	26394.9	26404.5	26414.1	26423.7	4	3.8
.4	26433.3	26442.8	26452.4	26462.0	26471.5	26481.1	26490.6	26500.2	26509.7	26519.3	5	4.8
											6	5.7
.5	26528.8	26538.3	26547.9	26557.4	26566.9	26576.4	26586.0	26595.5	26605.0	26614.5	7	6.7
.6	26624.0	26633.5	26643.0	26652.5	26662.0	26671.5	26680.9	26690.4	26699.9	26709.4	8	7.6
.7	26718.8	26728.3	26737.8	26747.2	26756.7	26766.1	26775.6	26785.0	26794.5	26803.9	9	8.6
.8	26813.3	26822.8	26832.2	26841.6	26851.0	26860.5	26869.9	26879.3	26888.7	26898.1		
.9	26907.5	26916.9	26926.3	26935.7	26945.1	26954.5	26963.8	26973.2	26982.6	26992.0		
28.0	27001.3	27010.7	27020.0	27029.4	27038.8	27048.1	27057.5	27066.8	27076.1	27085.5	9.2	
.1	27094.8	27104.1	27113.5	27122.8	27132.1	27141.4	27150.8	27160.1	27169.4	27178.7		
.2	27188.0	27197.3	27206.6	27215.9	27225.1	27234.4	27243.7	27253.0	27262.3	27271.5	1	0.9
.3	27280.8	27290.1	27299.3	27308.6	27317.8	27327.1	27336.3	27345.6	27354.8	27364.1	2	1.8
.4	27373.3	27382.5	27391.8	27401.0	27410.2	27419.4	27428.7	27437.9	27447.1	27456.3	3	2.8
											4	3.7
.5	27465.5	27474.7	27483.9	27493.1	27502.3	27511.5	27520.6	27529.8	27539.0	27548.2	5	4.6
.6	27557.3	27566.5	27575.7	27584.8	27594.0	27603.2	27612.3	27621.5	27630.6	27639.7	6	5.5
.7	27648.9	27658.0	27667.2	27676.3	27685.4	27694.5	27703.6	27712.8	27721.9	27731.0	7	6.4
.8	27740.1	27749.2	27758.3	27767.4	27776.5	27785.6	27794.7	27803.8	27812.8	27821.9	8	7.4
.9	27831.0	27840.1	27849.1	27858.2	27867.3	27876.3	27885.4	27894.4	27903.5	27912.5	9	8.3

Table II—Continued.

$$D_1 = 60384.3 \times \log H \text{ or } h.$$

Barometer in English inches.	Hundredths of an inch. 0.00	0.01	0.02	0.03	0.04	0.05	0.06	0.07	0.08	0.09	Thousandths of an inch.	8.9
	Feet.	*Feet.*	*Feet.*	*Feet.*	*Feet.*	*Feet.*	*Feet.*	*Feet.*	*Feet.*	*Feet.*		
29.0	27921.6	27930.6	27939.6	27948.7	27957.7	27966.8	27975.8	27984.8	27993.8	28002.8	1	0.9
.1	28011.9	28020.9	28029.9	28038.9	28047.9	28056.9	28065.9	28074.9	28083.8	28092.8	2	1.8
.2	28101.8	28110.8	28119.8	28128.7	28137.7	28146.7	28155.6	28164.6	28173.6	28182.5	3	2.7
.3	28191.5	28200.4	28209.4	28218.3	28227.3	28236.2	28245.1	28254.1	28263.0	28271.9	4	3.6
.4	28280.8	28289.7	28298.7	28307.6	28316.5	28325.4	28334.3	28343.2	28352.1	28361.0	5	4.4
.5	28369.9	28378.8	28387.6	28396.5	28405.4	28414.3	28423.2	28432.0	28440.9	28449.8	6	5.3
.6	28458.6	28467.5	28476.3	28485.2	28494.0	28502.9	28511.7	28520.6	28529.4	28538.2	7	6.2
.7	28547.1	28555.9	28564.7	28573.5	28582.4	28591.2	28600.0	28608.8	28617.6	28626.4	8	7.1
.8	28635.2	28644.0	28652.8	28661.6	28670.4	28679.2	28688.0	28696.7	28705.5	28714.3	9	8.0
.9	28723.1	28731.8	28740.6	28749.4	28758.1	28766.9	28775.6	28784.4	28793.1	28801.9		
30.0	28810.6	28819.4	28828.1	28836.8	28845.6	28854.3	28863.0	28871.8	28880.5	28889.2		8.6
.1	28897.9	28906.6	28915.3	28924.0	28932.7	28941.4	28950.1	28958.8	28967.5	28976.2	1	0.9
.2	28984.9	28993.6	29002.2	29010.9	29019.6	29028.3	29036.9	29045.6	29054.3	29062.9	2	1.7
.3	29071.6	29080.2	29088.9	29097.5	29106.2	29114.8	29123.5	29132.1	29140.7	29149.4	3	2.6
.4	29158.0	29166.6	29175.2	29183.8	29192.5	29201.1	29209.7	29218.3	29226.9	29235.5	4	3.4
.5	29244.1	29252.7	29261.3	29269.9	29278.5	29287.1	29295.6	29304.2	29312.8	29321.4	5	4.3
.6	29329.9	29338.5	29347.1	29355.6	29364.2	29372.8	29381.3	29389.9	29398.4	29407.0	6	5.2
.7	29415.5	29424.1	29432.6	29441.1	29449.7	29458.2	29466.7	29475.2	29483.7	29492.3	7	6.0
.8	29500.8	29509.3	29517.8	29526.3	29534.8	29543.3	29551.8	29560.3	29568.8	29577.3	8	6.9
.9	29585.8	29594.3	29602.8	29611.2	29619.7	29628.2	29636.7	29645.1	29653.6	29662.1	9	7.7

TABLE III.—*Correction for Temperature.*

$$D_{II} = A \log \frac{h}{H} \times B \text{ or } C.$$

$t + t'$	B	C	$t + t'$	B	C
	$\frac{t + t' - 64}{982.2647}$	$\frac{t + t' - 64}{900}$		$\frac{t + t' - 64}{982.2647}$	$\frac{t + t' - 64}{900}$
°			°		
30	−0.0346	−0.0378	70	+0.0061	+0.0067
31	.0336	.0367	71	.0071	.0078
32	.0326	.0356	72	.0081	.0089
33	.0316	.0344	73	.0092	.0100
34	.0305	.0333	74	.0102	.0111
35	.0295	.0322	75	.0112	.0122
36	.0285	.0311	76	.0122	.0133
37	.0275	.0300	77	.0132	.0144
38	.0265	.0289	78	.0142	.0156
39	.0255	.0278	79	.0153	.0167
40	.0244	.0267	80	.0163	.0178
41	.0234	.0256	81	.0173	.0189
42	.0224	.0244	82	.0183	.0200
43	.0214	.0233	83	.0193	.0211
44	.0204	.0222	84	.0204	.0222
45	.0193	.0211	85	.0214	.0233
46	.0183	.0200	86	.0224	.0244
47	.0173	.0189	87	.0234	.0256
48	.0163	.0178	88	.0244	.0267
49	.0153	.0167	89	.0255	.0278
50	.0143	.0156	90	.0265	.0289
51	.0132	.0144	91	.0275	.0300
52	.0122	.0133	92	.0285	.0311
53	.0112	.0122	93	.0295	.0322
54	.0102	.0111	94	.0305	.0333
55	.0092	.0100	95	.0316	.0344
56	.0081	.0089	96	.0326	.0356
57	.0071	.0078	97	.0336	.0367
58	.0061	.0067	98	.0346	.0378
59	.0051	.0056	99	.0356	.0389
60	.0041	.0044	100	.0366	.0400
61	.0030	.0033	101	.0377	.0411
62	.0020	.0022	102	.0387	.0422
63	−0.0010	−0.0011	103	.0397	.0433
64	.0000	.0000	104	.0407	.0444
65	+0.0010	+0.0011	105	.0417	.0456
66	.0020	.0022	106	.0428	.0467
67	.0030	.0033	107	.0438	.0478
68	.0041	.0044	108	.0448	.0489
69	+0.0051	+0.0056	109	+0.0458	+0.0500

TABLE III.—*Correction for Temperature*—Continued.

$$D_{II} = A \log \frac{h}{H} \times B \text{ or } C.$$

$t + t'$	B	C	$t + t'$	B	C
	$\frac{t + t' - 64}{982.2647}$	$\frac{t + t' - 64}{900}$		$\frac{t + t' - 64}{982.2647}$	$\frac{t + t' - 64}{900}$
°			°		
110	+0.0468	+0.0511	150	+0.0875	+0.0958
111	.0478	.0522	151	.0886	.0967
112	.0489	.0533	152	.0896	.0978
113	.0499	.0544	153	.0906	.0989
114	.0509	.0556	154	.0916	.1000
115	.0519	.0567	155	.0926	.1011
116	.0529	.0578	156	.0937	.1022
117	.0540	.0589	157	.0947	.1033
118	.0550	.0600	158	.0957	.1044
119	.0560	.0611	159	.9967	.1056
120	.0570	.0622	160	.0977	.1067
121	.0580	.0633	161	.0987	.1078
122	.0590	.0644	162	.0998	.1089
123	.0601	.0656	163	.1008	.1100
124	.0611	.0667	164	.1018	.1111
125	.0621	.0678	165	.1028	.1122
126	.0631	.0689	166	.1038	.1133
127	.0641	.0700	167	.1049	.1144
128	.0651	.0711	168	.1059	.1156
129	.0662	.0722	169	.1069	.1167
130	.0672	.0733	170	.1079	.1178
131	.0682	.0744	171	.1089	.1189
132	.0692	.0756	172	.1099	.1200
133	.0702	.0767	173	.1109	.1211
134	.0713	.0778	174	.1120	.1222
135	.0723	.0789	175	.1130	.1233
136	.0733	.0800	176	.1140	.1244
137	.0743	.0811	177	.1150	.1256
138	.0753	.0822	178	.1160	.1267
139	.0763	.0833	179	.1171	.1278
140	.0774	.0844	180	.1181	.1289
141	.0784	.0856	181	.1191	.1300
142	.0794	.0867	182	.1201	.1311
143	.0804	.0878	183	.1211	.1322
144	.0814	.0889	184	.1222	.1333
145	.0825	.0900	185	.1232	.1344
146	.0835	.0911	186	.1242	.1356
147	.0845	.0922	187	.1252	.1367
148	.0855	.0933	188	.1262	.1378
149	+0.0865	+0.0944	189	+0.1272	+0.1389

TABLE IV.—*Correction for the Difference of Gravity in Various Latitudes.*

$$D_{III} = 60384.3 \log \frac{h}{H} (0.002657 \times \cos 2 \text{ lat.})$$

Correction, additive from 0° to 45°; substractive from 45° to 90°.

Approx. dif. of level.	0°	2°	4°	6°	8°	10°	12°	14°	16°	18°	20°	22°	24°	26°	28°	30°	32°	34°	36°	38°	40°	42°	44°	45°
	90°	88°	86°	84°	82°	80°	78°	76°	74°	72°	70°	68°	66°	64°	62°	60°	58°	56°	54°	52°	50°	48°	46°	
	Feet.	*Feet.*	*Feet.*	*Feet.*	*Feet.*	*Feet.*	*Feet.*	*Feet.*	*Feet.*	*Feet.*	*Feet.*	*Feet.*	*Feet.*	*Feet.*	*Feet.*	*Feet.*	*Feet.*	*Feet.*	*Feet.*	*Feet.*	*Feet.*	*Feet.*	*Feet.*	
1000	2.6	2.6	2.6	2.6	2.5	2.5	2.4	2.3	2.2	2.1	2.0	1.9	1.8	1.6	1.5	1.3	1.2	1.0	0.8	0.6	0.5	0.3	0.1	0
2000	5.3	5.2	5.2	5.1	5.0	4.9	4.8	4.6	4.5	4.2	4.0	3.8	3.5	3.2	2.9	2.6	2.3	2.0	1.6	1.3	0.9	0.5	0.2	0
3000	7.9	7.9	7.8	7.7	7.6	7.4	7.2	7.0	6.7	6.4	6.0	5.7	5.3	4.8	4.4	3.9	3.5	3.0	2.4	1.9	1.4	0.8	0.3	0
4000	10.5	10.5	10.4	10.3	10.1	9.9	9.6	9.3	8.9	8.5	8.0	7.6	7.0	6.5	5.9	5.3	4.6	3.9	3.2	2.5	1.8	1.1	0.4	0
5000	13.1	13.1	13.0	12.8	12.6	12.3	12.0	11.6	11.1	10.6	10.1	9.4	8.8	8.1	7.3	6.6	5.8	4.9	4.1	3.2	2.3	1.4	0.5	0
6000	15.8	15.7	15.6	15.4	15.1	14.8	14.4	13.9	13.4	12.7	12.1	11.3	10.5	9.7	8.8	7.9	6.9	5.9	4.9	3.8	2.7	1.6	0.5	0
7000	18.4	18.3	18.2	18.0	17.7	17.3	16.8	16.2	15.6	14.9	14.1	13.2	12.3	11.3	10.3	9.2	8.1	6.9	5.7	4.4	3.2	1.9	0.6	0
8000	21.0	21.0	20.8	20.5	20.2	19.7	19.2	18.5	17.8	17.0	16.1	15.1	14.1	12.9	11.7	10.5	9.2	7.9	6.5	5.1	3.6	2.2	0.7	0
9000	23.6	23.6	23.4	23.1	22.7	22.2	21.6	20.9	20.0	19.1	18.1	17.0	15.8	14.5	13.2	11.8	10.4	8.9	7.3	5.7	4.1	2.5	0.8	0
10000	26.3	26.2	26.0	25.7	25.2	24.7	24.0	23.2	22.3	21.2	20.1	18.9	17.6	16.2	14.7	13.1	11.5	9.8	8.1	6.4	4.6	2.7	0.9	0
11000	28.9	28.8	28.6	28.2	27.8	27.1	26.4	25.5	24.5	23.4	22.1	20.8	19.3	17.8	16.2	14.4	12.7	10.8	8.9	7.0	5.0	3.0	1.0	0
12000	31.5	31.4	31.2	30.8	30.3	29.6	28.8	27.8	26.7	25.5	24.1	22.7	21.1	19.4	17.6	15.8	13.8	11.8	9.7	7.6	5.5	3.3	1.1	0
13000	34.1	34.1	33.8	33.4	32.8	32.1	31.2	30.1	28.9	27.6	26.1	24.6	22.8	21.0	19.1	17.1	15.0	12.8	10.5	8.3	5.9	3.6	1.2	0
14000	36.8	36.7	36.4	36.0	35.3	34.5	33.6	32.5	31.2	29.7	28.2	26.4	24.6	22.6	20.6	18.4	16.1	13.8	11.4	8.9	6.4	3.8	1.3	0
15000	39.4	39.3	39.0	38.5	37.9	37.0	36.0	34.8	33.4	31.9	30.2	28.3	26.3	24.2	22.0	19.7	17.3	14.8	12.2	9.5	6.8	4.1	1.4	0
16000	42.0	41.9	41.6	41.1	40.4	39.5	38.4	37.1	35.6	34.0	32.2	30.2	28.1	25.9	23.5	21.0	18.4	15.7	13.0	10.2	7.3	4.4	1.5	0
17000	44.6	44.5	44.2	43.7	42.9	41.9	40.8	39.4	37.9	36.1	34.2	32.1	29.9	27.5	25.0	22.3	19.6	16.7	13.8	10.8	7.8	4.7	1.6	0
18000	47.3	47.1	46.8	46.2	45.4	44.4	43.2	41.7	40.1	38.2	36.2	34.0	31.6	29.1	26.4	23.6	20.7	17.7	14.6	11.4	8.2	4.9	1.6	0
19000	49.9	49.8	49.4	48.8	48.0	46.9	45.6	44.0	42.3	40.4	38.2	35.9	33.4	30.7	27.9	24.9	21.9	18.7	15.4	12.1	8.7	5.2	1.7	0
20000	52.5	52.4	52.0	51.4	50.5	49.3	48.0	46.4	44.5	42.5	40.2	37.8	35.1	32.3	29.4	26.3	23.0	19.7	16.2	12.7	9.1	5.5	1.8	0
21000	55.1	55.0	54.6	53.9	53.0	51.8	50.4	48.7	46.8	44.6	42.2	39.7	36.9	33.9	30.8	27.6	24.2	20.7	17.0	13.3	9.6	5.8	1.9	0
22000	57.8	57.6	57.2	56.5	55.5	54.3	52.8	51.0	49.0	46.7	44.3	41.6	38.7	35.6	32.3	28.9	25.3	21.6	17.9	14.0	10.0	6.0	2.0	0
23000	60.4	60.2	59.8	59.1	58.1	56.7	55.2	53.3	51.2	48.9	46.3	43.4	40.4	37.2	33.8	30.2	26.5	22.6	18.7	14.6	10.5	6.3	2.1	0
24000	63.0	62.9	62.4	61.6	60.6	59.2	57.6	55.6	53.4	51.0	48.3	45.3	42.2	38.8	35.2	31.5	27.6	23.6	19.5	15.2	10.9	6.6	2.2	0

TABLE V.—*Correction for Decrease of Gravity on a Vertical acting on the Density of the Mercurial Column.*

$$D_{IV} = 60384.3 \log \frac{h}{H} + \frac{60384.3 \log \frac{h}{H} + 52252}{20886860}$$

Approx. diff. of alt.	000	100	200	300	400	500	600	700	800	900
	Feet.	*Feet.*	*Feet.*	*Feet.*	*Feet.*	*Feet.*	*Feet.*	*Feet.*	*Feet.*	*Feet.*
1000	2.5	2.8	3.1	3.3	3.6	3.9	4.1	4.4	4.7	4.9
2000	5.2	5.5	5.7	6.0	6.3	6.6	6.8	7.1	7.4	7.7
3000	7.9	8.2	8.5	8.8	9.1	9.3	9.6	9.9	10.2	10.5
4000	10.8	11.1	11.4	11.6	11.9	12.2	12.5	12.8	13.1	13.4
5000	13.7	14.0	14.3	14.6	14.9	15.2	15.5	15.8	16.1	16.4
6000	16.7	17.0	17.4	17.7	18.0	18.3	18.6	18.9	19.2	19.5
7000	19.9	20.2	20.5	20.8	21.1	21.5	21.8	22.1	22.4	22.8
8000	23.1	23.4	23.7	24.1	24.4	24.7	25.1	25.4	25.7	26.1
9000	26.4	26.7	27.1	27.4	27.7	28.1	28.4	28.8	29.1	29.5
10000	29.8	30.2	30.5	30.8	31.2	31.5	31.9	32.2	32.6	33.0
11000	33.3	33.7	34.0	34.4	34.7	35.1	35.5	35.8	36.2	36.5
12000	36.9	37.3	37.6	38.0	38.4	38.8	39.1	39.5	39.9	40.2
13000	40.6	41.0	41.4	41.7	42.1	42.5	42.8	43.3	43.6	44.0
14000	44.4	44.8	45.2	45.6	46.0	46.3	46.7	47.1	47.5	47.9
15000	48.3	48.7	49.1	49.5	49.9	50.3	50.7	51.1	51.5	51.9
16000	52.3	52.7	53.1	53.5	53.9	54.3	54.7	55.1	55.5	56.0
17000	56.4	56.8	57.2	57.6	58.0	58.4	58.9	59.3	59.7	60.1
18000	60.5	61.0	61.4	61.8	62.2	62.7	63.1	63.5	64.0	64.4
19000	64.8	65.2	65.7	66.1	66.6	67.0	67.4	67.9	68.3	68.7
20000	69.2	69.6	70.1	70.5	71.0	71.4	71.9	72.3	72.7	73.2
21000	73.6	74.1	74.6	75.0	75.5	75.9	76.4	76.8	77.3	77.7
22000	78.2	78.7	79.1	79.6	80.1	80.5	81.0	81.5	81.9	82.4
23000	82.9	83.3	83.8	84.3	84.8	85.2	85.7	86.2	86.7	87.1
24000	87.6	88.1	88.6	89.1	89.5	90.0	90.5	91.0	91.5	92.0

NOTE.—In Table I the scale of the barometer is supposed to be of brass, extending from the cistern to the top of the mercurial column, and the difference of expansion of brass and of mercury is taken into account. The standard temperature of the yard being 62° Fahrenheit and not 32°, the difference of expansion of the scale and of the mercurial column carries the point of no correction down to 29° Fahrenheit.

TABLE VI.—*Correction for Decrease of Gravity on a Vertical acting on the Density of the Air.*

$$D_v = 60384.3 \log \frac{h}{H} \times \frac{s}{10443430}$$

Approx. diff. of alt.	Height of the barometer in English inches at the lower station.																	
	12	13	14	15	16	17	18	19	20	21	22	23	24	25	26	27	28	29
1000	2.3	2.1	1.9	1.7	1.6	1.4	1.3	1.1	1.0	0.9	0.8	0.7	0.6	0.5	0.4	0.3	0.2	0.1
2000	4.6	4.2	3.8	3.5	3.1	2.8	2.6	2.3	2.0	1.8	1.5	1.3	1.1	0.9	0.7	0.5	0.3	0.2
3000	6.9	6.3	5.7	5.2	4.7	4.3	3.8	3.4	3.0	2.7	2.3	2.0	1.7	1.4	1.1	0.8	0.5	0.2
4000	9.2	8.4	7.6	6.9	6.3	5.7	5.1	4.6	4.0	3.6	3.1	2.6	2.2	1.8	1.4	1.0	0.7	0.3
5000	11.5	10.5	9.5	8.7	7.9	7.1	6.4	5.7	5.1	4.4	3.9	3.3	2.8	2.3	1.8	1.3	0.8	0.4
6000	13.8	12.6	11.4	10.4	9.4	8.5	7.7	6.8	6.1	5.3	4.6	4.0	3.3	2.7	2.1	1.5	1.0	0.5
7000	16.1	14.7	13.4	12.1	11.0	9.9	8.9	8.0	7.1	6.2	5.4	4.6	3.9	3.2	2.5	1.8	1.2	0.6
8000	18.4	16.7	15.3	13.9	12.6	11.4	10.2	9.1	8.1	7.1	6.2	5.3	4.4	3.6	2.8	2.1	1.3	0.6
9000	20.6	18.8	17.2	15.6	14.1	12.8	11.5	10.3	9.1	8.0	7.0	5.9	5.0	4.1	3.2	2.3	1.5	0.7
10000	22.9	20.9	19.1	17.3	15.7	14.2	12.8	11.4	10.1	8.9	7.7	6.6	5.5	4.5	3.5	2.6	1.7	0.8
11000	25.2	23.0	21.0	19.1	17.3	15.6	14.0	12.5	11.1	9.8	8.5	7.3	6.1	5.0	3.9	2.8	1.8	0.9
12000	27.5	25.1	22.9	20.8	18.9	17.0	15.3	13.7	12.1	10.7	9.3	7.9	6.6	5.4	4.2	3.1	2.0	0.9
13000	29.8	27.2	24.8	22.5	20.4	18.5	16.6	14.8	13.2	11.6	10.0	8.6	7.2	5.9	4.6	3.4	2.2	1.0
14000	32.1	29.3	26.7	24.3	22.0	19.9	17.9	16.0	14.2	12.4	10.8	9.2	7.8	6.3	4.9	3.6	2.3	1.1
15000	34.4	31.4	28.6	26.0	23.6	21.3	19.2	17.1	15.2	13.3	11.6	9.9	8.3	6.8	5.3	3.9	2.5	1.2
16000	36.7	33.5	30.5	27.7	25.2	22.7	20.4	18.2	16.2	14.2	12.4	10.6	8.9	7.2	5.6	4.1	2.7	1.3
17000	39.0	35.6	32.4	29.5	26.7	24.1	21.7	19.4	17.2	15.1	13.1	11.2	9.4	7.7	6.0	4.4	2.8	1.3
18000	41.3	37.7	34.3	31.2	28.3	25.6	23.0	20.5	18.2	16.0	13.9	11.9	10.0	8.1	6.4	4.6	3.0	1.4
19000	43.6	39.8	36.2	32.9	29.9	27.0	24.2	21.7	19.2	16.9	14.7	12.6	10.5	8.6	6.7	4.9	3.2	1.5
20000	45.9	41.9	38.1	34.7	31.4	28.4	25.5	22.8	20.2	17.8	15.4	13.2	11.1	9.0	7.1	5.2	3.3	1.6
21000	48.2	44.0	40.1	36.4	33.0	29.8	26.8	23.9	21.2	18.7	16.2	13.9	11.6	9.5	7.4	5.4	3.5	1.7
22000	50.5	46.1	42.0	38.1	34.6	31.2	28.1	25.1	22.3	19.6	17.0	14.5	12.2	9.9	7.8	5.7	3.7	1.7
23000	52.8	48.1	43.9	39.9	36.2	32.7	29.4	26.2	23.3	20.4	17.8	15.2	12.7	10.4	8.1	5.9	3.8	1.8
24000	55.1	50.2	45.8	41.6	37.7	34.1	30.6	27.4	24.3	21.3	18.5	15.9	13.3	10.8	8.5	6.2	4.0	1.9

TABLE VII.—*Giving the Relative Humidities in Fractions of Unity.*

Diff. of wet and dry bulb thermometers.	Wet bulb.						
	10°	15°	20°	22°	24°	26°	28°
° 0.0	1.000	1.000	1.000	1.000	1.000	1.000	1.000
.2	.925	.926	.929	.935	.941	.949	.958
.4	.854	.856	.864	.874	.887	.902	.919
.6	.788	.791	.804	.817	.837	.858	.883
.8	.727	.731	.749	.764	.790	.818	.850
1.0	.671	.675	.699	.717	.747	.781	.819
.2	.619	.624	.653	.674	.707	.747	.791
.4	.570	.577	.610	.635	.670	.716	.765
.6	.525	.534	.571	.599	.636	.687	.741
.8	.483	.494	.535	.566	.605	.660	.719
2.0	.445	.456	.502	.535	.577	.635	.698
.2	.410	.421	.471	.506	.552	.612	.679
.4	.377	.389	.443	.479	.529	.591	.662
.6	.346	.360	.417	.454	.507	.572	.646
.8	.318	.333	.393	.432	.487	.554	.631
3.0	.292	.308	.371	.412	.469	.538	.618
.2	.268	.285	.351	.394	.453	.524	.605
.4	.245	.264	.332	.377	.439	.511	.593
.6	.224	.245	.314	.361	.426	.499	.582
.8	.205	.227	.298	.346	.413	.488	.572
4.0	.187	.211	.283	.332	.401	.478	.562
.2	.170	.196	.269	.320	.390	.469	.553
.4	.154	.182	.257	.309	.379	.461	.544
.6	.140	.169	.246	.299	.370	.453	.536
.8	.128	.157	.236	.290	.362	.446	.528
5.0	.117	.145	.226	.281	.355	.440	.521
.2			.217	.273	.348	.434	.514
.4			.209	.266	.342	.429	.507
.6			.201	.259	.336	.424	.501
.8			.194	.253	.331	.419	.495
6.0			.188	.248	.327	.415	.489
.2			.182	.243	.324	.411	.482
.4			.176	.239	.321	.407	.476
.6			.171	.236	.319	.403	.469
.8			.167	.233	.317	.399	.463
7.0			.164	.231	.315	.396	.457

TABLE VII.—*Giving the Relative Humidities, &c.*—Continued.

Diff. of wet and dry bulb thermometers.	Wet bulb.						
	30°	32°	34°	36°	38°	40°	42°
°							
0	1.000	1.000	1.000	1.000	1.000	1.000	1.000
1	.856	.885	.903	.909	.914	.917	.920
2	.756	.799	.821	.831	.837	.843	.847
3	.684	.728	.750	.761	.768	.776	.781
4	.628	.665	.686	.697	.707	.715	.721
5	.579	.609	.628	.640	.652	.660	.667
6	.533	.558	.575	.589	.601	.610	.618
7	.490	.512	.528	.543	.554	.565	.573
8	.450	.470	.486	.500	.512	.523	.532
9	.414	.432	.448	.461	.473	.485	.495
10	.381	.398	.413	.426	.438	.451	.462
11	.351	.367	.381	.394	.406	.420	.432
12	.324	.339	.352	.365	.378	.391	.404
13	.299	.313	.326	.339	.352	.365	.378
14	.277	.289	.302	.315	.328	.341	.354
15	.257	.268	.280	.293	.306	.320	.332
16	.238	.249	.261	.274	.286	.300	.312
17	.221	.232	.244	.256	.268	.281	.293
18	.206	.216	.228	.240	.252	.263	.275
19	.192	.202	.213	.225	.237	.246	.258
20	.180	.189	.200	.211	.223	.231	.242
21	.168	.178	.189	.199	.209	.217	.226
22	.158	.168	.178	.187	.195	.204	.213
23	.149	.158	.167	.175	.183	.192	.200
24	.140	.149	.157	.164	.172	.181	.188
25	.132	.141	.148	.154	.162	.170	.177
26						.160	.168
27						.151	.159
28						.143	.151
29						.135	.143
30						.128	.135

TABLE VII.—*Giving the Relative Humidities, &c.*—Continued.

Diff. of wet and dry bulb thermometers.	Wet bulb.						
	44°	46°	48°	50°	52°	54°	56°
°							
0	1.000	1.000	1.000	1.000	1.000	1.000	1.000
1	.922	.923	.925	.928	.930	.931	.933
2	.851	.853	.858	.862	.866	.869	.872
3	.786	.791	.796	.802	.807	.811	.816
4	.727	.734	.740	.747	.753	.758	.764
5	.674	.682	.689	.697	.704	.709	.716
6	.626	.635	.643	.651	.658	.664	.671
7	.582	.592	.601	.609	.616	.622	.629
8	.542	.552	.562	.570	.577	.583	.590
9	.506	.516	.526	.534	.541	.547	.554
10	.473	.483	.493	.501	.507	.513	.521
11	.443	.453	.462	.470	.476	.482	.491
12	.415	.425	.433	.441	.448	.454	.463
13	.389	.398	.406	.414	.422	.428	.437
14	.365	.373	.381	.389	.397	.404	.413
15	.343	.350	.358	.366	.374	.382	.391
16	.322	.329	.337	.345	.353	.361	.370
17	.302	.310	.318	.326	.333	.341	.350
18	.284	.292	.300	.308	.315	.322	.331
19	.267	.275	.283	.291	.298	.305	.313
20	.251	.259	.267	.275	.282	.289	.296
21	.236	.244	.252	.260	.267	.274	.280
22	.222	.230	.238	.246	.253	.260	.265
23	.209	.217	.225	.233	.240	.246	.251
24	.197	.205	.213	.221	.227	.233	.238
25	.186	.194	.202	.209	.215	.221	.226
26	.176	.184	.192	.198	.204	.209	.214
27	.167	.174	.182	.188	.193	.198	.203
28	.158	.165	.172	.178	.183	.188	.193
29	.150	.157	.163	.169	.174	.178	.183
30	.142	.148	.154	.160	.165	.169	.174
31				.151	.156	.161	.166
32				.143	.148	.153	.158
33				.136	.141	.146	.150
34				.129	.134	.139	.143
35				.123	.127	.132	.136

TABLE VII.—*Giving the Relative Humidities, &c.*—Continued.

Diff. of wet and dry bulb thermometers.	Wet bulb.						
	58°	60°	62°	64°	66°	67°	68°
°							
0	1.000	1.000	1.000	1.000	1.000	1.000	1.000
1	.935	.936	.937	.938	.940	.940	.941
2	.875	.877	.879	.881	.884	.885	.886
3	.819	.822	.825	.828	.832	.834	.835
4	.767	.771	.775	.779	.783	.786	.788
5	.719	.724	.729	.734	.739	.741	.743
6	.675	.681	.686	.692	.697	.699	.702
7	.634	.641	.647	.653	.658	.660	.663
8	.597	.603	.610	.616	.621	.624	.627
9	.562	.568	.575	.582	.587	.590	.593
10	.529	.536	.543	.549	.555	.558	.561
11	.499	.506	.513	.519	.526	.528	.531
12	.471	.478	.485	.491	.497	.500	.503
13	.445	.452	.459	.465	.471	.474	.476
14	.420	.428	.434	.440	.446	.449	.451
15	.397	.405	.411	.417	.422	.426	.428
16	.376	.383	.389	.395	.400	.403	.406
17	.356	.362	.368	.374	.379	.382	.385
18	.337	.343	.348	.354	.360	.363	.366
19	.319	.325	.330	.336	.341	.344	.447
20	.302	.308	.313	.319	.324	.327	.330
21	.286	.292	.297	.303	.308	.311	.314
22	.271	.277	.282	.288	.293	.296	.299
23	.257	.263	.268	.273	.279	.282	.284
24	.244	.249	.255	.260	.265	.268	.271
25	.231	.236	.242	.247	.252	.255	.258
26	.219	.224	.230	.235	.240	.243	.246
27	.208	.213	.219	.224	.229	.232	.235
28	.198	.203	.208	.213	.219	.221	.224
29	.188	.193	.198	.203	.209	.211	.214
30	.179	.184	.189	.194	.200	.202	.204
31	.170	.175	.180	.185	.190	.193	.195
32	.162	.167	.172	.177	.182	.184	.186
33	.154	.159	.164	.169	.173	.176	.178
34	.147	.152	.157	.161	.166	.168	.170
35	.141	.145	.150	.154	.159	.161	.163

TABLE VII.—*Giving the Relative Humidities, &c.*—Continued.

Diff. of wet and dry bulb thermometers.	Wet bulb.						
	69°	70°	71°	73°	75°	77°	79°
°							
0	1.000	1.000	1.000	1.000	1.000	1.000	1.000
1	.941	.942	.942	.943	.945	.946	.946
2	.887	.888	.889	.891	.893	.895	.896
3	.836	.838	.839	.842	.845	.847	.849
4	.789	.791	.793	.796	.799	.801	.804
5	.745	.747	.749	.753	.756	.759	.762
6	.704	.706	.708	.712	.716	.719	.723
7	.665	.668	.670	.674	.678	.682	.686
8	.629	.632	.634	.638	.642	.647	.651
9	.595	.598	.600	.605	.609	.614	.618
10	.563	.566	.569	.573	.578	.583	.587
11	.533	.536	.539	.544	.549	.553	.558
12	.505	.508	.511	.516	.521	.526	.531
13	.479	.482	.484	.490	.495	.500	.505
14	.454	.457	.459	.465	.470	.476	.481
15	.431	.434	.436	.442	.447	.453	.458
16	.409	.412	.414	.420	.425	.431	.436
17	.388	.391	.393	.399	.405	.411	.416
18	.368	.371	.374	.380	.386	.392	.397
19	.350	.353	.356	.361	.367	.373	.379
20	.333	.336	.339	.344	.350	.356	.361
21	.317	.320	.323	.328	.334	.340	.345
22	.301	.304	.307	.313	.319	.324	.330
23	.287	.290	.293	.299	.304	.310	.315
24	.274	.276	.279	.285	.290	.296	.301
25	.261	.263	.266	.272	.277	.283	.288
26	.249	.252	.254	.260	.265	.270	.275
27	.238	.240	.243	.248	.253	.258	.263
28	.227	.229	.232	.237	.242	.247	.251
29	.217	.219	.222	.227	.231	.236	.240
30	.207	.209	.212	.217	.221	.225	.229
31	.198	.200	.202	.207	.211	.215	.219
32	.189	.191	.193	.198	.202	.205	.209
33	.181	.183	.185	.189	.193	.196	.200
34	.173	.175	.177	.181	.184	.187	.191
35	.165	.167	.169	.173	.176	.179	.183

TABLE VIII.—*First Part of Correction for Atmospheric Humidity.*

$$D_{VI} = \frac{\log \frac{h}{H}}{\sqrt{hH}} V_{t+t'} = 180^\circ \text{ F.}^{W}\ t + t' = 180^\circ \text{ F.}$$

Barometer at upper station.	Height of the barometer at lower station in English inches.													
	31	30	29	28	27	26	25	24	23	22	21	20	19	18
Eng. in.	*Feet.*	*Feet.*	*Feet.*	*Feet.*	*Feet.*	*Feet.*	*Feet.*	*Feet.*	*Feet.*	*Feet.*	*Feet.*	*Feet.*	*Feet.*	*Feet.*
17.0	207	199	190	181	171	160	148	135	121	105	88	70	49	26
.5	194	186	177	168	158	147	135	122	108	92	75	56	36	13
18.0	182	174	165	155	145	134	122	109	95	80	63	44	23	
.5	170	162	153	144	134	123	111	98	83	68	51	32	11	
19.0	159	151	142	133	123	111	99	86	72	57	40	21		
.5	149	141	132	122	112	101	89	76	62	46	29	10		
20.0	139	131	122	112	102	91	79	66	51	36	19			
.5	130	121	112	103	92	81	69	56	42	26	9			
21.0	121	112	103	94	83	72	60	47	33	17				
.5	112	104	95	85	75	63	51	38	24	8				
22.0	104	95	86	77	66	55	43	30	16					
.5	96	87	78	69	58	47	35	22	8					
23.0	88	80	71	61	51	40	27	14						
.5	81	73	64	54	44	32	20	7						
24.0	74	66	57	47	37	25	13							
.5	67	59	50	40	30	19	6							
25.0	61	53	44	34	23	12								
.5	55	46	37	28	17	6								
26.0	49	40	31	22	11									
.5	43	35	26	16	6									
27.0	38	29	20	10										
.5	32	24	15	5										
28.0	27	19	10											
.5	22	14	5											
29.0	18	9												
.5	13	4												
30.0	8													
.5	4													

TABLE IX.—*Second Part of Correc*

$$D_{VII} = \frac{\log \frac{h}{H}}{\sqrt{h H}} V. W.$$

(t + t')	(t + t')						
180° F.	170° F.	160° F.	150° F.	140° F.	130° F.	120° F.	110° F.
Feet.	*Feet.*	*Feet.*	*Feet.*	*Feet.*	*Feet.*	*Feet.*	*Feet.*
10	8.5	7.1	6.0	5.0	4.2	3.5	2.9
20	16.9	14.3	12.0	10.0	8.4	6.9	5.8
30	25.4	21.4	18.0	15.0	12.5	10.4	8.6
40	33.8	28.5	23.9	20.0	16.7	13.9	11.5
50	42.3	35.6	29.9	25.1	20.9	17.4	14.4
60	50.7	42.8	35.9	30.1	25.1	20.8	17.3
70	59.2	49.9	41.9	35.1	29.3	24.3	20.1
80	67.6	57.0	47.9	40.1	33.4	27.8	23.0
90	76.1	64.1	53.9	45.1	37.6	31.3	25.9
100	84.5	71.3	59.9	50.1	41.8	34.7	28.8
110	93.0	78.4	65.8	55.1	46.0	38.2	31.6
120	101.5	85.5	71.8	60.1	50.1	41.7	34.5
130	109.9	92.6	77.8	65.1	54.3	45.1	37.4
140	118.4	99.8	83.8	70.1	58.5	48.6	40.3
150	126.8	106.9	89.8	75.2	62.7	52.1	43.1
160	135.3	114.0	95.8	80.2	66.9	55.6	46.0
170	143.7	121.1	101.8	85.2	71.0	59.0	48.9
180	152.2	128.3	107.7	90.2	75.2	62.5	51.8
190	160.6	135.4	113.7	95.2	79.4	66.0	54.6
200	169.1	142.5	119.7	100.2	83.6	69.5	57.5
210	177.6	149.6	125.7	105.2	87.8	72.9	60.4
220	186.0	156.8	131.7	110.2	91.9	76.4	63.3
230	194.5	163.9	137.7	115.2	96.1	79.9	66.1
240	202.9	171.0	143.7	120.2	100.3	83.3	69.0
250	211.4	178.1	149.6	125.3	104.5	86.8	71.9
260	219.8	185.3	155.6	130.3	108.6	90.3	74.8
270	228.3	192.4	161.6	135.3	112.8	93.8	77.6
280	236.7	199.5	167.6	140.3	117.0	97.2	80.5
290	245.2	206.7	173.6	145.3	121.2	100.7	83.4
300	253.6	213.8	179.6	150.3	125.4	104.2	86.3
310	262.1	220.9	185.6	155.3	129.5	107.6	89.1
320	270.6	228.0	191.5	160.3	133.7	111.1	92.0
330	279.0	235.2	197.5	165.3	137.9	114.6	94.9
340	287.5	242.3	203.5	170.3	142.1	118.1	97.8
350	295.9	249.4	209.5	175.4	146.3	121.5	100.6
360	304.4	256.5	215.5	180.4	150.4	125.0	103.5
370	312.8	263.7	221.5	185.4	154.6	128.5	106.4
380	321.3	270.8	227.5	190.4	158.8	132.0	109.3

tion for Atmospheric Humidity.

$$D_{VII} = \frac{\log \frac{h}{H}}{\sqrt{h H}} \text{ V. W.}$$

(*t* + *t*′)

100° F.	90° F.	80° F.	70° F.	60° F.	50° F.	40° F.	30° F.	20° F.
Feet.	*Feet.*	*Feet.*	*Feet.*	*Feet.*	*Feet.*	*Feet.*	*Feet.*	*Feet.*
2.4	1.9	1.6	1.3	1.1	0.8	0.7	0.5	0.4
4.7	3.9	3.2	2.6	2.1	1.7	1.3	1.0	0.8
7.1	5.8	4.8	3.9	3.2	2.5	2.0	1.6	1.2
9.5	7.8	6.4	5.2	4.2	3.4	2.7	2.1	1.7
11.9	9.7	8.0	6.5	5.3	4.2	3.3	2.6	2.1
14.2	11.7	9.6	7.8	6.3	5.0	4.0	3.1	2.5
16.6	13.6	11.2	9.1	7.4	5.9	5.7	3.7	2.9
19.0	15.6	12.8	10.4	8.4	6.7	5.3	4.2	3.3
21.3	17.5	14.4	11.7	9.5	7.6	6.0	4.7	3.7
23.7	19.5	16.0	13.0	10.5	8.4	6.7	5.2	4.1
26.1	21.4	17.5	14.3	11.6	9.2	7.3	5.8	4.5
28.5	23.4	19.1	15.6	12.6	10.1	8.0	6.3	5.0
30.8	25.3	20.7	16.9	13.7	10.9	8.7	6.8	5.4
33.2	27.3	22.3	18.2	14.7	11.8	9.3	7.3	5.8
35.6	29.2	23.9	19.5	15.8	12.6	10.0	7.9	6.2
38.0	31.2	25.5	20.8	16.8	13.4	10.7	8.4	6.6
40.3	33.1	27.1	22.1	17.9	14.3	11.3	8.9	7.0
42.7	35.1	28.7	23.4	18.9	15.1	12.0	9.4	7.4
45.1	37.0	30.3	24.7	20.0	16.0	12.7	10.0	7.8
47.4	39.0	31.9	26.0	21.0	16.8	13.3	10.5	8.3
49.8	40.9	33.5	27.3	22.1	17.7	14.0	11.0	8.7
52.2	42.9	35.1	28.6	23.1	18.5	14.7	11.5	9.1
54.6	44.8	36.7	29.9	24.2	19.3	15.3	12.1	9.5
56.9	46.8	38.3	31.2	25.2	20.2	16.0	12.6	9.9
59.3	48.7	39.9	32.5	26.3	21.0	16.7	13.1	10.3
61.7	50.7	41.5	33.8	27.4	21.9	17.3	13.6	10.7
64.0	52.6	43.1	35.1	28.4	22.7	18.0	14.2	11.1
66.4	54.6	44.7	36.4	29.5	23.5	18.7	14.7	11.6
68.8	56.5	46.3	37.7	30.5	24.4	19.3	15.2	12.0
71.2	58.5	47.9	39.0	31.6	25.2	20.0	15.7	12.4
73.5	60.4	49.5	40.3	32.6	26.1	20.7	16.3	12.8
75.9	62.4	51.0	41.6	33.7	26.9	21.3	16.8	13.2
78.3	64.3	52.6	42.9	34.7	27.7	22.0	17.3	13.6
80.7	66.3	54.2	44.2	35.8	28.6	22.7	17.8	14.0
83.0	68.2	55.8	45.5	36.8	29.4	23.3	18.4	14.4
85.4	70.2	57.4	46.8	37.9	30.3	24.0	18.9	14.9
87.8	72.1	59.0	48.1	38.9	31.1	24.7	19.4	15.3
90.1	74.1	60.6	49.4	40.0	31.9	25.3	19.9	15.7

Table to facilitate the Calculation of the Reduction to Level, by showing at what Hours 0.001 *is to be added on Account of the Fractional Part of the Correction for one Hour.*

Hour.	No.	When the fractional part of the correction for one hour is—																							No.	Hour.
		$\frac{1}{24}$	$\frac{2}{24}$	$\frac{3}{24}$	$\frac{4}{24}$	$\frac{5}{24}$	$\frac{6}{24}$	$\frac{7}{24}$	$\frac{8}{24}$	$\frac{9}{24}$	$\frac{10}{24}$	$\frac{11}{24}$	$\frac{12}{24}$	$\frac{13}{24}$	$\frac{14}{24}$	$\frac{15}{24}$	$\frac{16}{24}$	$\frac{17}{24}$	$\frac{18}{24}$	$\frac{19}{24}$	$\frac{20}{24}$	$\frac{21}{24}$	$\frac{22}{24}$	$\frac{23}{24}$		
		Add.	Add.	Add.	Add.	Add.	Add.	Add.	Add.	Add.	Add.	Add.	Add.	Add.	Add.	Add.	Add.	Add.	Add.	Add.	Add.	Add.	Add.	Add.		
7 a.m.	0	.000	.000	.000	.000	.000	.000	.000	.000	.000	.000	.000	.000	.000	.000	.000	.000	.000	.000	.000	.000	.000	.000	.000	0	7 a.m.
8	1	0	0	0	0	0	0	0	0	0	0	0	0	.001	.001	.001	.001	.001	.001	.001	.001	.001	.001	.001	1	8
9	2	0	0	0	0	0	0	.001	.001	.001	.001	.001	.001	0	0	0	0	0	0	.001	.001	.001	.001	.001	2	9
10	3	0	0	0	0	.001	.001	0	0	0	0	0	0	.001	.001	.001	.001	.001	.001	0	0	.001	.001	.001	3	10
11	4	0	0	0	.001	0	0	0	0	0	.001	.001	.001	0	0	0	.001	.001	.001	.001	.001	0	.001	.001	4	11
Noon.	5	0	0	.001	0	0	0	0	.001	.001	0	0	0	.001	.001	.001	0	.001	.001	.001	.001	.001	.001	.001	5	Noon.
1 p.m.	6	0	0	0	0	0	0	.001	0	0	0	.001	.001	0	0	.001	.001	0	0	.001	.001	.001	.001	.001	6	1 p.m.
2	7	0	.001	0	0	0	.001	0	0	.001	.001	0	0	.001	.001	0	.001	.001	.001	.001	.001	.001	0	.001	7	2
3	8	0	0	0	0	.001	0	0	.001	0	0	.001	.001	0	.001	.001	0	.001	.001	0	.001	.001	.001	.001	8	3
4	9	0	0	0	0	0	0	.001	0	0	.001	0	0	.001	0	.001	.001	0	.001	.001	0	.001	.001	.001	9	4
5	10	0	0	0	.001	0	0	0	0	.001	0	.001	.001	0	.001	0	.001	.001	0	.001	.001	.001	.001	.001	10	5
6	11	0	0	0	0	0	.001	0	.001	0	.001	0	0	.001	0	.001	0	.001	.001	.001	.001	.001	.001	.001	11	6
7	12	0	0	0	0	0	0	0	0	0	0	0	.001	.001	.001	0	.001	0	.001	0	.001	0	.001	0	12	7
8	13	.001	0	.001	0	.001	0	.001	0	.001	0	.001	0	0	.001	.001	.001	.001	.001	.001	.001	.001	.001	.001	13	8
9	14	0	0	0	0	0	0	0	.001	0	.001	0	.001	.001	0	.001	0	.001	0	.001	.001	.001	.001	.001	14	9
10	15	0	0	0	0	0	.001	0	0	.001	0	.001	0	0	.001	0	.001	.001	.001	.001	0	.001	.001	.001	15	10
11	16	0	0	0	.001	0	0	.001	0	0	.001	0	.001	.001	0	.001	.001	0	.001	.001	.001	.001	.001	.001	16	11
Midnight.	17	0	0	0	0	.001	0	0	.001	0	0	.001	0	0	.001	.001	0	.001	.001	0	.001	.001	.001	.001	17	Midnight.
1 a.m.	18	0	0	0	0	0	0	0	0	.001	0	0	.001	.001	0	0	.001	.001	0	.001	.001	.001	.001	.001	18	1 a.m.
2	19	0	.001	0	0	0	.001	.001	0	0	.001	.001	0	0	.001	.001	.001	0	.001	.001	.001	.001	0	.001	19	2
3	20	0	0	.001	0	0	0	0	.001	0	0	0	.001	.001	.001	0	0	.001	.001	.001	.001	0	.001	.001	20	3
4	21	0	0	0	0	0	0	0	0	.001	.001	.001	0	0	0	.001	.001	.001	.001	.001	0	.001	.001	.001	21	4
5	22	0	0	0	.001	.001	0	0	0	0	0	0	.001	.001	.001	.001	.001	.001	0	0	.001	.001	.001	.001	22	5
6	23	0	0	0	0	0	.001	.001	.001	.001	.001	.001	0	0	0	0	0	0	.001	.001	.001	.001	.001	.001	23	6
7	24	.000	.000	.000	.000	.000	.000	.000	.000	.000	.000	.000	.001	.001	.001	.001	.001	.001	.001	.001	.001	.001	.001	.001	24	7

XLVI.—*Thermometrical Measurement of Heights.*

BAROMETRIC PRESSURES CORRESPONDING TO TEMPERATURES OF THE BOILING-POINT OF WATER.

Degrees of Fahrenheit.	Tenths of a degree of Fahrenheit.				
	0	2	4	6	8
185	17.048	17.122	17.197	17.272	17.348
186	.423	.499	.575	.652	.729
187	.806	.883	.961	18.039	18.117
188	18.195	18.274	18.353	.432	.512
189	.592	.672	.753	.833	.914
190	.996	19.077	19.159	19.241	19.324
191	19.407	.490	.573	.657	.741
192	.825	.910	.995	20.080	20.166
193	20.251	20.338	20.424	.511	.598
194	.685	.773	.861	.949	21.038
195	21.126	21.216	21.305	21.395	.485
196	.576	.666	.758	.849	.941
197	22.033	22.125	22.218	22.311	22.404
198	.498	.592	.686	.781	.876
199	.971	23.067	23.163	23.259	23.356
200	23.453	.550	.648	.746	.845
201	.943	24.042	24.142	24.241	24.341
202	24.442	.542	.644	.745	.847
203	.949	25.051	25.154	25.257	25.361
204	25.465	.569	.674	.779	.884
205	.990	26.096	26.202	26.309	26.416
206	26.523	.631	.740	.848	.957
207	27.066	27.176	27.286	27.397	27.507
208	.618	.730	.842	.954	28.067
209	28.180	28.293	28.407	28.521	.636
210	.751	.866	.982	29.098	29.215
211	29.331	29.449	29.566	.684	.803
212	.922	30.041	30.161	30.281	30.401

Table of Comparison of Fahrenheit's Thermometer with Reaumur's and the Centesimal.

Fah.	Reaum.	Centes.	Fah.	Reaum.	Centes.	Fah.	Reaum.	Centes.
°	°	°	°	°	°	°	°	°
			33	+ 0.4	+ 0.6	67	+15.6	+19.4
0	—14.2	—17.8	34	0.9	1.1	68	16.0	20.0
1	13.8	17.2	35	1.3	1.7	69	16.4	20.6
2	13.3	16.7	36	1.8	2.2	70	16.9	21.1
3	12.9	16.1	37	2.2	2.8	71	17.3	21.7
4	12.4	15.6	38	2.7	3.3	72	17.8	22.2
5	12.0	15.0	39	3.1	3.9	73	18.2	22.8
6	11.6	14.4	40	3.6	4.4	74	18.7	23.3
7	11.1	13.9	41	4.0	5.0	75	19.1	23.9
8	10.7	13.3	42	4.4	5.6	76	19.6	24.4
9	10.2	12.8	43	4.9	6.1	77	20.0	25.0
10	9.8	12.2	44	5.3	6.7	78	20.4	25.6
11	9.3	11.7	45	5.8	7.2	79	20.9	26.1
12	8.9	11.1	46	6.2	7.8	80	21.3	26.7
13	8.4	10.6	47	6.7	8.3	81	21.8	27.2
14	8.0	10.0	48	7.1	8.9	82	22.2	27.8
15	7.6	9.4	49	7.6	9.4	83	22.7	28.3
16	7.1	8.9	50	8.0	10.0	84	23.1	28.9
17	6.7	8.3	51	8.4	10.6	85	23.6	29.4
18	6.2	7.8	52	8.9	11.1	86	24.0	30.0
19	5.8	7.2	53	9.3	11.7	87	24.4	30.6
20	5.3	6.7	54	9.8	12.2	88	24.9	31.1
21	4.9	6.1	55	10.2	12.8	89	25.3	31.7
22	4.4	5.6	56	10.7	13.3	90	25.8	32.2
23	4.0	5.0	57	11.1	13.9	91	26.2	32.8
24	3.6	4.4	58	11.6	14.4	92	26.7	33.3
25	3.1	3.9	59	12.0	15.0	93	27.1	33.9
26	2.7	3.3	60	12.4	15.6	94	27.6	34.4
27	2.2	2.8	61	12.9	16.1	95	28.0	35.0
28	1.8	2.2	62	13.3	16.7	96	28.4	35.6
29	1.3	1.7	63	13.8	17.2	97	28.9	36.1
30	0.9	1.1	64	14.2	17.8	98	29.3	36.7
31	— 0.4	— 0.6	65	14.7	18.3	99	29.8	37.2
32	0.0	0.0	66	+15.1	+18.9	100	+30.2	+37.8

x° Reaumur $= (32^\circ + \frac{9}{4}x^\circ)$ Fah. $= \frac{5}{4}x^\circ$ Centes.

x° Centes. $= (32^\circ + \frac{9}{5}x^\circ)$ Fah. $= \frac{4}{5}x^\circ$ Reaum.

x° Fah. $= \frac{4}{9}(x^\circ - 32^\circ)$ Reau. $= \frac{5}{9}(x^\circ - 32^\circ)$ Cen.

TABLES AND FORMULÆ.

PART III.

ASTRONOMY.

ASTRONOMY.

XLVII.—*Of Sidereal and Solar Time.*

True or *apparent solar time* is that deduced from observations of the sun, or is the same as that shown by a well-adjusted sun-dial.

Mean solar time is derived from the time employed by the earth in revolving on its axis, as compared with the sun, supposed to move at a mean rate in its orbit, and to make 365.242218 revolutions in a mean Gregorian year.

It cannot be immediately obtained from observation, but is always deduced from apparent time by the aid of the equation of time, which is the angular distance, in time, between the mean and true sun; or, mean solar time = apparent solar time ± equation of time.

Sidereal time is the portion of a sidereal day which has elapsed since the transit of the first point of Aries.

Its point of origin cannot be determined by observation, but it is known at any moment by the right ascension of whatever object may be then in the meridian; or,

Sidereal time of a star's culmination = AR. of ✱;

Sidereal time at mean noon = AR. mean ☉ at mean noon; and, generally,

Sidereal time = sidereal time at mean noon ± solar time from mean noon, (expressed in sidereal intervals;)

Solar time = sidereal time − sidereal time at mean noon, (the difference being reduced to a solar interval.)

XLVII.—*Of Sidereal and Solar Time*—Continued.

EXAMPLE.

To find the mean solar time of the passage of Altair over the meridian of Washington, on the 10th July, 1849:

		h. m. s.
AR. Altair, July 10, 1849....................	=	19 43 27.39
Sidereal time at mean noon at Washington	=	7 14 00.96
Sidereal interval past Washington mean noon...	=	12 29 26.43
Retardation of mean on sidereal time..........	=	— 02 02.77
Corresponding mean time interval past mean noon or mean time of culmination	=	12 27 23.66

The nautical almanacs give the sidereal time at mean noon for each day of the year for a certain meridian.

If the sidereal day be taken equal to 24 sidereal hours, the mean solar day will be equal to $24^h\ 3^m\ 56^s.55$ of those sidereal hours; or the daily acceleration of sidereal on mean solar time (which is the mean motion of the earth in a mean solar day) is $3^m\ 56^s.5554$ of sidereal time; hence the sidereal time at mean noon under any meridian other than that of the nautical almanac used will be found by allowing the proportion of this quantity due to the difference of longitude of the two places.

If the mean solar day be taken equal to 24 mean solar hours, the sidereal day will be equal to $23^h\ 56^m\ 4^s.09$ of those solar hours, or the daily retardation of mean solar on sidereal time is $3^m\ 55^s.9093$ of solar time.

The astronomical day begins at noon. In the civil or common method of reckoning, the day is supposed to commence at the *preceding* midnight. The civil reckoning is therefore 12 hours in advance of the astronomical reckoning, and in the above example, July 10th, $12^h\ 27^m\ 23^s.66$ astronomical time, corresponds to July 11th, $0^h\ 27^m\ 23^s.66$ a. m. civil time.

XLVIII.—*To find the Time by an Altitude of the Sun or a Star.*

Sidereal time = AR. ✱ ± ✱'s hour-angle.
Solar time = 24^h ± ☉'s hour-angle.

$$2m = L + \triangle + A$$

$$\sin^2 \tfrac{1}{2} p = \frac{\cos m \,.\, \sin (m - A)}{\cos L \,.\, \sin \triangle}$$

where—

L = the latitude of the place of observation;

△ = the north polar distance of the sun or the star;

A = the corrected altitude of the sun or star
= observed altitude — (refraction — parallax) ± semi-diameter; and

p = the hour-angle of the sun or star.

The formula gives the arc in degrees, which must be converted into time, as in one of the following four cases:

1. When we have the *corrected* altitude of the sun's *center*, the hour-angle, p, in time, is the *apparent* time when the sun is in the west, or the complement of 24 hours when in the east. To reduce it to *mean* time apply the equation of time.

2. But should the *sidereal* time be required, transform the mean time thus obtained to sidereal time, as previously explained.

3. When the altitude is that of a star, the sidereal time is at once deduced from the hour-angle, p.

4. And if, in this last instance, solar time should be required, convert this sidereal time into solar time by means of the equation—

Solar time = AR. ✱ — AR. ☉ ± p

in which the sign + is used if the star is observed in the west, and the sign — if in the east; or,

Mean solar time = the mean solar equivalent of (sidereal time of observation — sidereal time of preceding mean noon at place.)

XLIX.—*Sun-Dial.*

The most common dial is that in which the plane of the dial is horizontal, and the *style*, placed in the meridian, is inclined to the plane of the dial at an angle equal to the latitude of the place.

Hour-lines are drawn from the center, or point where the style intersects the plane, to the exterior limit of the surface of the dial. Their positions are calculated from the formula—

$$\tan x = \tan p \sin L$$

in which—

x = hour-angle on the horizontal plane;

p = 15°, 30°, 45°, etc., the hour-angle on the equatorial plane; and

L = latitude of the place.

The geometrical determination of these lines will be readily seen from the following figure.

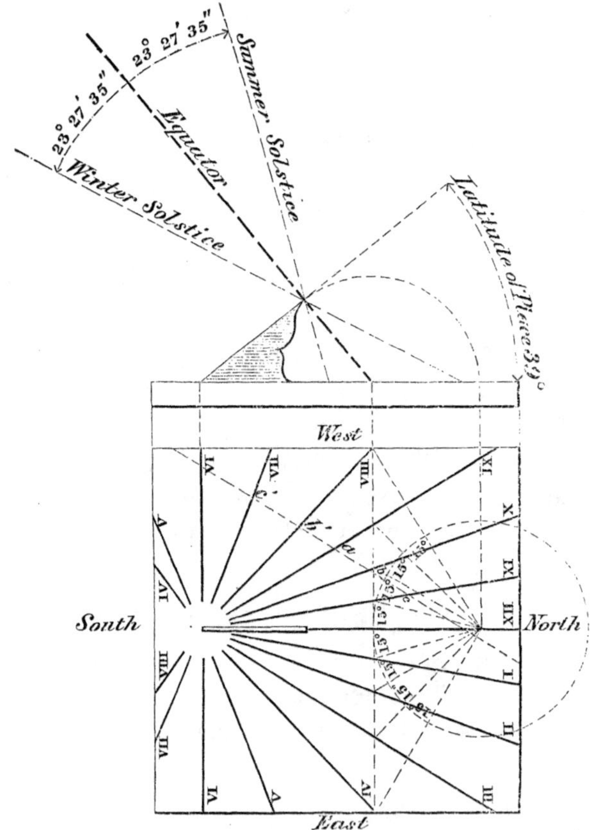

XLIX.—*Sun-Dial*—Continued.

As the lines of IV, V, VIII, and VII cannot, generally, be directly drawn, owing to want of space upon the surface of the dial, draw, from any point of the line of IX hours, a line parallel to that of III hours, and take $ab' = ab$, and $ac' = ac$; b' and c' will be points in the lines VIII and VII. The lines IV and V will make the same angles on the opposite side. The line of VI is perpendicular to that of XII.

To determine the Meridian Line.

Take a point in the plane of the dial through which it is intended the meridian plane shall pass. With this point as a center describe several concentric circles. Fix a straight pin in the center, perpendicular to the plane of the dial, of such a length that the extremity of the shadow cast by it shall fall within the circles at XII M. Mark the points where the extremity of the shadow passes over these circles in the forenoon and again the same in the afternoon. The line drawn from the middle of these arcs, contained between the points of passage, to the center of the circles will be the meridian.

SUN-DIAL CORRECTION.

Mean Time at Apparent Noon.

Day.	January.		February.		March.		April.		May.		June.	
	h.	*m.*	*h.*	*m.*	*h.*	*m.*	*h.*	*m.*	*h.*	*m.*	*h.*	*m.*
1	12	4	12	14	12	12	12	4	11	57	11	58
8	12	7	12	14	12	11	12	2	11	56	11	59
16	12	10	12	14	12	9	12	0	11	56	12	0
24	12	12	12	13	12	6	11	58	11	57	12	2

Day.	July.		August.		September.		October.		November.		December.	
	h.	*m.*	*h.*	*m.*	*h.*	*m.*	*h.*	*m.*	*h.*	*m.*	*h.*	*m.*
1	12	3	12	6	12	0	11	50	11	44	11	50
8	12	5	12	5	11	58	11	48	11	44	11	53
16	12	6	12	4	11	55	11	46	11	45	11	56
24	12	6	12	2	11	52	11	45	11	47	12	0

For converting Intervals of SIDEREAL *into Corresponding Intervals of* MEAN SOLAR *Time.*

Hours.		Minutes.				Seconds.			
h.	*m. s.*	*m.*	*s.*	*m.*	*s.*	*s.*	*s.*	*s.*	*s.*
1	0 9.830	1	0.164	31	5.079	1	0.003	31	0.085
2	0 19.659	2	0.328	32	5.242	2	0.005	32	0.087
3	0 29.489	3	0.491	33	5.406	3	0.008	33	0.090
4	0 39.318	4	0.655	34	5.570	4	0.011	34	0.093
5	0 49.148	5	0.819	35	5.734	5	0.014	35	0.096
6	0 58.977	6	0.983	36	5.898	6	0.016	36	0.098
7	1 8.807	7	1.147	37	6.062	7	0.019	37	0.101
8	1 18.636	8	1.311	38	6.225	8	0.022	38	0.104
9	1 28.466	9	1.474	39	6.389	9	0.025	39	0.106
10	1 38.296	10	1.638	40	6.553	10	0.027	40	0.109
11	1 48.125	11	1.802	41	6.717	11	0.030	41	0.112
12	1 57.955	12	1.966	42	6.881	12	0.033	42	0.115
13	2 7.784	13	2.130	43	7.044	13	0.036	43	0.118
14	2 17.614	14	2.294	44	7.208	14	0.038	44	0.120
15	2 27.443	15	2.457	45	7.372	15	0.041	45	0.123
16	2 37.273	16	2.621	46	7.536	16	0.044	46	0.126
17	2 47.103	17	2.785	47	7.700	17	0.047	47	0.128
18	2 56.932	18	2.949	48	7.864	18	0.049	48	0.131
19	3 6.762	19	3.113	49	8.027	19	0.052	49	0.134
20	3 16.591	20	3.277	50	8.191	20	0.055	50	0.137
21	3 26.421	21	3.440	51	8.355	21	0.057	51	0.140
22	3 36.250	22	3.604	52	8.519	22	0.060	52	0.142
23	3 46.080	23	3.768	53	8.683	23	0.063	53	0.145
24	3 55.909	24	3.932	54	8.847	24	0.066	54	0.148
		25	4.096	55	9.010	25	0.068	55	0.150
		26	4.259	56	9.174	26	0.071	56	0.153
		27	4.423	57	9.338	27	0.074	57	0.156
		28	4.587	58	9.502	28	0.076	58	0.159
		29	4.751	59	9.666	29	0.079	59	0.161
		30	4.915	60	9.830	30	0.082	60	0.164

The quantities taken from this table must be *subtracted* from a sidereal interval to obtain the corresponding interval in mean solar time.

For converting Intervals of Mean Solar *Time into Corresponding Intervals of* Sidereal *Time.*

Hours.		Minutes.				Seconds.			
h.	*m. s.*	*m.*	*s.*	*m.*	*s.*	*s.*	*s.*	*s.*	*s.*
1	0 9.856	1	0.164	31	5.092	1	0.003	31	0.085
2	0 19.713	2	0.329	32	5.257	2	0.005	32	0.088
3	0 29.569	3	0.493	33	5.421	3	0.008	33	0.090
4	0 39.426	4	0.657	34	5.585	4	0.011	34	0.093
5	0 49.282	5	0.821	35	5.750	5	0.014	35	0.096
6	0 59.139	6	0.986	36	5.914	6	0.016	36	0.098
7	1 8.995	7	1.150	37	6.078	7	0.019	37	0.101
8	1 18.852	8	1.314	38	6.242	8	0.022	38	0.104
9	1 28.708	9	1.478	39	6.407	9	0.025	39	0.106
10	1 38.565	10	1.643	40	6.571	10	0.027	40	0.109
11	1 48.421	11	1.807	41	6.735	11	0.030	41	0.112
12	1 58.278	12	1.971	42	6.900	12	0.033	42	0.115
13	2 8.134	13	2.136	43	7.064	13	0.036	43	0.118
14	2 17.991	14	2.300	44	7.228	14	0.038	44	0.120
15	2 27.847	15	2.464	45	7.392	15	0.041	45	0.123
16	2 37.704	16	2.628	46	7.557	16	0.044	46	0.126
17	2 47.560	17	2.793	47	7.721	17	0.047	47	0.129
18	2 57.416	18	2.957	48	7.885	18	0.049	48	0.131
19	3 7.273	19	3.121	49	8.050	19	0.052	49	0.134
20	3 17.129	20	3.285	50	8.214	20	0.055	50	0.137
21	3 26.986	21	3.450	51	8.378	21	0.057	51	0.140
22	3 36.842	22	3.614	52	8.542	22	0.060	52	0.142
23	3 46.699	23	3.778	53	8.707	23	0.063	53	0.145
24	3 56.555	24	3.943	54	8.871	24	0.066	54	0.148
		25	4.107	55	9.035	25	0.068	55	0.151
		26	4.271	56	9.199	26	0.071	56	0.153
		27	4.436	57	9.364	27	0.074	57	0.156
		28	4.600	58	9.528	28	0.077	58	0.159
		29	4.764	59	9.692	29	0.079	59	0.161
		30	4.928	60	9.856	30	0.082	60	0.164

The quantities taken from this table must be *added* to a mean interval to obtain the corresponding interval in sidereal time.

To convert Parts of the Equator in Arc into Sidereal Time, or to convert Terrestrial Longitude in Arc into Time.

DEGREES.

Arc.	Time.		Arc.	Time.		Arc.	Time.		Arc.	Time.	
°	*h.*	*m.*	°	*h.*	*m.*	°	*h.*	*m.*	°	*h.*	*m.*
1	0	4	31	2	4	61	4	4	91	6	4
2	0	8	32	2	8	62	4	8	92	6	8
3	0	12	33	2	12	63	4	12	93	6	12
4	0	16	34	2	16	64	4	16	94	6	16
5	0	20	35	2	20	65	4	20	95	6	20
6	0	24	36	2	24	66	4	24	96	6	24
7	0	28	37	2	28	67	4	28	97	6	28
8	0	32	38	2	32	68	4	32	98	6	32
9	0	36	39	2	36	69	4	36	99	6	36
10	0	40	40	2	40	70	4	40	100	6	40
11	0	44	41	2	44	71	4	44	101	6	44
12	0	48	42	2	48	72	4	48	102	6	48
13	0	52	43	2	52	73	4	52	103	6	52
14	0	56	44	2	56	74	4	56	104	6	56
15	1	0	45	3	0	75	5	0	105	7	0
16	1	4	46	3	4	76	5	4	106	7	4
17	1	8	47	3	8	77	5	8	107	7	8
18	1	12	48	3	12	78	5	12	108	7	12
19	1	16	49	3	16	79	5	16	109	7	16
20	1	20	50	3	20	80	5	20	110	7	20
21	1	24	51	3	24	81	5	24	111	7	24
22	1	28	52	3	28	82	5	28	112	7	28
23	1	32	53	3	32	83	5	32	113	7	32
24	1	36	54	3	36	84	5	36	114	7	36
25	1	40	55	3	40	85	5	40	115	7	40
26	1	44	56	3	44	86	5	44	116	7	44
27	1	48	57	3	48	87	5	48	117	7	48
28	1	52	58	3	52	88	5	52	118	7	52
29	1	56	59	3	56	89	5	56	119	7	56
30	2	0	60	4	0	90	6	0	120	8	0

To convert Parts of the Equator in Arc into Sidereal Time, or to convert Terrestrial Longitude in Arc into Time—Continued.

DEGREES.

Arc.	Time.		Arc.	Time.		Arc.	Time.		Arc.	Time.	
°	*h.*	*m.*	°	*h.*	*m.*	°	*h.*	*m.*	°	*h.*	*m.*
121	8	4	151	10	4	181	12	4	211	14	4
122	8	8	152	10	8	182	12	8	212	14	8
123	8	12	153	10	12	183	12	12	213	14	12
124	8	16	154	10	16	184	12	16	214	14	16
125	8	20	155	10	20	185	12	20	215	14	20
126	8	24	156	10	24	186	12	24	216	14	24
127	8	28	157	10	28	187	12	28	217	14	28
128	8	32	158	10	32	188	12	32	218	14	32
129	8	36	159	10	36	189	12	36	219	14	36
130	8	40	160	10	40	190	12	40	220	14	40
131	8	44	161	10	44	191	12	44	221	14	44
132	8	48	162	10	48	192	12	48	222	14	48
133	8	52	163	10	52	193	12	52	223	14	52
134	8	56	164	10	56	194	12	56	224	14	56
135	9	0	165	11	0	195	13	0	225	15	0
136	9	4	166	11	4	196	13	4	226	15	4
137	9	8	167	11	8	197	13	8	227	15	8
138	9	12	168	11	12	198	13	12	228	15	12
139	9	16	169	11	16	199	13	16	229	15	16
140	9	20	170	11	20	200	13	20	230	15	20
141	9	24	171	11	24	201	13	24	231	15	24
142	9	28	172	11	28	202	13	28	232	15	28
143	9	32	173	11	32	203	13	32	233	15	32
144	9	36	174	11	36	204	13	36	234	15	36
145	9	40	175	11	40	205	13	40	235	15	40
146	9	44	176	11	44	206	13	44	236	15	44
147	9	48	177	11	48	207	13	48	237	15	48
148	9	52	178	11	52	208	13	52	238	15	52
149	9	56	179	11	56	209	13	56	239	15	56
150	10	0	180	12	0	210	14	0	240	16	0

To convert Parts of the Equator in Arc into Sidereal Time, or to convert Terrestrial Longitude in Arc into Time—Continued.

DEGREES.

Arc.	Time.		Arc.	Time.		Arc.	Time.		Arc.	Time.	
°	*h.*	*m.*	°	*h.*	*m.*	°	*h.*	*m.*	°	*h.*	*m.*
241	16	4	271	18	4	301	20	4	331	22	4
242	16	8	272	18	8	302	20	8	332	22	8
243	16	12	273	18	12	303	20	12	333	22	12
244	16	16	274	18	16	304	20	16	334	22	16
245	16	20	275	18	20	305	20	20	335	22	20
246	16	24	276	18	24	306	20	24	336	22	24
247	16	28	277	18	28	307	20	28	337	22	28
248	16	32	278	18	32	308	20	32	338	22	32
249	16	36	279	18	36	309	20	36	339	22	36
250	16	40	280	18	40	310	20	40	340	22	40
251	16	44	281	18	44	311	20	44	341	22	44
252	16	48	282	18	48	312	20	48	342	22	48
253	16	52	283	18	52	313	20	52	343	22	52
254	16	56	284	18	56	314	20	56	344	22	56
255	17	0	285	19	0	315	21	0	345	23	0
256	17	4	286	19	4	316	21	4	346	23	4
257	17	8	287	19	8	317	21	8	347	23	8
258	17	12	288	19	12	318	21	12	348	23	12
259	17	16	289	19	16	319	21	16	349	23	16
260	17	20	290	19	20	320	21	20	350	23	20
261	17	24	291	19	24	321	21	24	351	23	24
262	17	28	292	19	28	322	21	28	352	23	28
263	17	32	293	19	32	323	21	32	353	23	32
264	17	36	294	19	36	324	21	36	354	23	36
265	17	40	295	19	40	325	21	40	355	23	40
266	17	44	296	19	44	326	21	44	356	23	44
267	17	48	297	19	48	327	21	48	357	23	48
268	17	52	298	19	52	328	21	52	358	23	52
269	17	56	299	19	56	329	21	56	359	23	56
270	18	0	300	20	0	330	22	0	360	24	0

To convert Parts of the Equator in Arc into Sidereal Time, or to convert Terrestrial Longitude in Arc into Time—Continued.

MINUTES.						SECONDS.			
Arc.	Time.		Arc.	Time.		Arc.	Time.	Arc.	Time.
′	*m.*	*s.*	′	*m.*	*s.*	″	*s.*	″	*s.*
1	0	4	31	2	4	1	0.067	31	2.067
2	0	8	32	2	8	2	0.133	32	2.133
3	0	12	33	2	12	3	0.200	33	2.200
4	0	16	34	2	16	4	0.267	34	2.267
5	0	20	35	2	20	5	0.333	35	2.333
6	0	24	36	2	24	6	0.400	36	2.400
7	0	28	37	2	28	7	0.467	37	2.467
8	0	32	38	2	32	8	0.533	38	2.533
9	0	36	39	2	36	9	0.600	39	2.600
10	0	40	40	2	40	10	0.667	40	2.667
11	0	44	41	2	44	11	0.733	41	2.733
12	0	48	42	2	48	12	0.800	42	2.800
13	0	52	43	2	52	13	0.867	43	2.867
14	0	56	44	2	56	14	0.933	44	2.933
15	1	0	45	3	0	15	1.000	45	3.000
16	1	4	46	3	4	16	1.067	46	3.067
17	1	8	47	3	8	17	1.133	47	3.133
18	1	12	48	3	12	18	1.200	48	3.200
19	1	16	49	3	16	19	1.267	49	3.267
20	1	20	50	3	20	20	1.333	50	3.333
21	1	24	51	3	24	21	1.400	51	3.400
22	1	28	52	3	28	22	1.467	52	3.467
23	1	32	53	3	32	23	1.533	53	3.533
24	1	36	54	3	36	24	1.600	54	3.600
25	1	40	55	3	40	25	1.667	55	3.667
26	1	44	56	3	44	26	1.733	56	3.733
27	1	48	57	3	48	27	1.800	57	3.800
28	1	52	58	3	52	28	1.867	58	3.867
29	1	56	59	3	56	29	1.933	59	3.933
30	2	0	60	4	0	30	2.000	60	4.000

To convert Sidereal Time into Parts of the Equator in Arc, or to convert Time into Terrestrial Longitude in Arc.

HOURS.		MINUTES.				SECONDS.			
Time.	Arc.	Time.	Arc.	Time.	Arc.	Time.	Arc.	Time.	Arc.
h.	°	*m.*	° ′	*m.*	° ′	*s.*	′ ″	*s.*	′ ″
1	15	1	0 15	31	7 45	1	0 15	31	7 45
2	30	2	0 30	32	8 0	2	0 30	32	8 0
3	45	3	0 45	33	8 15	3	0 45	33	8 15
4	60	4	1 0	34	8 30	4	1 0	34	8 30
5	75	5	1 15	35	8 45	5	1 15	35	8 45
6	90	6	1 30	36	9 0	6	1 30	36	9 0
7	105	7	1 45	37	9 15	7	1 45	37	9 15
8	120	8	2 0	38	9 30	8	2 0	38	9 30
9	135	9	2 15	39	9 45	9	2 15	39	9 45
10	150	10	2 30	40	10 0	10	2 30	40	10 0
11	165	11	2 45	41	10 15	11	2 45	41	10 15
12	180	12	3 0	42	10 30	12	3 0	42	10 30
13	195	13	3 15	43	10 45	13	3 15	43	10 45
14	210	14	3 30	44	11 0	14	3 30	44	11 0
15	225	15	3 45	45	11 15	15	3 45	45	11 15
16	240	16	4 0	46	11 30	16	4 0	46	11 30
17	255	17	4 15	47	11 45	17	4 15	47	11 45
18	270	18	4 30	48	12 0	18	4 30	48	12 0
19	285	19	4 45	49	12 15	19	4 45	49	12 15
20	300	20	5 0	50	12 30	20	5 0	50	12 30
21	315	21	5 15	51	12 45	21	5 15	51	12 45
22	330	22	5 30	52	13 0	22	5 30	52	13 0
23	345	23	5 45	53	13 15	23	5 45	53	13 15
24	360	24	6 0	54	13 30	24	6 0	54	13 30
		25	6 15	55	13 45	25	6 15	55	13 45
		26	6 30	56	14 0	26	6 30	56	14 0
		27	6 45	57	14 15	27	6 45	57	14 15
		28	7 0	58	14 30	28	7 0	58	14 30
		29	7 15	59	14 45	29	7 15	59	14 45
		30	7 30	60	15 0	30	7 30	60	15 0

To convert Sidereal Time into Parts of the Equator in Arc, or to convert Time into Terrestrial Longitude in Arc—Continued.

TENTHS OF SECONDS.

Time.	Arc.	Time.	Arc.	Time.	Arc.	Time.	Arc.
s.	"	*s.*	"	*s.*	"	*s.*	"
0.01	0.15	0.31	4.65	0.61	9.15	0.91	13.65
0.02	0.30	0.32	4.80	0.62	9.30	0.92	13.80
0.03	0.45	0.33	4.95	0.63	9.45	0.93	13.95
0.04	0.60	0.34	5.10	0.64	9.60	0.94	14.10
0.05	0.75	0.35	5.25	0.65	9.75	0.95	14.25
0.06	0.90	0.36	5.40	0.66	9.90	0.96	14.40
0.07	1.05	0.37	5.55	0.67	10.05	0.97	14.55
0.08	1.20	0.38	5.70	0.68	10.20	0.98	14.70
0.09	1.35	0.39	5.85	0.69	10.35	0.99	14.85
0.10	1.50	0.40	6.00	0.70	10.50	1.00	15.00
0.11	1.65	0.41	6.15	0.71	10.65	Thousandths of seconds of time.	Arc.
0.12	1.80	0.42	6.30	0.72	10.80		
0.13	1.95	0.43	6.45	0.73	10.95		
0.14	2.10	0.44	6.60	0.74	11.10		
0.15	2.25	0.45	6.75	0.75	11.25		
0.16	2.40	0.46	6.90	0.76	11.40		
0.17	2.55	0.47	7.05	0.77	11.55		
0.18	2.70	0.48	7.20	0.78	11.70		
0.19	2.85	0.49	7.35	0.79	11.85		
0.20	3.00	0.50	7.50	0.80	12.00	*s.*	"
0.21	3.15	0.51	7.65	0.81	12.15	0.001	0.015
0.22	3.30	0.52	7.80	0.82	12.30	0.002	0.030
0.23	3.45	0.53	7.95	0.83	12.45	0.003	0.045
0.24	3.60	0.54	8.10	0.84	12.60	0.004	0.060
0.25	3.75	0.55	8.25	0.85	12.75	0.005	0.075
0.26	3.90	0.56	8.40	0.86	12.90	0.006	0.090
0.27	4.05	0.57	8.55	0.87	13.05	0.007	0.105
0.28	4.20	6.58	8.70	0.88	13.20	0.008	0.120
0.29	4.35	0.59	8.85	0.89	13.35	0.009	0.135
0.30	4.50	0.60	9.00	0.90	13.50	0.010	0.150

To convert Right Ascension in Arc into Mean Time.

DEGREES.

AR. in arc.	Mean time.			AR. in arc.	Mean time.			AR. in arc.	Mean time.		
°	*h.*	*m.*	*s.*	°	*h.*	*m.*	*s.*	°	*h.*	*m.*	*s.*
1	0	3	59.345	31	2	3	39.686	61	4	3	20.027
2	0	7	58.689	32	2	7	39.030	62	4	7	19.371
3	0	11	58.034	33	2	11	38.375	63	4	11	18.716
4	0	15	57.379	34	2	15	37.720	64	4	15	18.061
5	0	19	56.724	35	2	19	37.064	65	4	19	17.405
6	0	23	56.068	36	2	23	36.409	66	4	23	16.750
7	0	27	55.413	37	2	27	35.754	67	4	27	16.095
8	0	31	54.758	38	2	31	35.099	68	4	31	15.639
9	0	35	54.102	39	2	35	34.443	69	4	35	14.784
10	0	39	53.447	40	2	39	33.788	70	4	39	14.129
11	0	43	52.792	41	2	43	33.133	71	4	43	13.474
12	0	47	52.136	42	2	47	32.477	72	4	47	12.818
13	0	51	51.481	43	2	51	31.822	73	4	51	12.163
14	0	55	50.826	44	2	55	31.167	74	4	55	11.508
15	0	59	50.170	45	2	59	30.511	75	4	59	10.852
16	1	3	49.515	46	3	3	29.856	76	5	3	10.197
17	1	7	48.860	47	3	7	29.201	77	5	7	9.542
18	1	11	48.205	48	3	11	28.545	78	5	11	8.886
19	1	15	47.549	49	3	15	27.890	79	5	15	8.231
20	1	19	46.894	50	3	19	27.235	80	5	19	7.576
21	1	23	46.239	51	3	23	26.580	81	5	23	6.920
22	1	27	45.583	52	3	27	25.924	82	5	27	6.265
23	1	31	44.928	53	3	31	25.269	83	5	31	5.610
24	1	35	44.273	54	3	35	24.614	84	5	35	4.955
25	1	39	43.617	55	3	39	23.958	85	5	39	4.299
26	1	43	42.962	56	3	43	23.303	86	5	43	3.644
27	1	47	42.307	57	3	47	22.648	87	5	47	2.989
28	1	51	41.652	58	3	51	21.992	88	5	51	2.333
29	1	55	40.996	59	3	55	21.337	89	5	55	1.678
30	1	59	40.341	60	3	59	20.682	90	5	59	1.023

To convert Right Ascension in Arc into Mean Time—Continued.

DEGREES.

AR. in arc.	Mean time.			AR. in arc.	Mean time.			AR. in arc.	Mean time.		
°	h.	m.	s.	°	h.	m.	s.	°	h.	m.	s.
91	6	3	0.367	121	8	2	40.708	151	10	2	21.049
92	6	6	59.712	122	8	6	40.053	152	10	6	20.394
93	6	10	59.057	123	8	10	39.398	153	10	10	19.738
94	6	14	58.401	124	8	14	38.742	154	10	14	19.083
95	6	18	57.746	125	8	18	38.087	155	10	18	18.428
96	6	22	57.091	126	8	22	37.432	156	10	22	17.773
97	6	26	56.436	127	8	26	36.776	157	10	26	17.117
98	6	30	55.780	128	8	30	36.121	158	10	30	16.462
99	6	34	55.125	129	8	34	35.466	159	10	34	15.807
100	6	38	54.470	130	8	38	34.810	160	10	38	15.151
101	6	42	53.814	131	8	42	34.155	161	10	42	14.496
102	6	46	53.159	132	8	46	33.500	162	10	46	13.841
103	6	50	52.504	133	8	50	32.845	163	10	50	13.185
104	6	54	51.848	134	8	54	32.189	164	10	54	12.530
105	6	58	51.193	135	8	58	31.534	165	10	58	11.875
106	7	2	50.538	136	9	2	30.879	166	11	2	11.220
107	7	6	49.883	137	9	6	30.223	167	11	6	10.564
108	7	10	49.227	138	9	10	29.568	168	11	10	9.909
109	7	14	48.572	139	9	14	28.913	169	11	14	9.254
110	7	18	47.917	140	9	18	28.257	170	11	18	8.598
111	7	22	47.261	141	9	22	27.602	171	11	22	7.943
112	7	26	46.606	142	9	26	26.947	172	11	26	7.288
113	7	30	45.951	143	9	30	26.292	173	11	30	6.632
114	7	34	45.295	144	9	34	25.636	174	11	34	5.977
115	7	38	44.640	145	9	38	24.981	175	11	38	5.322
116	7	42	43.985	146	9	42	24.326	176	11	42	4.666
117	7	46	43.329	147	9	46	23.670	177	11	46	4.011
118	7	50	42.674	148	9	50	23.015	178	11	50	3.356
119	7	54	42.019	149	9	54	22.360	179	11	54	2.701
120	7	58	41.364	150	9	58	21.704	180	11	58	2.045

To convert Right Ascension in Arc into Mean Time—Continued.

MINUTES.				SECONDS.			
AR. in arc.	Mean time.	AR. in arc.	Mean time.	AR. in arc.	Mean time.	AR. in arc.	Mean time.
′	*m. s.*	′	*m. s.*	″	*s.*	″	*s.*
1	0 3.989	31	2 3.661	1	0.066	31	2.061
2	0 7.978	32	2 7.650	2	0.133	32	2.128
3	0 11.969	33	2 11.640	3	0.199	33	2.194
4	0 15.956	34	2 15.629	4	0.266	34	2.261
5	0 19.945	35	2 19.618	5	0.332	35	2.327
6	0 23.935	36	2 23.607	6	0.399	36	2.393
7	0 27.924	37	2 27.596	7	0.465	37	2.460
8	0 31.913	38	2 31.585	8	0.532	38	2.526
9	0 35.902	39	2 35.574	9	0.598	39	2.593
10	0 39.891	40	2 39.563	10	0.665	40	2.659
11	0 43.880	41	2 43.552	11	0.731	41	2.726
12	0 47.869	42	2 47.541	12	0.798	42	2.792
13	0 51.858	43	2 51.530	13	0.864	43	2.859
14	0 55.847	44	2 55.519	14	0.931	44	2.925
15	0 59.836	45	2 59.509	15	0.997	45	2.992
16	1 3.825	46	3 3.498	16	1.064	46	3.058
17	1 7.814	47	3 7.487	17	1.130	47	3.125
18	1 11.803	48	3 11.476	18	1.197	48	3.191
19	1 15.793	49	3 15.465	19	1.263	49	3.258
20	1 19.782	50	3 19.454	20	1.330	50	3.324
21	1 23.771	51	3 23.443	21	1.396	51	3.391
22	1 27.760	52	3 27.432	22	1.463	52	3.457
23	1 31.749	53	3 31.421	23	1.529	53	3.524
24	1 35.738	54	3 35.410	24	1.596	54	3.590
25	1 39.727	55	3 39.399	25	1.662	55	3.657
26	1 43.716	56	3 43.388	26	1.729	56	3.723
27	1 47.705	57	3 47.377	27	1.795	57	3.790
28	1 51.694	58	3 51.367	28	1.862	58	3.856
29	1 55.683	59	3 55.356	29	1.928	59	3.923
30	1 59.672	60	3 59.345	30	1.995	60	3.989

To convert Mean Time into Right Ascension in Arc.

HOURS.				MINUTES.							
Mean time.	AR. in arc.			Mean time.	AR. in arc.			Mean time.	AR. in arc.		
h.	°	′	″	*m.*	°	′	″	*m.*	°	′	″
1	15	2	27.85	1	0	15	2.46	31	7	46	16.39
2	30	4	52.69	2	0	30	4.93	32	8	1	18.85
3	45	7	23.54	3	0	45	30.39	33	8	16	21.31
4	60	9	51.39	4	1	0	9.86	34	8	31	23.78
5	75	12	19.24	5	1	15	12.32	35	8	46	26.24
6	90	14	47.08	6	1	30	14.79	36	9	1	28.71
7	105	17	14.93	7	1	45	17.25	37	9	16	31.17
8	120	19	42.78	8	2	0	19.71	38	9	31	33.64
9	135	22	10.62	9	2	15	22.18	39	9	46	36.10
10	150	24	38.47	10	2	30	24.64	40	10	1	38.57
11	165	27	6.32	11	2	45	27.11	41	10	16	41.03
12	180	29	34.16	12	3	0	29.57	42	10	31	43.39
13	195	32	2.01	13	3	15	32.03	43	10	46	45.96
14	210	34	29.86	14	3	30	34.50	44	11	1	48.42
15	225	36	57.70	15	3	45	36.96	45	11	16	50.89
16	240	39	25.55	16	4	0	39.43	46	11	31	53.35
17	255	41	53.40	17	4	15	41.89	47	11	46	55.81
18	270	44	21.24	18	4	30	44.35	48	12	1	58.38
19	285	46	49.09	19	4	45	46.82	49	12	17	0.74
20	300	49	16.94	20	5	0	49.28	50	12	32	3.21
21	315	51	44.78	21	5	15	51.75	51	12	47	5.57
22	330	54	12.63	22	5	30	54.21	52	13	2	8.13
23	345	56	40.48	23	5	45	56.67	53	13	17	10.60
24	360	59	8.33	24	6	0	59.14	54	13	32	13.06
				25	6	16	1.60	55	13	47	15.53
				26	6	31	4.07	56	14	2	17.99
				27	6	46	6.53	57	14	17	20.45
				28	7	1	9.00	58	14	32	22.92
				29	7	16	11.46	59	14	47	25.38
				30	7	31	13.92	60	15	2	27.85

To convert Mean Time into Right Ascension in Arc—Continued.

SECONDS.						TENTHS OF SECONDS.			
Mean time.	AR. in arc.		Mean time.	AR. in arc.		Mean time.	AR. in arc.	Mean time.	AR. in arc.
s.	′	″	s.	′	″	s.	″	s.	″
1	0	15.04	31	7	46.27	0.01	0.15	0.31	4.66
2	0	30.08	32	8	1.31	0.02	0.30	0.32	4.81
3	0	45.12	33	8	16.36	0.03	0.45	0.33	4.96
4	1	0.16	34	8	31.40	0.04	0.60	0.34	5.12
5	1	15.21	35	8	46.44	0.05	0.75	0.35	5.27
6	1	30.25	36	9	1.48	0.06	0.90	0.36	5.42
7	1	45.29	37	9	16.52	0.07	1.05	0.37	5.57
8	2	0.33	38	9	31.56	0.08	1.20	0.38	5.72
9	2	15.37	39	9	46.60	0.09	1.35	0.39	5.87
10	2	30.41	40	10	1.64	0.10	1.50	0.40	6.02
11	2	45.45	41	10	16.68	0.11	1.65	0.41	6.17
12	3	0.49	42	10	31.73	0.12	1.81	0.42	6.32
13	3	15.53	43	10	46.77	0.13	1.96	0.43	6.47
14	3	30.58	44	11	1.81	0.14	2.11	0.44	6.62
15	3	45.62	45	11	16.85	0.15	2.26	0.45	6.77
16	4	0.66	46	11	31.89	0.16	2.41	0.46	6.92
17	4	15.70	47	11	46.93	0.17	2.56	0.47	7.07
18	4	30.74	48	12	1.97	0.18	2.71	0.48	7.22
19	4	45.78	49	12	17.01	0.19	2.86	0.49	7.37
20	5	0.82	50	12	32.05	0.20	3.01	0.50	7.52
21	5	15.86	51	12	47.09	0.21	3.16	0.51	7.67
22	5	30.90	52	13	2.14	0.22	3.31	0.52	7.82
23	5	45.94	53	13	17.18	0.23	3.46	0.53	7.97
24	6	1.00	54	13	32.22	0.24	3.61	0.54	8.12
25	6	16.03	55	13	47.26	0.25	3.76	0.55	8.27
26	6	31.07	56	14	2.30	0.26	3.91	0.56	8.43
27	6	46.11	57	14	17.34	0.27	4.06	0.57	8.58
28	7	1.15	58	14	32.38	0.28	4.21	0.58	8.73
29	7	16.19	59	14	47.42	0.29	4.36	0.59	8.88
30	7	31.23	60	15	2.46	0.30	4.51	0.60	9.03

To convert Mean Time into Right Ascension in Arc—Continued.

TENTHS OF SECONDS.						THOUSANDTHS OF SECONDS.	
Mean time.	AR. in arc.	Mean time.	AR. in arc.	Mean time.	AR. in arc.	Mean time.	AR. in arc.
s.	"	s.	"	s.	"	s.	"
0.61	9.18	0.76	11.43	0.91	13.69	0.001	0.02
0.62	9.33	0.77	11.58	0.92	13.84	0.002	0.03
0.63	9.48	0.78	11.74	0.93	13.99	0.003	0.05
0.64	9.63	0.79	11.89	0.94	14.14	0.004	0.06
0.65	9.78	0.80	12.04	0.95	14.29	0.005	0.08
0.66	9.93	0.81	12.19	0.96	14.44	0.006	0.09
0.67	10.08	0.82	12.34	0.97	14.59	0.007	0.11
0.68	10.23	0.83	12.49	0.98	14.74	0.008	0.12
0.69	10.38	0.84	12.64	0.99	14.89	0.009	0.14
0.70	10.53	0.85	12.79	1.00	15.05	0.010	0.15
0.71	10.68	0.86	12.94				
0.72	10.83	0.87	13.09				
0.73	10.98	0.88	13.24				
0.74	11.13	0.89	13.39				
0.75	11.28	0.90	13.54				

CONSTANT LOGARITHMS.

			Logarithms.
12 hours, expressed in seconds	=	43200.	4.6354837
Complement to the same	=	.00002315	5.3645163
24 hours, expressed in seconds	=	86400.	4.9365137
Complement to the same	=	.00001157	5.0634863
360 degrees, expressed in seconds	=	1296000.	6.1126050
To convert sidereal time into mean solar time			9.9988126

FORM FOR

SURVEY OF .... DETERMINATION OF TIME,

DATE AND STATION.—1843, *October* 13.—*Mouth of the Big Black River*,

INSTRUMENTS ... { Sextant No. 2197, by *Troughton & Simms*, and *Mean Solar* Chronometer No. 76, by *Charles*

Names of stars.	Observed double altitudes of star with sextant.	True altitudes of star affected by corrections for refraction and errors of sextant, = A.	*Mean solar* time of observation deduced.	Time of observation noted by chronometer.
	° ′ ″	° ′ ″	*h. m. s.*	*h. m. s.*
	91 43 40	45 52 58.8	7 05 47.69	6 57 02.4
	92 18 00	46 10 09.3	7 07 28.67	6 58 43.2
	92 41 15	46 21 47.3	7 08 37.15	6 59 52.8
a Andromedæ,	93 04 05	46 33 12.6	7 09 44.37	7 00 59.6
(*east.*)	93 45 20	46 53 50.8	7 11 45.92	7 03 01.2
	94 13 45	47 08 03.7	7 13 09.73	7 04 25.6
	94 40 50	47 21 36.6	7 14 29.64	7 05 45.
	95 07 25	47 34 54.5	7 15 48.14	7 07 03.6

Mean result of 8 observations on *a Andromedæ*, in the *east* ....

	° ′ ″	° ′ ″	*h. m. s.*	*h. m. s.*
	95 20 05	47 41 14.7	8 55 32.36	8 46 49.2
	95 00 00	47 31 11.6	8 56 32.06	8 47 50.4
	94 30 40	47 16 31.2	8 57 59.42	8 49 16.
a Lyræ	94 12 20	47 07 21.	8 58 54.	8 50 10.8
(*west.*)	93 53 45	46 58 03.1	8 59 49.4	8 51 06.9
	93 29 20	46 45 50.2	9 01 02.1	8 52 19.4
	93 07 35	46 34 57.3	9 02 07.	8 53 24.8
	92 46 50	46 24 34.5	9 03 09.	8 54 26.
	92 28 45	46 15 31.7	9 04 02.96	8 55 21.2

Mean result of *nine* observations on the star *a Lyræ*, in the *west*...........
Mean result of *eight* observations on the star *a Andromedæ*, in the east, as above.

CHRONOMETER ERROR.—*Slow* of *mean solar* time *at* 8ʰ *p. m.*, by a mean of these results from east and west stars...............................

RECORD AND COMPUTATION.

by Observed Double Altitudes of East and West Stars.

a tributary to the river Saint John, Maine.
artificial horizon of Mercury.

Young.

Chronometer (*C. Y.* 76) slow of mean solar time by each observation.	Remarks.
h. m. s.	Index error of sextant .... $=+2' 40''$
0 08 45.29	Error of eccentricity of sextant ...... $=+1\ 32$
8 45.47	Thermometer, 31°.5 Fahrenheit.
8 44.35	Barometer, 29.14 inches.
8 44.77	Apparent right ascension of star ..... $= 0^h 00^m 21^s.72$
8 44.56	Apparent declination of star .... $=28° 13' 59''.5$ N.
8 44.13	Apparent north polar distance of star . $=61\ 46\ 00\ .5 = \triangle$
8 44.64	Approximate latitude of this station .. $=46\ 57\ 00$ N. $=$ L
8 44.54	Approximate longitude of this station . $= 4^h 37^m 47^s$
	Sidereal time of mean noon at station . $=13\ 26\ 20\ .83$
$0^h 08^m 44^s.74$	
h. m. s.	
0 08 43.16	Thermometer, 29° Fahrenheit.
8 41.66	Barometer, 29.14 inches.
8 43.42	Apparent right ascension of star ..... $=18^h 31^m 39^s.16$
8 43.20	Apparent declination of star north.... $=38° 38' 46''.5$
8 42.50	Apparent north polar distance of star . $=51\ 21\ 13\ .5 = \triangle$
8 42.70	Index error of sextant...... .... $=+2' 40''$
8 42.20	Error of eccentricity of sextant....... $=+1\ 32$
8 43.00	
8 41.76	
$0^h 08^m 42^s.6$	
0 08 44.7	
$0^h 08^m 43^s.6$	Observer, *Major J. D. Graham.* Computer, *Private F. Herbst.*

Computation of the Fifth of the Preceding Altitudes of α Andromeda, (*formula, page* 187.)

Observed double altitude................	=	93° 45′ 20″
Index error, sextant.....................	= +	02 40
Eccentricity, sextant.....................	= +	01 32
Double altitude corrected................	=	93 49 32
Altitude	=	46 54 46
Refraction, (thermom., 31°.5; barom., 29°.1)..	= −	56 .6
True altitude of ✱ = A..................	=	46 53 49 .4

$2m = L + \varDelta + A$

L = 46° 57′	cos =	9.8341894
Δ = 61 46 00″.5	sin =	9.9449899
A = 46 53 49.4	cos L sin Δ =	9.7791793
2 m = 155 36 49.9		
m = 77 48 24.4	cos =	9.3247127
(m − A) = 30 54 35.0	sin =	9.7106984
	cos m sin (m − A) =	9.0354111
	$\sin^2 \frac{1}{2} p = \dfrac{\cos m \sin (m - A)}{\cos L \sin \varDelta}$ =	19.2562318
	sin ½ p =	9.6281159
	½ p =	25° 08′ 01″.5
	p in arc =	50 16 03 .0
	(page 192) p in time = −	3 21 04 .20
	AR. ✱ =	24 00 21 .72

Sidereal time of observation... = AR. ± p	=	20 39 17 .52
Sidereal time, mean noon, at place, (nautical almanac)	=	13 26 20 .83
Sidereal interval past mean noon	=	7 12 56 .69
Retardation of mean on sidereal interval, (page 190)	= −	01 10 .93
Mean solar interval past mean noon, or mean time p. m. of observation.........		7 11 45 .76
Time of observation by chronometer		7 03 01 .20
Chronometer slow......................		8 44 .56

L.—*To Find the Time by Equal Altitudes of the Sun.*

Correction in time, to be applied as an equation to the mean of the times of observed equal altitudes of the sun, in order to obtain the time of its meridional passage:

$$x = \delta \tan D \frac{T}{30 \tan 7\frac{1}{2} T} - \delta \tan L \frac{T}{30 \sin 7\frac{1}{2} T}$$

Make—

$$\frac{T}{30 \sin 7\frac{1}{2} T} = A ; \frac{T}{30 \tan 7\frac{1}{2} T} = B$$

$$x = \mp A \delta \tan L + B \delta \tan D$$

$$\text{Apparent noon} = \tfrac{1}{2} (t + t') + x$$

t, t' = the times of observations;

$T = (t' - t)$ = the interval of time between the observations, expressed in hours and decimals;

L = the latitude of the place of observation, (*minus* when south;)

D = the sun's declination at apparent noon on the given day, (*minus* when south;)

δ = the hourly variation in the declination at noon, (*minus* when the sun is proceeding toward the south;) and

x = required correction in *seconds,* where A is to be *minus* where the time of noon is required and *plus* where the time of midnight is required, *i. e.*, when the first observation is made in the afternoon and the corresponding one the morning following.

Logarithmic values of A and B are given in the following tables.

Equations to Equal Altitudes.

Interval.	Log A.	Log B.	Interval.	Log A.	Log B.
h. *m.*			*h.* *m.*		
2 0	9.4109	9.3959	3 0	9.4172	9.3828
2	.4111	.3955	2	.4174	.3822
4	.4113	.3952	4	.4177	.3817
6	.4114	.3948	6	.4179	.3811
8	.4116	.3944	8	.4182	.3806
10	.4118	.3941	10	.4184	.3800
12	.4120	.3937	12	.4187	.3794
14	.4121	.3933	14	.4190	.3789
16	.4123	.3929	16	.4193	.3783
18	.4125	.3925	18	.4195	.3777
20	.4127	.3921	20	.4198	.3771
22	.4129	.3917	22	.4201	.3765
24	.4131	.3913	24	.4204	.3759
26	.4133	.3909	26	.4207	.3752
28	.4135	.3905	28	.4209	.3746
30	.4137	.3900	30	.4212	.3740
32	.4139	.3896	32	.4215	.3733
34	.4141	.3892	34	.4218	.3727
36	.4144	.3887	36	.4221	.3720
38	.4146	.3882	38	.4224	.3713
40	.4148	.3878	40	.4227	.3707
42	.4150	.3873	42	.4231	.3700
44	.4152	.3868	44	.4234	.3693
46	.4155	.3863	46	.4237	.3686
48	.4157	.3859	48	.4240	.3679
50	.4159	.3854	50	.4243	.3672
52	.4162	.3849	52	.4246	.3665
54	.4164	.3843	54	.4250	.3657
56	.4167	.3838	56	.4253	.3650
2 58	9.4169	9.3833	3 58	9.4256	9.3643

$$x = \mp A\,\delta \tan L + B\,\delta \tan D$$

Equations to Equal Altitudes—Continued.

Interval.	Log A.	Log B.	Interval.	Log A.	Log B.
h. *m.*			*h.* *m.*		
4 0	9.4260	9.3635	5 0	9.4374	9.3369
2	.4263	.3627	2	.4378	.3358
4	.4266	.3620	4	.4383	.3348
6	.4270	.3612	6	.4387	.3337
8	.4273	.3604	8	.4391	.3327
10	.4277	.3596	10	.4396	.3316
12	.4280	.3588	12	.4400	.3305
14	.4284	.3580	14	.4405	.3294
16	.4288	.3572	16	.4409	.3283
18	.4291	.3564	18	.4414	.3272
20	.4295	.3555	20	.4418	.3261
22	.4299	.3547	22	.4423	.3249
24	.4302	.3538	24	.4427	.3238
26	.4306	.3530	26	.4432	.3226
28	.4310	.3521	28	.4437	.3214
30	.4314	.3512	30	.4441	.3203
32	.4317	.3503	32	.4446	.3191
34	.4321	.3494	34	.4451	.3178
36	.4325	.3485	36	.4456	.3166
38	.4329	.3476	38	.4460	.3154
40	.4333	.3467	40	.4465	.3142
42	.4337	.3457	42	.4470	.3129
44	.4341	.3448	44	.4475	.3116
46	.4345	.3438	46	.4480	.3103
48	.4349	.3429	48	.4485	.3091
50	.4353	.3419	50	.4490	.3078
52	.4357	.3409	52	.4494	.3064
54	.4361	.3399	54	.4500	.3051
56	.4366	.3389	56	.4505	.3038
4 58	9.4370	9.3379	5 58	9.4510	9.3024

$$x = \mp \text{A}\ \delta \tan \text{L} + \text{B}\ \delta \tan \text{D}$$

Equations to Equal Altitudes—Continued.

Interval.		Log A.	Log B.	Interval.		Log A.	Log B.
h.	*m.*			*h.*	*m.*		
6	0	9.4515	9.3010	7	0	9.4685	9.2530
	2	.4521	.2996		2	.4691	.2511
	4	.4526	.2982		4	.4697	.2492
	6	.4531	.2968		6	.4704	.2473
	8	.4536	.2954		8	.4710	.2454
	10	.4542	.2940		10	.4716	.2434
	12	.4547	.2925		12	.4723	.2415
	14	.4552	.2911		14	.4729	.2395
	16	.4558	.2896		16	.4735	.2375
	18	.4563	.2881		18	.4742	.2355
	20	.4569	.2866		20	.4748	.2334
	22	.4574	.2850		22	.4755	.2313
	24	.4580	.2835		24	.4761	.2292
	26	.4585	.2819		26	.4768	.2271
	28	.4591	.2804		28	.4774	.2250
	30	.4597	.2788		30	.4781	.2228
	32	.4602	.2772		32	.4788	.2206
	34	.4608	.2756		34	.4794	.2184
	36	.4614	.2739		36	.4801	.2162
	38	.4620	.2723		38	.4808	.2140
	40	.4625	.2706		40	.4815	.2117
	42	.4631	.2689		42	.4821	.2094
	44	.4637	.2672		44	.4828	.2070
	46	.4643	.2655		46	.4835	.2047
	48	.4649	.2638		48	.4842	.2023
	50	.4655	.2620		50	.4849	.1999
	52	.4661	.2602		52	.4856	.1974
	54	.4667	.2584		54	.4863	.1950
	56	.4673	.2566		56	.4870	.1925
6	58	9.4679	9.2548	7	58	9.4877	9.1900

$$x = \mp A\,\delta \tan L + B\,\delta \tan D$$

Equations to Equal Altitudes—Continued.

Interval.	Log A.	Log B.	Interval.	Log A.	Log B.
h. m.			*h. m.*		
8 0	9.4884	9.1874	9 0	9.5115	9.0943
2	.4892	.1848	2	.5123	.0906
4	.4899	.1822	4	.5132	.0867
6	.4906	.1796	6	.5140	.0828
8	.4913	.1769	8	.5148	.0789
10	.4921	.1742	10	.5157	.0749
12	.4928	.1715	12	.5165	.0708
14	.4935	.1687	14	.5174	.0667
16	.4943	.1659	16	.5182	.0625
18	.4950	.1630	18	.5191	.0583
20	.4958	.1602	20	.5199	.0540
22	.4965	.1573	22	.5208	.0496
24	.4973	.1543	24	.5217	.0452
26	.4980	.1513	26	.5225	.0406
28	.4988	.1483	28	.5234	.0360
30	.4996	.1453	30	.5243	.0314
32	.5003	.1422	32	.5252	.0266
34	.5011	.1390	34	.5261	.0218
36	.5019	.1359	36	.5269	.0169
38	.5027	.1327	38	.5278	.0119
40	.5035	.1294	40	.5287	.0069
42	.5042	.1261	42	.5296	.0017
44	.5050	.1228	44	.5305	8.9965
46	.5058	.1194	46	.5315	.9911
48	.5066	.1159	48	.5324	.9857
50	.5074	.1125	50	.5333	.9802
52	.5082	.1089	52	.5342	.9745
54	.5091	.1054	54	.5351	.9688
56	.5099	.1017	56	.5361	.9630
8 58	9.5107	9.0981	9 58	9.5370	8.9570

$$x = \mp A \delta \tan L + B \delta \tan D$$

Equations to Equal Altitudes—Continued.

Interval.	Log A.	Log B.	Interval.	Log A.	Log B.
h. m.			*h. m.*		
14 0	9.6841	−9.0971	15 0	9.7333	−9.3162
2	.6856	.1057	2	.7351	.3225
4	.6872	.1141	4	.7369	.3287
6	.6887	.1224	6	.7386	.3350
8	.6903	.1306	8	.7404	.3411
10	.6919	.1387	10	.7422	.3472
12	.6934	.1468	12	.7440	.3533
14	.6950	.1547	14	.7458	.3593
16	.6966	.1625	16	.7476	.3653
18	.6982	.1703	18	.7494	.3713
20	.6998	.1779	20	.7512	.3772
22	.7014	.1855	22	.7531	.3831
24	.7030	.1930	24	.7549	.3889
26	.7047	.2004	26	.7568	.3947
28	.7063	.2078	28	.7586	.4005
30	.7079	.2150	30	.7605	.4062
32	.7096	.2222	32	.7624	.4119
34	.7112	.2293	34	.7642	.4175
36	.7129	.2364	36	.7661	.4232
38	.7146	.2434	38	.7680	.4288
40	.7162	.2503	40	.7699	.4343
42	.7179	.2571	42	.7718	.4399
44	.7196	.2639	44	.7738	.4454
46	.7213	.2706	46	.7757	.4509
48	.7230	.2773	48	.7776	.4563
50	.7247	.2839	50	.7796	.4617
52	.7264	.2905	52	.7815	.4671
54	.7281	.2970	54	.7835	.4725
56	.7299	.3034	56	.7855	.4778
14 58	9.7316	−9.3098	15 58	9.7875	−9.4831

$$x = \mp A\,\delta \tan L + B\,\delta \tan D$$

Equations to Equal Altitudes—Continued.

Interval.	Log A.	Log B.	Interval.	Log A.	Log B.
h. m.			*h. m.*		
16 0	9.7895	−9.4884	17 0	9.8539	−9.6383
2	.7915	.4937	2	.8562	.6431
4	.7935	.4990	4	.8585	.6478
6	.7955	.5042	6	.8608	.6526
8	.7975	.5094	8	.8632	.6573
10	.7996	.5146	10	.8655	.6621
12	.8016	.5197	12	.8679	.6668
14	.8037	.5248	14	.8703	.6715
16	.8058	.5300	16	.8727	.6762
18	.8078	.5351	18	.8751	.6809
20	.8099	.5401	20	.8775	.6856
22	.8120	.5452	22	.8799	.6903
24	.8141	.5502	24	.8824	.6949
26	.8162	.5553	26	.8848	.6996
28	.8184	.5603	28	.8873	.7043
30	.8205	.5653	30	.8898	.7089
32	.8227	.5702	32	.8923	.7136
34	.8248	.5752	34	.8948	.7182
36	.8270	.5801	36	.8973	.7228
38	.8292	.5850	38	.8999	.7275
40	.8314	.5900	40	.9024	.7321
42	.8336	.5948	42	.9050	.7367
44	.8358	.5997	44	.9075	.7413
46	.8380	.6046	46	.9101	.7459
48	.8402	.6094	48	.9127	.7505
50	.8425	.6143	50	.9154	.7552
52	.8447	.6191	52	.9180	.7598
54	.8470	.6239	54	.9206	.7644
56	.8493	.6287	56	.9233	.7690
16 58	9.8516	−9.6335	17 58	9.9260	−9.7736

$$x = \mp A\,\delta \tan L + B\,\delta \tan D$$

Equations to Equal Altitudes—Continued.

Interval.		Log A.	Log B.	Interval.		Log A.	Log B.
h.	*m.*			*h.*	*m.*		
18	0	9.9287	−9.7782	19	0	0.0172	−9.9167
	2	.9314	.7827		2	.0204	.9213
	4	.9341	.7873		4	.0237	.9260
	6	.9368	.7919		6	.0270	.9307
	8	.9396	.7965		8	.0303	.9355
	10	.9424	.8011		10	.0336	.9402
	12	.9451	.8057		12	.0370	.9449
	14	.9479	.8103		14	.0403	.9497
	16	.9508	.8149		16	.0437	.9544
	18	.9536	.8195		18	.0472	.9592
	20	.9564	.8241		20	.0506	.9640
	22	.9593	.8287		22	.0541	.9687
	24	.9622	.8333		24	.0576	.9735
	26	.9651	.8379		26	.0611	.9784
	28	.9680	.8425		28	.0646	.9832
	30	.9709	.8471		30	.0682	.9880
	32	.9739	.8517		32	.0718	.9929
	34	.9769	.8563		34	.0754	−9.9977
	36	.9798	.8609		36	.0790	−0.0026
	38	.9829	.8655		38	.0827	.0075
	40	.9859	.8701		40	.0864	.0124
	.42	.9889	.8748		42	.0901	.0173
	44	.9920	.8794		44	.0939	.0223
	46	.9951	.8840		46	.0976	.0272
	48	9.9982	.8887		48	.1015	.0322
	50	0.0013	.8933		50	.1053	.0372
	52	.0044	.8980		52	.1092	.0422
	54	.0076	.9026		54	.1131	.0473
	56	.0108	.9073		56	.1170	.0523
18	58	0.0140	−9.9120	19	58	0.1209	−0.0574

$$x = \mp A\,\delta \tan L + B\,\delta \tan D$$

Equations to Equal Altitudes—Continued.

Interval.	Log A.	Log B.	Interval.	Log A.	Log B.
h. m.			*h. m.*		
20 0	0.1249	—0.0625	21 0	0.2623	—0.2279
2	.1290	.0676	2	.2676	.2339
4	.1330	.0727	4	.2729	.2401
6	.1371	.0779	6	.2783	.2462
8	.1412	.0830	8	.2838	.2524
10	.1454	.0882	10	.2893	.2587
12	.1496	.0935	12	.2949	.2650
14	.1538	.0987	14	.3005	.2714
16	.1581	.1040	16	.3063	.2778
18	.1623	.1093	18	.3120	.2843
20	.1667	.1146	20	.3179	.2909
22	.1711	.1200	22	.3238	.2975
24	.1755	.1253	24	.3298	.3041
26	.1799	.1308	26	.3359	.3109
28	.1844	.1362	28	.3420	.3177
30	.1889	.1417	30	.3482	.3245
32	.1935	.1472	32	.3545	.3315
34	.1981	.1527	34	.3609	.3385
36	.2028	.1582	36	.3674	.3456
38	.2075	.1638	38	.3739	.3527
40	.2122	.1695	40	.3805	.3599
42	.2170	.1751	42	.3873	.3673
44	.2218	.1808	44	.3941	.3747
46	.2267	.1866	46	.4010	.3822
48	.2316	.1924	48	.4080	.3897
50	.2366	.1982	50	.4151	.3974
52	.2416	.2040	52	.4223	.4052
54	.2467	.2099	54	.4297	.4130
56	.2518	.2159	56	.4371	.4210
20 58	0.2570	—0.2219	21 58	0.4446	—0.4291

$$x = \mp A\,\delta \tan L + B\,\delta \tan D$$

Equations to Equal Altitudes—Continued.

Interval.	Log A.	Log B.	Interval.	Log A.	Log B.
h. *m.*			*h.* *m.*		
22 0	0.4523	—0.4372	23 0	0.7689	—0.7652
2	.4601	.4455	2	.7842	.7807
4	.4680	.4540	4	.8000	.7967
6	.4761	.4625	6	.8163	.8133
8	.4842	.4711	8	.8333	.8305
10	.4926	.4799	10	.8508	.8483
12	.5010	.4889	12	.8691	.8667
14	.5097	.4980	14	.8882	.8860
16	.5184	.5072	16	.9080	.9060
18	.5274	.5165	18	.9288	.9270
20	.5365	.5261	20	.9506	.9489
22	.5458	.5358	22	.9734	.9719
24	.5553	.5457	24	0.9975	—0.9961
26	.5649	.5557	26	1.0228	—1.0216
28	.5748	.5660	28	.0497	.0487
30	.5848	.5764	30	.0783	.0774
32	.5951	.5871	32	.1089	.1081
34	.6056	.5979	34	.1416	.1409
36	.6164	.6090	36	.1770	.1764
38	.6273	.6204	38	.2154	.2149
40	.6386	.6319	40	.2573	.2569
42	.6501	.6438	42	.3037	.3033
44	.6619	.6559	44	.3554	.3552
46	.6740	.6684	46	.4140	.4138
48	.6865	.6811	48	.4815	.4814
50	.6993	.6942	50	.5613	.5612
52	.7124	.7076	52	.6588	.6587
54	.7259	.7214	54	.7844	.7843
56	.7398	.7355	56	1.9610	—1.9610
22 58	0.7541	—0.7501	23 58	2.2627	—2.2627

$$x = \mp A\ \delta \tan L + B\ \delta \tan D$$

Computation of the Equation of Equal Altitudes to correct the Chronometer for Noon, August 9, 1844, *by the First of the following Equal Altitudes of the Sun's Limbs.*

$$x = (- \mathrm{A}\,\delta \tan \mathrm{L}) + (\mathrm{B}\,\delta \tan \mathrm{D})$$

$\mathrm{T} = 6^h\ 33^m$	log A =	− 9.4605	log B =	9.2764
$\delta = 43''.63$	log δ =	− 1.6397	log δ =	− 1.6397
$\mathrm{L} = 45^\circ\ 48'$	log tan L =	0.0121	log tan D =	9.4493
1st term	$12^s.95$ =	1.1123	$- 2^s.32$ =	− 0.3654
2d term	− 2 .32			
	+ 10 .63 = equation of equal altitudes.			

Computation of the First Two of the following Pairs of Equal Altitudes of the Sun's Limbs.

		1st pair.	2d pair.
A. M.	= t =	$1^h\ 28^m\ 23^s.0$	$1^h\ 29^m\ 52^s.8$
P. M.	= t' =	8 03 16 .5	8 01 46 .5
	$t + t'$ =	9 31 39 .5	9 31 39 .3
	$\frac{t + t'}{2}$ =	4 45 49 .75	4 45 49 .65
Equat'n of equal alts. =	x =	+ 10 .63	10 .63
Time by chron. of appt. noon	=	4 46 00 .38	4 46 00 .28
Correct mean time at apparent noon (Naut. Alm.)	=	0 05 09 .09	0 05 09 .09
Chron. fast of mean time at app't noon, August 9, 1844	=	4 40 51 .29	4 40 51 .19

SURVEY OF .... DETERMINATION OF THE TIME,

Chronometer

DATE AND STATION.—1844, *August* 9—*American Camp, near Tasche*

INSTRUMENTS { Sextant No. 2197, by *Troughton &*
Mean Solar Chronometer, No. 2440,

Observed double altitudes of the sun's upper and lower limbs.	Times, by chronometer, of observed equal altitudes. August 9th. A. M. = t	Times, by chronometer, of observed equal altitudes. August 9th. P. M. = t'	$t' - t =$ the elapsed time, = T.	Equation of equal altitudes = x.
Upper Limb.	*h. m. s.*	*h. m. s.*	*h. m.*	*s.*
78° 50′ 00″	1 28 23	8 03 16.5	6 33	+10.63
79 9 30	1 29 52.8	8 01 46.5		
Lower Limb.				
83° 10′ 00″	1 45 01	7 46 40.5		
83 40 00	1 46 34.5	7 45 06.2	5 59½	+10.24
84 00 00	1 47 38	7 44 04		
Upper Limb.				
85° 36′ 00″	1 49 23	7 42 18	5 48	+10.1
87 02 10	1 53 55.5	7 37 46.2		

CHRONOMETER ERROR.—*Fast* of mean solar time at apparent noon of *August* 9, 1844, by a mean of 7 pairs of equal altitudes of the sun..........

by Observed Equal Altitudes of the Sun's Limbs, to Correct the at Noon.

reau's house, on the highland boundary between Maine and Canada.

Simms, and Artificial Horizon of Mercury.
by *Parkinson & Frodsham.*

Chronometer *No.* 2440 *fast* of mean time at apparent noon by each pair of equal altitudes.	Remarks.
h. m. s.	
4 40 51.29	Index error of sextant ....
4 40 51.19	Error of eccentricity of sextant..........
	Thermometer (a. m.) 70° Fahr. barom...
	Thermometer (p. m.) 69° Fahr. barom...
4 40 51.9	Sun's appar't declination at appar't noon (D)= 15° 43′ 12″ N.
4 40 51.5	Hourly variation of sun's declination...(δ) = 43″.63
4 40 52.15	Equation of time at apparent noon....... + 5^{m} $09^{s}.09$
	Latitude of station (approximate) .... + 45° 48′ = (L.)
4 40 51.51	
4 40 51.86	
	Observer, *Major J. D. Graham.*
	Computer, *Do.*
4 40 51.6	

Sun's Parallax in Altitude.

Sun's altitude.	Sun's horizontal parallax.				Sun's altitude.	Sun's horizontal parallax.			
	8″.7	8″.8	8″.9	9″.0		8″.7	8″.8	8″.9	9″.0
°	″	″	″	″	°	″	″	″	″
0	8.70	8.80	8.90	9.00	45	6.15	6.22	6.29	6.36
5	8.67	8.77	8.87	8.97	50	5.59	5.66	5.72	5.79
10	8.57	8.67	8.76	8.86	55	4.99	5.05	5.12	5.16
15	8.40	8.50	8.60	8.70	60	4.35	4.40	4.45	4.50
20	8.18	8.27	8.36	8.46	65	3.68	3.72	3.76	3.80
25	7.88	7.98	8.06	8.16	70	2.98	3.01	3.04	3.08
30	7.53	7.62	7.70	7.79	75	2.25	2.28	2.31	2.33
35	7.13	7.21	7.29	7.37	80	1.51	1.53	1.55	1.56
40	6.66	6.74	6.82	6.90	85	0.76	0.77	0.77	0.78
45	6.15	6.22	6.29	6.36	90	0.00	0.00	0.00	0.00

Parallax in altitude = horizontal parallax × cosine of altitude.

Decimals of an Hour.

Minutes.						Seconds.					
m.	*decm.*	*m.*	*decm.*	*m.*	*decm.*	*s.*	*decm.*	*s.*	*decm.*	*s.*	*decm.*
1	.01667	21	.35000	41	.68333	1	.00028	21	.00583	41	.01139
2	.03333	22	.36667	42	.70000	2	.00056	22	.00611	42	.01167
3	.05000	23	.38333	43	.71667	3	.00083	23	.00639	43	.01194
4	.06667	24	.40000	44	.73333	4	.00111	24	.00667	44	.01222
5	.08333	25	.41667	45	.75000	5	.00139	25	.00694	45	.01250
6	.10000	26	.43333	46	.76667	6	.00167	26	.00722	46	.01278
7	.11667	27	.45000	47	.78333	7	.00194	27	.00750	47	.01306
8	.13333	28	.46667	48	.80000	8	.00222	28	.00778	48	.01333
9	.15000	29	.48333	49	.81667	9	.00250	29	.00806	49	.01361
10	.16667	30	.50000	50	.83333	10	.00278	30	.00833	50	.01389
11	.18333	31	.51667	51	.85000	11	.00306	31	.00861	51	.01417
12	.20000	32	.53333	52	.86667	12	.00333	32	.00889	52	.01444
13	.21667	33	.55000	53	.88333	13	.00361	33	.00917	53	.01472
14	.23333	34	.56667	54	.90000	14	.00389	34	.00944	54	.01500
15	.25000	35	.58333	55	.91667	15	.00417	35	.00972	55	.01528
16	.26667	36	.60000	56	.93333	16	.00444	36	.01000	56	.01556
17	.28333	37	.61667	57	.95000	17	.00472	37	.01028	57	.01583
18	.30000	38	.63333	58	.96667	18	.00500	38	.01056	58	.01611
19	.31667	39	.65000	59	.98333	19	.00528	39	.01083	59	.01639
20	.33333	40	.66667	60	1.00000	20	.00556	40	.01111	60	.01667

LI.—*The Transit Instrument.*

Knowing the apparent right ascension of a star, to compute the corrections to its observed transit on account of the three principal errors of the transit instrument—in azimuth, in the inclination of the axis, and in collimation—in order to obtain the correct clock error:

$$E = T + a\frac{\sin(L - D)}{\cos D} + b\frac{\cos(L - D)}{\cos D} + \frac{c}{\cos D} - AR.$$

where—

E denotes the error of the clock, *minus* when slow;

T, the observed time of transit;

L, the latitude of the place;

D, the declination of the star, *plus* when north, and *minus* when south, for the upper culminations, and *vice versa* for the lower culminations;

a, the deviation in the telescope in azimuth, *plus* when (pointing to the south) the vertical which it describes falls to the east, and *minus* when it falls to the west, and *vice versa* when pointing to the north;

b, the bias or inclination of the axis of the telescope, *plus* when the west end of the axis is too high;

c, the error in collimation, *plus* when the circle described by the line of collimation of the telescope falls to the east, and *minus* when it falls to the west, for upper culminations, and *vice versa* for lower culminations; and

AR., the right ascension of the star.

When the clock marks mean solar time, the mean time of transit of the object over the meridian must be substituted for AR.

LI.—*The Transit Instrument*—Continued.

1. To determine the value (*in time*) of the co-efficients a, b, c, in the preceding formula:

For inclination of the axis of the telescope:

$$b = \frac{d}{60}\left\{(w + w') - (e + e')\right\}$$

where—

w' and e' denote respectively the values of w and e after *reversing* the *level*;

d, the value of each division of the level in seconds of space;

w, the inclination of the level to the west; and

e, the inclination of the level to the east.

For collimation:

$$c = \tfrac{1}{2}(t' - t)\cos D + \tfrac{1}{2}(b' - b)\cos(L - D)$$

where—

t' and b' denote respectively the values of t and b, after *reversing* the *instrument;*

D, the declination of a circumpolar star; and

t, the time of the transit of the circumpolar star deduced from an observation at a given *side* wire of the instrument.

For the deviation in azimuth:

By observations of a circumpolar star:

$$a = \frac{12^{h} - (T' - T)}{2\cos L\tan D} + \frac{b\cos(L - D) - b'\cos(L + D) + 2c}{2\cos L\sin D}$$

where—

T′ and b' denote respectively the values of T and b at the *lower* culmination.

LI.—*The Transit Instrument*—Continued.

Deviation in azimuth by transits of a high and low star:

$$a = \left\{ (\text{AR.}' - \text{AR.}) - (\text{T}' - \text{T}) \right\} \times \frac{\cos \text{D}' \cos \text{D}}{\cos \text{L} \sin (\text{D} - \text{D}')}$$

where—

T′, AR.′, and D′ denote respectively the values of T, AR., and D of the *second* star observed.

Or—

$$a = \frac{(\text{AR.}' - \text{AR.}) - (\text{T}' - \text{T})}{\cos \text{L} (\tan \text{D} - \tan \text{D}')}$$

If one of the stars is observed at its lower culmination, use $180° - \text{D}'$ and $12^{\text{h}} + \text{AR.}'$ for its declination and right ascension.

Or make—

$$\frac{\sin (\text{L} - \text{D})}{\cos \text{D}} \text{ for the } \textit{first} \text{ star} = n$$

and—

$$\frac{\sin (\text{L} - \text{D}')}{\cos \text{D}'} \text{ for the } \textit{second} \text{ star} = n$$

then—

$$a = \frac{(\text{AR.}' - \text{AR.}) - (\text{T}' - \text{T})}{n' - n}$$

n is negative for a star north of the zenith.

2. To find the equatorial interval of each wire from the central wire, observe the transit of a star of any declination D; then—

$$\text{Equatorial interval} = \text{observed interval} \times \cos \text{D}.$$

3. When the intervals on each side of the central wire are equal, the mean of the times of transit over each wire will denote the transit over the middle wire. But should they not be equal, a correction must be applied to obtain a correct mean.

Call I. II; IV. V, the equatorial intervals of each wire from the central wire, the instrument having say 5 wires; then—

$$\text{Reduction to middle wire} = \frac{(\text{I} + \text{II}) - (\text{IV} + \text{V})}{5 \cos \text{D}}$$

FORM FOR RECORD AND COMPUTATION.

SURVEY OF .... STATION.

Transits of Stars .... *with* .... *Inch transit No.* ....;
Sidereal Chronometer Hardy No. 50.

Illuminated end of axis, west.

Date (1847)	October 6th.	October 6th.	October 6th.
Observer	T. J. L.	T. J. L.	T. J. L.
Object	π Capricorni.	14 Capricorni.	a Cygni.
Level	E. 32.2 W. 33.0	E. 32.7 W. 32.5	E. 32.7 W. 32.5
Value of 1 division of scale = 7″.5	E. 32.2 W. 33.0	E. 32.5 W. 33.3	E. 33.0 W. 32.5
	h. m. s.	h. m. s.	h. m. s.
Wires I	20 17 33.0	20 29 43.7	20 35 00.0
II	17 53.5	30 02.7	35 26.0
III	18 12.7	30 22.0	35 52.0
IV	18 32.7	30 41.7	36 18.7
V	20 18 52.5	20 31 00.7	20 36 45.5
Sum	184.4	110.8	142.2
Mean	20 18 12.88	20 30 22.16	20 35 52.44
Reduc'n to middle wire	— .07	— .07	— .10
Transit on instrument.	12.81	22.09	52.39
Corr'ns in time: for collimation.			
Corr'ns in time: for level	+ .10	+ .04	— .12
Corr'ns in time: for dev'n in az'h	+ .17	+ .18	— .01
Transit by chronom'r.	20 18 13.08	20 30 22.31	20 35 52.21
AR. of star	20 18 36.66	20 30 45.89	20 36 15.80
Error of chronometer.	23.58	23.58	23.59

Chronometer .... slow of .... time at .... p. m., October 6th, 1847.

Computation of the Corrections a and b, in the Preceding Transits.

Declination of π Capricorni = 18° 42 S.
14 Capricorni = 15° 29 S.
α Cygni = 44° 44 N.
Latitude of Station = L = 43°13′

Level Correction of π Capricorni.

$$L = 43^\circ\ 13'$$
$$D = -\ 18^\circ\ 42'$$
$$(L - D) = 61^\circ\ 55'$$
$$\cos\frac{(L - D)}{\cos D} = 0.50$$

	E.	W.
	32.2	33
	32.2	33
	64.4	66

66 − 64.4 = 1.6

$$b = \frac{7.5}{60} \times 1.6 = 0^s.20$$

$$\text{Level correction} = b\,\frac{\cos (L - D)}{\cos D} = 0^s.20 \times 0.50 = 0^s.10$$

Deviation in Azimuth.

$$a = \frac{(AR.' - AR.) - (T' - T)}{n' - n}$$

T′ and T being the times of transit corrected for level and collimation.

Combining π Capricorni and α Cygni,

	h.	m.	s.		h.	m.	s.
AR.′ =	20	36	15.80	T′ =	20	35	52.22
AR. =	20	18	36.66	T =	20	18	12.91
		17	39.14			17	39.31
		17	39.31				

$$(AR.' - AR.) - (T' - T) = -\ 0.17$$

$$n' = \frac{(\sin L - D')}{\cos D'} = \frac{\sin(-1^\circ\ 31')}{\cos 44^\circ\ 44'} = -\ 0.03$$

$$n = \frac{\sin (L - D)}{\cos D} = \frac{\sin 61^\circ\ 55'}{\cos 18^\circ\ 42'} = +\ 0.93$$

$$a = \frac{-\ 0^s.17}{-\ 0.03 - 0.93} = \frac{0.17}{0.96} = +\ 0^s.18$$

Combining 14 Capricorni and α Cygni, $a = +\ 0^s.19$

Correction for deviation in azimuth of π Capricorni,

$$= a\,\frac{\sin (L - D)}{\cos D} = 0^s.18 \times 0.93 = 0^s.17$$

Numerical Values of Factors $\frac{\sin (L - D)}{\cos D}$, $\frac{\cos (L - D)}{\cos D}$,

For deviation.	Star's declination = ± D							For level.
Star's Z.-D. =(L—D)	0°	10°	20°	25°	30°	35°	40°	Star's Z.-D. =(L—D)
°								°
1	.02	.02	.02	.02	.02	.02	.02	89
2	.04	.04	.04	.04	.04	.04	.05	88
4	.07	.07	.07	.08	.08	.08	.09	86
6	.11	.11	.11	.11	.12	.13	.14	84
8	.14	.14	.15	.15	.16	.17	.18	82
10	.17	.18	.19	.19	.20	.21	.23	80
12	.21	.21	.22	.23	.24	.25	.27	78
14	.24	.25	.26	.27	.28	.29	.32	76
16	.28	.28	.29	.30	.32	.34	.36	74
18	.31	.31	.33	.34	.36	.38	.40	72
20	.34	.35	.36	.38	.40	.42	.45	70
22	.37	.38	.40	.42	.44	.46	.49	68
24	.41	.41	.43	.45	.47	.49	.53	66
26	.44	.45	.47	.49	.51	.54	.57	64
28	.47	.48	.50	.52	.54	.57	.61	62
30	.50	.51	.53	.55	.58	.61	.65	60
32	.53	.54	.56	.58	.61	.65	.69	58
34	.56	.57	.59	.61	.65	.69	.73	56
36	.59	.60	.63	.65	.68	.72	.77	54
38	.62	.63	.66	.68	.71	.75	.80	52
40	.64	.65	.68	.71	.74	.78	.84	50
45	.71	.72	.75	.78	.82	.86	.92	45
50	.77	.78	.82	.84	.89	.93	1.00	40
55	.82	.83	.87	.90	.95	.98	1.07	35
60	.87	.88	.92	.95	1.00	1.06	1.13	30
65	.91	.92	.96	1.00	1.05	1.10	1.18	25
70	.94	.95	1.00	1.04	1.09	1.14	1.23	20
75	.97	.98	1.03	1.07	1.12	1.17	1.26	15
80	.98	1.00	1.05	1.09	1.14	1.20	1.29	10
89	1.00	1.02	1.06	1.10	1.15	1.22	1.31	1
For collimation ..	1.00	1.02	1.06	1.10	1.15	1.22	1.31	$= \frac{1}{\cos D}$

$\frac{1}{\cos D}$, *for facilitating the Reduction of Transit Observations.*

For deviation.	Star's declination = ± D							For level.
Star's Z.-D. =(L—D)	45°	50°	55°	60°	65°	70°	75°	Star's Z.-D. =(L—D)
°								°
1	.02	.03	.03	.03	.04	.05	.07	89
2	.05	.05	.06	.07	.08	.10	.13	88
4	.10	.11	.12	.14	.17	.20	.27	86
6	.15	.16	.18	.21	.25	.31	.40	84
8	.20	.22	.24	.28	.33	.41	.54	82
10	.25	.27	.30	.35	.41	.51	.67	80
12	.29	.32	.36	.42	.49	.61	.80	78
14	.34	.38	.42	.48	.57	.71	.94	76
16	.39	.43	.48	.55	.65	.81	1.06	74
18	.44	.48	.54	.62	.73	.90	1.19	72
20	.48	.53	.60	.68	.81	1.00	1.32	70
22	.53	.58	.65	.75	.89	1.09	1.45	68
24	.58	.63	.71	.81	.96	1.19	1.57	66
26	.62	.68	.76	.88	1.04	1.28	1.69	64
28	.66	.73	.82	.94	1.11	1.37	1.81	62
30	.71	.78	.87	1.00	1.18	1.46	1.93	60
32	.75	.82	.92	1.06	1.25	1.55	2.05	58
34	.79	.87	.97	1.12	1.32	1.65	2.16	56
36	.83	.91	1.03	1.18	1.39	1.74	2.27	54
38	.87	.96	1.07	1.23	1.46	1.80	2.38	52
40	.91	1.00	1.12	1.29	1.52	1.88	2.48	50
45	1.00	1.10	1.23	1.41	1.67	2.07	2.73	45
50	1.08	1.19	1.34	1.53	1.81	2.24	2.96	40
55	1.16	1.27	1.43	1.64	1.94	2.40	3.16	35
60	1.22	1.35	1.51	1.73	2.05	2.53	3.35	30
65	1.28	1.41	1.58	1.81	2.14	2.65	3.50	25
70	1.33	1.46	1.64	1.88	2.22	2,75	3.63	20
75	1.37	1.50	1.68	1.93	2.29	2.82	3.67	15
80	1.39	1.53	1.72	1.97	2.33	2.88	3.81	10
89	1.41	1.56	1.74	2.00	2.37	2.92	3.86	1
For collimation ..	1.41	1.56	1.74	2.00	2.37	2.92	3.86	$=\frac{1}{\cos D}$

LII.—*Reduction of Transits by Least Squares.*

Let—

E be the error of chronometer at an assumed time T;

t_1, t_2, t_3, &c., the observed times of transit (corrected for rate and level error) of stars having the right ascensions AR.$_1$, AR.$_2$, AR.$_3$, &c.;

a and c, the errors of azimuth and collimation; and

A_1, A_2, A_3, &c., C_1, C_2, C_3, &c., the factors of azimuth and of collimation for the several stars;

then—

$$t_1 + E + A_1\, a + C_1\, c = AR_1$$
$$t_2 + E + A_2\, a + C_2\, c = AR_2$$
$$t_3 + E + A_3\, a + C_3\, c = AR_3$$

&c., &c.

Let—

$$E = \mathit{E} + \varepsilon$$

where ε is the unknown correction to an assumed chronometer error E;

and let, also,

$$AR_1 - t_1 = e_1$$
$$AR_2 - t_2 = e_2$$
$$AR_3 - t_3 = e_3$$

&c., &c.

then—

$$\mathit{E} + \varepsilon + A_1\, a + C_1\, c = e_1$$
$$\mathit{E} + \varepsilon + A_2\, a + C_2\, c = e_2$$
$$\mathit{E} + \varepsilon + A_3\, a + C_3\, c = e_3$$

&c., &c.

Let now—

$$e_1 - \mathit{E} = n_1$$
$$e_2 - \mathit{E} = n_2$$
$$e_3 - \mathit{E} = n_3$$

&c., &c.

then—

$$\varepsilon + A_1\, a + C_1\, c = n_1$$
$$\varepsilon + A_2\, a + C_2\, c = n_2$$
$$\varepsilon + A_3\, a + C_3\, c = n_3$$

&c., &c.

From which form the normal equations—

$$\Sigma\, \varepsilon + \Sigma \cdot A\, a + \Sigma\, C\, c = n$$
$$\Sigma\, A\, \varepsilon + \Sigma\, A^2\, a + \Sigma\, A\, C\, c = A\, n \qquad (1)$$
$$\Sigma\, C\, \varepsilon + \Sigma\, A\, C\, a + \Sigma\, C_2\, c = C\, n$$

from which ε, a, and c can be obtained.

LII.—*Reduction of Transits, &c.*—Continued.

If the errors of collimation are known, and the times t_1, t_2, t_3, &c., corrected for it, the azimuthal deviation and correction to assumed chronometer-error may be deduced from the equations—

$$\begin{aligned} \Sigma \varepsilon + \Sigma A\, a &= \Sigma n \\ \Sigma A\, \varepsilon + \Sigma A^2\, a &= \Sigma A\, n \end{aligned} \qquad (2)$$

Equations (1) cannot be advantageously employed unless the nstrument be reversed.

Example of the Computation of Equations (1).

Latitude, 36° 38′ N.—April 11, 1852—Assumed time, T = 11^h sidereal—Chronometer losing $1^s.83$ daily—Assume $E = + 3^h\ 14^m\ 30^s.0$.

	Illum'n star.	*e*	*n*	A	C	A*n*	C*n*	A²	AC	C²
		h. m. s.								
E.	α Urs. Maj.......	3 14 29.77	−0.23	−0.95	+2.17	+0.22	−0.50	+0.90	− 2.06	+4.71
	δ Leonis.........	29.91	−0.09	+0.27	+1.07	−0.02	−0.10	0.07	+ 0.29	1.14
	δ Hydra.........	29.87	−0.13	+0.80	+1.04	−0.10	−0.14	0.64	+ 0.83	1.08
W.	γ Cephei sub. polo	28.50	−1.50	+4.05	+4.39	−6.07	−6.58	16.40	−17.78	19.27
	γ Urs. Maj.......	30.56	+0.56	−0.55	−1.72	−0.31	−0.96	0.30	+ 0.95	2.96
	4052 B. A. C.....	30.19	+0.19	+0.46	−1.01	+0.09	−0.19	0.21	− 0.46	1.02
	4072 B. A. C.....	30.06	+0.06	+0.46	−1.02	+0.03	−0.06	0.21	− 0.47	1.04
			−1.14	+4.54	+4.92	−6.16	−8.53	+18.73	+16.86	+31.22

Normal Equations.

$$\begin{aligned} 7\,\varepsilon + 4.54\,a + 4.92\,c &= -\ 1.14 \\ 4.54\,\varepsilon + 18.73\,a + 16.86\,c &= -\ 6.16 \\ 4.92\,\varepsilon + 16.86\,a + 31.22\,c &= -\ 8.53 \end{aligned}$$

from which—

$$\begin{aligned} \varepsilon &= +\ 0^s.09 \\ a &= -\ 0^s.18 \\ c &= -\ 0^s.19 \end{aligned}$$

hence,

Azimuthal deviation of the instrument .... = $0^s.18$ W. of S.
Error of collimation of mean of wires, illumination east .... = $0^s.19$ W.
Error of chronometer, (slow) .... = $3^h\ 14^m\ 30^s.06$

LIII.—*Tables of Refraction.*

Table I gives the refraction when the barometer stands at 30 inches and the Fahrenheit thermometer at 50°.

Table II, to be used when greater accuracy is desired, gives the correction of the mean refraction depending upon the observed height of the barometer and thermometer.

In column A of this table, the refraction is regarded as a function of the *apparent* zenith-distance Z. The adopted form of this function is—

$$r = \alpha\beta^{A}\,\gamma^{\lambda}\tan Z$$

in which α varies slowly with the zenith-distance, and its logarithm is therefore readily taken from the table with the argument Z. The exponents A and λ differ sensibly from unity only for great zenith-distances, and also vary slowly; their values are therefore readily found from the table.

The factor β depends upon the barometer. The actual pressure indicated by the barometer depends not only upon the height of the column, but also upon its temperature. It is therefore put under the form—

$$\beta = \mathrm{B\,T}$$

and log B and log T are given in the supplementary tables with the arguments "Height of the barometer" and "Height of the attached thermometer," respectively; so that—

$$\log\beta = \log \mathrm{B} + \log \mathrm{T}$$

Finally, log γ is given directly in the supplementary table with the argument "External thermometer." This thermometer should be so exposed as to indicate truly the temperature of the atmosphere at the place of observation.

LIII.—*Tables of Refraction*—Continued.

Example.—Given the apparent zenith distance, Z = 78°30′0″; barometer, 29.770 inches; attached thermometer, − 0°.4 F.; external thermometer, − 2°.0 F.

From table II for 78°30′; log α = 1.74981
A = 1.0032; λ = 1.0328

and from the tables for barometer and thermometer—

$$\log B = +\ 0.00253$$
$$\log T = +\ 0.00127 \qquad \log \gamma = +\ 0.04545$$
$$\log \beta = +\ 0.00380$$

Hence the refraction is computed as follows:

$$\log \alpha = 1.74981$$
$$A \log \beta = \log \beta^A = +\ 0.00381$$
$$\lambda \log \gamma = \log \gamma^\lambda = +\ 0.04694$$
$$\log \tan Z = 0.69154$$
$$r = 5'\ 10''.53 = 310''.53;\ \log r = 2.49210$$

The true zenith-distance is, therefore,

$$78°\ 30' + 5'\ 10''.53 = 78°\ 35'\ 10''.53$$

TABLE I.—*Mean Refraction.*

Barometer, 30 inches—Fahrenheit thermometer, 50°.

Apparent altitude.	Mean refraction.	Apparent altitude.	Mean refraction.	Apparent altitude.	Mean refraction.
° ′	′ ″	° ′	′ ″	° ′	′ ″
		5 30	9 7.0	6 30	7 53.9
0 0	36 29	35	9 0.1	35	7 48.7
1 0	24 54	40	8 53.4	40	7 43.5
2 0	18 26	45	8 46.8	45	7 38.4
3 0	14 25	50	8 40.4	50	7 33.5
4 0	11 44	55	8 34.2	55	7 28.6
5 0	9 52.0	6 0	8 28.0	7 0	7 23.8
5	9 44.0	5	8 22.1	5	7 19.2
10	9 36.2	10	8 16.2	10	7 14.6
15	9 28.6	15	8 10.5	15	7 10.1
20	9 21.2	20	8 4.8	20	7 5.7
25	9 14.0	25	7 59.3	25	7 1.4

TABLE I.—*Mean Refraction*—Continued.

Barometer, 30 inches—Fahrenheit thermometer, 50°.

Apparent altitude.	Mean refraction.	Apparent altitude.	Mean refraction.	Apparent altitude.	Mean refraction.
° ′	′ ″	° ′	′ ″	° ′	′ ″
7 30	6 57.1	10 30	5 4.6	13 30	3 58.1
35	6 53.0	35	5 2.3	35	3 56.6
40	6 48.9	40	5 0.0	40	3 55.2
45	6 44.9	45	4 57.8	45	3 53.7
50	6 41.0	50	4 55.6	50	3 52.3
55	6 37.1	55	4 53.4	55	3 50.9
8 0	6 33.3	11 0	4 51.2	14 0	3 49.5
5	6 29.6	5	4 49.1	5	3 48.1
10	6 25.9	10	4 47.0	10	3 46.8
15	6 22.3	15	4 44.9	15	3 45.5
20	6 18.8	20	4 42.9	20	3 44.2
25	6 15.3	25	4 40.9	25	3 42.9
30	6 11.9	30	4 38.9	30	3 41.6
35	6 8.5	35	4 36.9	35	3 40.3
40	6 5.2	40	4 35.0	40	3 39.0
45	6 2.0	45	4 33.1	45	3 37.7
50	5 58.8	50	4 31.2	50	3 36.5
55	5 55.7	55	4 29.4	55	3 35.3
9 0	5 52.6	12 0	4 27.5	15 0	3 34.1
5	5 49.6	5	4 25.7	5	3 32.9
10	5 46.6	10	4 23.9	10	3 31.7
15	5 43.6	15	4 22.2	15	3 30.5
20	5 40.7	20	4 20.4	20	3 29.4
25	5 37.9	25	4 18.7	25	3 28.2
30	5 35.1	30	4 17.0	30	3 27.1
35	5 32.4	35	4 15.3	35	3 25.9
40	5 29.6	40	4 13.6	40	3 24.8
45	5 27.0	45	4 12.0	45	3 23.7
50	5 24.3	50	4 10.4	50	3 22.6
55	5 21.7	55	4 8.8	55	3 21.5
10 0	5 19.2	13 0	4 7.2	16 0	3 20.5
5	5 16.7	5	4 5.6	5	3 19.4
10	5 14.2	10	4 4.1	10	3 18.4
15	5 11.7	15	4 2.6	15	3 17.3
20	5 9.3	20	4 1.0	20	3 16.3
25	5 6.9	25	3 59.6	25	3 15.2

TABLE I.—*Mean Refraction*—Continued.

Barometer, 30 inches—Fahrenheit thermometer, 50°.

Apparent altitude.		Mean refraction.		Apparent altitude.		Mean refraction.		Apparent altitude.		Mean refraction.	
°	′	′	″	°	′	′	″	°	′	′	″
16	30	3	14.2	20	0	2	38.8	26	0	1	58.9
	35	3	13.2		10	2	37.4		10	1	58.1
	40	3	12.2		20	2	36.0		20	1	57.2
	45	3	11.2		30	2	34.6		30	1	56.4
	50	3	10.3		40	2	33.3		40	1	55.5
	55	3	9.3		50	2	32.0		50	1	54.7
17	0	3	8.3	21	0	2	30.7	27	0	1	53.9
	5	3	7.3		10	2	29.4		10	1	53.1
	10	3	6.4		20	2	28.1		20	1	52.3
	15	3	5.5		30	2	26.9		30	1	51.5
	20	3	4.6		40	2	25.7		40	1	50.7
	25	3	3.7		50	2	24.5		50	1	50.0
	30	3	2.8	22	0	2	23.3	28	0	1	49.2
	35	3	1.9		10	2	22.1		10	1	48.4
	40	3	1.0		20	2	20.9		20	1	47.7
	45	3	0.1		30	2	19.8		30	1	46.9
	50	2	59.2		40	2	18.7		40	1	46.2
	55	2	58.3		50	2	17.5		50	1	45.5
18	0	2	57.5	23	0	2	16.4	29	0	1	44.8
	5	2	56.6		10	2	15.4		20	1	43.4
	10	2	55.8		20	2	14.3		40	1	42.0
	15	2	54.9		30	2	13.3	30	0	1	40.6
	20	2	54.1		40	2	12.2		20	1	39.3
	25	2	53.2		50	2	11.2		40	1	38.0
	30	2	52.4	24	0	2	10.2	31	0	1	36.7
	35	2	51.6		10	2	9.2		20	1	35.5
	40	2	50.8		20	2	8.2		40	1	34.2
	45	2	50.0		30	2	7.2	32	0	1	33.0
	50	2	49.2		40	2	6.2		20	1	31.8
	55	2	48.4		50	2	5.3		40	1	30.7
19	0	2	47.7	25	0	2	4.4	33	0	1	29.5
	10	2	46.1		10	2	3.4		20	1	28.4
	20	2	44.6		20	2	2.5		40	1	27.3
	30	2	43.1		30	2	1.6	34	0	1	26.2
	40	2	41.6		40	2	0.7		20	1	25.1
	50	2	40.2		50	1	59.8		40	1	24.1

TABLE I.—*Mean Refraction*—Continued.

Barometer, 30 inches—Fahrenheit thermometer, 50°.

Apparent altitude.	Mean refraction.	Apparent altitude.	Mean refraction.	Apparent altitude.	Mean refraction.
° ′	′ ″	° ′	′ ″	° ′	′ ″
35 0	1 23.1	47 0	0 54.3	59 0	0 35.0
20	1 22.0	20	0 53.7	20	0 34.5
40	1 21.0	40	0 53.1	40	0 34.1
36 0	1 20.1	48 0	0 52.5	60 0	0 33.6
20	1 19.1	20	0 51.9	20	0 33.2
40	1 18.2	40	0 51.2	40	0 32.7
37 0	1 17.2	49 0	0 50.6	61 0	0 32.3
20	1 16.3	20	0 50.0	62 0	0 31.0
40	1 15.4	40	0 49.4	63 0	0 29.7
38 0	1 14.5	50 0	0 48.9	64 0	0 28.4
20	1 13.6	20	0 48.3	65 0	0 27.2
40	1 12.7	40	0 47.8	66 0	0 25.9
39 0	1 11.9	51 0	0 47.2	67 0	0 24.7
20	1 11.0	20	0 46.6	68 0	0 23.6
40	1 10.2	40	0 46.1	69 0	0 22.4
40 0	1 9.4	52 0	0 45.5	70 0	0 21.2
20	1 8.6	20	0 45.0	71 0	0 20.1
40	1 7.8	40	0 44.4	72 0	0 18.9
41 0	1 7.0	53 0	0 43.9	73 0	0 17.8
20	1 6.2	20	0 43.4	74 0	0 16.7
40	1 5.4	40	0 42.8	75 0	0 15.6
42 0	1 4.7	54 0	0 42.3	76 0	0 14.5
20	1 3.9	20	0 41.8	77 0	0 13.5
40	1 3.2	40	0 41.3	78 0	0 12.4
43 0	1 2.4	55 0	0 40.8	79 0	0 11.3
20	1 1.7	20	0 40.3	80 0	0 10.3
40	1 1.0	40	0 39.8	81 0	0 9.2
44 0	1 0.3	56 0	0 39.3	82 0	0 8.2
20	0 59.6	20	0 38.8	83 0	0 7.2
40	0 58.9	40	0 38.3	84 0	0 6.1
45 0	0 58.2	57 0	0 37.8	85 0	0 5.1
20	0 57.6	20	0 37.3	86 0	0 4.1
40	0 56.9	40	0 36.9	87 0	0 3.1
46 0	0 56.2	58 0	0 36.4	88 0	0 2.0
20	0 55.6	20	0 35.9	89 0	0 1.0
40	0 55.0	40	0 35.5	90 0	0 0.0

Table II.—*Bessel's Refraction-Table.*

Zenith-distance. °	′	Arg. app. zen.-dist. Log α.		A	λ
0	0	1.76156	2		
10	0	1.76154	5		
20	0	1.76149	10		
30	0	1.76139	9		
35	0	1.76130	11		
40	0	1.76119	15		
45	0	1.76104	4		1.0018
46	0	1.76100	4		1.0019
47	0	1.76096	4		1.0019
48	0	1.76092	5		1.0020
49	0	1.76087	5		1.0021
50	0	1.76082	5		1.0023
51	0	1.76077	6		1.0025
52	0	1.76071	6		1.0026
53	0	1.76065	7		1.0027
54	0	1.76058	8		1.0029
55	0	1.76050	8		1.0031
56	0	1.76042	9		1.0034
57	0	1.76033	10		1.0037
58	0	1.76023	11		1.0040
59	0	1.76012	11		1.0043
60	0	1.76001	13		1.0046
61	0	1.75988	15		1.0049
62	0	1.75973	16		1.0054
63	0	1.75957	18		1.0058
64	0	1.75939	20		1.0063
65	0	1.75919	22		1.0068
66	0	1.75897	26		1.0075
67	0	1.75871	29		1.0083
68	0	1.75842	33		1.0092
69	0	1.75809	38		1.0101
70	0	1.75771	45		1.0111
71	0	1.75726	51		1.0124
72	0	1.75675	60		1.0139
73	0	1.75615	72		1.0156
74	0	1.75543	86		1.0175
75	0	1.75457	16		1.0197
	10	1.75441	16		1.0200
	20	1.75425	17		1.0204
	30	1.75408	17		1.0208
	40	1.75391	18		1.0212
	50	1.75373	18		1.0216
76	0	1.75355	19		1.0220
	10	1.75336	20		1.0225
	20	1.75316	21		1.0230
	30	1.75295	21		1.0235
	40	1.75274	22		1.0241
	50	1.75252	23		1.0246
77	0	1.75229		1.0026	1.0252

Zenith-distance. °	′	Arg. app. zen.-dist. Log α.		A	λ
77	0	1.75229	24	1.0026	1.0252
	10	1.75205	25	1.0026	1.0258
	20	1.75180	25	1.0027	1.0264
	30	1.75155	26	1.0027	1.0272
	40	1.75129	28	1.0028	1.0281
	50	1.75101	29	1.0029	1.0290
78	0	1.75072	29	1.0030	1.0299
	10	1.75043	30	1.0030	1.0308
	20	1.75013	32	1.0031	1.0318
	30	1.74981	34	1.0032	1.0328
	40	1.74947	35	1.0033	1.0338
	50	1.74912	36	1.0034	1.0347
79	0	1.74876	37	1.0035	1.0357
	10	1.74839	40	1.0036	1.0367
	20	1.74799	42	1.0037	1.0377
	30	1.74757	43	1.0038	1.0387
	40	1.74714	44	1.0039	1.0398
	50	1.74670	47	1.0040	1.0409
80	0	1.74623	50	1.0041	1.0420
	10	1.74573	52	1.0042	1.0431
	20	1.74521	53	1.0043	1.0442
	30	1.74468	56	1.0045	1.0454
	40	1.74412	60	1.0046	1.0466
	50	1.74352	64	1.0047	1.0479
81	0	1.74288	65	1.0049	1.0493
	10	1.74223	68	1.0050	1.0508
	20	1.74155	72	1.0052	1.0523
	30	1.74083	76	1.0054	1.0540
	40	1.74007	79	1.0056	1.0559
	50	1.73928	83	1.0058	1.0579
82	0	1.73845	88	1.0060	1.0600
	10	1.73757	94	1.0062	1.0622
	20	1.73663	99	1.0065	1.0646
	30	1.73564	105	1.0067	1.0671
	40	1.73459	112	1.0070	1.0697
	50	1.73347	118	1.0073	1.0725
83	0	1.73229	124	1.0075	1.0754
	10	1.73105	131	1.0078	1.0784
	20	1.72974	142	1.0081	1.0815
	30	1.72832	151	1.0084	1.0846
	40	1.72681	162	1.0088	1.0879
	50	1.72519	173	1.0092	1.0914
84	0	1.72346	186	1.0096	1.0951
	10	1.72160	199	1.0100	1.0992
	20	1.71961	212	1.0105	1.1036
	30	1.71749	227	1.0110	1.1082
	40	1.71522	243	1.0115	1.1130
	50	1.71279	259	1.0121	1.1178
85	0	1.71020		1.0127	1.1229

TABLE II.—*Bessel's Refraction-Table.*

Factor depending upon the barometer.

Eng. ins.	Log B.
27.5	−0.03191
27.6	0.03033
27.7	0.02876
27.8	0.02720
27.9	0.02564
28.0	0.02409
28.1	0.02254
28.2	0.02099
28.3	0.01946
28.4	0.01793
28.5	0.01640
28.6	0.01488
28.7	0.01336
28.8	0.01185
28.9	0.01035
29.0	0.00885
29.1	0.00735
29.2	0.00586
29.3	0.00438
29.4	0.00290
29.5	−0.00142
29.6	+0.00005
29.7	0.00151
29.8	0.00297
29.9	0.00443
30.0	0.00588
30.1	0.00732
30.2	0.00876
30.3	0.01020
30.4	0.01163
30.5	0.01306
30.6	0.01448
30.7	0.01589
30.8	0.01731
30.9	0.01871
31.0	+0.02012

Factor depending upon the attached thermometer.

F.	Log T.
°	
−30	+0.00242
20	0.00203
−10	0.00164
0	0.00125
+10	0.00086
20	0.00047
30	+0.00008
40	−0.00031
50	0.00070
60	0.00109
70	0.00148
80	0.00186
90	0.00225
+100	−0.00264

$\text{Log } \beta = \log B + \log T.$

Factor depending upon the external thermometer.

F.	Log γ.	F.	Log γ.
°		°	
−20	+0.06279	+35	+0.01185
19	0.06181	36	0.01098
18	0.06083	37	0.01011
17	0.05985	38	0.00924
16	0.05887	39	0.00837
15	0.05790	40	0.00750
14	0.05693	41	0.00664
13	0.05596	42	0.00578
12	0.05500	43	0.00492
11	0.05403	44	0.00406
10	0.05307	45	0.00320
9	0.05211	46	0.00234
8	0.05115	47	0.00149
7	0.05020	48	+0.00064
6	0.04924	49	−0.00021
5	0.04829	50	0.00106
4	0.04734	51	0.00191
3	0.04640	52	0.00275
2	0.04545	53	0.00360
−1	0.04451	54	0.00444
0	0.04357	55	0.00528
+1	0.04263	56	0.00612
2	0.04169	57	0.00696
3	0.04076	58	0.00780
4	0.03982	59	0.00863
5	0.03889	60	0.00946
6	0.03796	61	0.01029
7	0.03704	62	0.01112
8	0.03611	63	0.01195
9	0.03519	64	0.01278
10	0.03427	65	0.01360
11	0.03335	66	0.01443
12	0.03243	67	0.01525
13	0.03152	68	0.01607
14	0.03060	69	0.01689
15	0.02969	70	0.01770
16	0.02878	71	0.01852
17	0.02787	72	0.01933
18	0.02697	73	0.02015
19	0.02606	74	0.02096
20	0.02516	75	0.02177
21	0.02426	76	0.02257
22	0.02336	77	0.02338
23	0.02247	78	0.02419
24	0.02157	79	0.02499
25	0.02068	80	0.02579
26	0.01979	81	0.02659
27	0.01890	82	0.02738
28	0.01801	83	0.02819
29	0.01713	84	0.02898
30	0.01624	85	0.02978
31	0.01536	86	0.03057
32	0.01448	87	0.03136
33	0.01360	88	0.03216
34	0.01273	89	0.03294
+35	+0.01185	+90	−0.03373

LIV.—*To determine the Latitude from the Meridional Altitude of an Object whose Declination is known.*

1. When the object observed is south of the zenith:

$$L = 90° + D - A = Z + D = 90° + Z - \triangle = 180° - (A + \triangle)$$

2. When the star is between the zenith and the pole:

$$L = A - \triangle = D - Z = 90° - (Z + \triangle) = A + D - 90°$$

3. When the star is between the pole and the horizon to the north:

$$L = A + \triangle = 90° + \triangle - Z = 90° + A - D = 180° - (Z + D)$$

where L = the latitude sought;

D = the declination of the object, *minus* when south;

$\triangle$ = its north-polar distance;

A = its meridional altitude; and

Z = its meridional zenith-distance.

A and Z must be corrected for refraction.

When the sun is the object observed,

A = observed altitude — (refraction — parallax) ± semi-diameter.

LV.—*Determination of the Latitude of a Place by the Method of Circum-Meridian Altitudes.*

Reduction to meridian =

$$x = k\left\{i\frac{\cos l \cos D}{\cos a}\right\} - m \tan a\left\{i\frac{\cos l \cos D}{\cos a}\right\}^2$$

$$k = \frac{2 \sin^2 \frac{1}{2} p}{\sin 1''}$$

$$m = \frac{2 \sin^4 \frac{1}{2} p}{\sin 1''}$$

$$a = 90° + D - l$$

Where—

$A = a + x =$ the meridional altitude of the object;

a = its observed altitude − (refraction − parallax) ± semi-diameter;

p = its correct hour-angle;

D = its declination;

l = the assumed latitude of the place; and

x = the required correction in seconds.

When a *star* is the object observed and the chronometer marks *mean* time—

$$i = 1.005473; \log i = 0.0023708$$

When the *sun* is observed and the chronometer marks *sidereal* time—

$$i = 0.99455418; \log i = 9.9976285$$

and, generally, when the chronometer has a large losing rate, x must be multiplied by $1 + 0.00002315\, r$; when it has a gaining rate it must be divided by $1 + 00002315\, r$; r being the rate in 24 hours, which must be assumed *minus* when *gaining*, and *plus* when *losing*.

LV.—*Determination of the Latitude, &c.*—Continued.

The values of k and m for each value of p are given in the following tables.

The meridian altitude,

$$A = a + x$$

for each observation; for any number of observations, n,

$$\frac{a' + a'' + \ldots\ldots}{n} + \frac{x' + x'' + \ldots\ldots}{n}$$

= the mean, a, of all the observed altitudes + the mean, x, of all the corrections. Consequently,

1. Measure several successive altitudes of the object both before and after its meridional passage.

2. Note the times of each observation, and compute the time of the object's culmination; the differences between this and the times of each successive observation are the values of p', p'', &c., in time, for which the corresponding values of k', k'', &c., and m', m'', &c., must be taken from the tables.

3. The means k and m of these results will be introduced into the equation for the value of the correction, x, to be applied to a to obtain the meridional altitude, A, of the object.

4. If the final latitude differ much from the assumed, the computation should be repeated with the new value for l.

5. It is not necessary that the time of the object's culmination should be known with great precision, provided an equal number of altitudes be taken upon each side of the meridian, and at nearly equal distances from it.

6. The second correction, m, is seldom necessary, unless great accuracy is desired, and the object is observed more than ten minutes of time from the meridian.

Reduction to the Meridian; Values of $k = \frac{2 \sin^2 \frac{1}{2} p}{\sin 1''}$

Sec.	0^m	1^m	2^m	3^m	4^m	5^m	6^m	7^m
	″	″	″	″	″	″	″	″
0	0.00	1.96	7.8	17.7	31.4	49.1	70.7	96.2
1	0.co	2.03	8.0	17.9	31.7	49.4	71.1	96.7
2	0.00	2.10	8.1	18.1	31.9	49.7	71.5	97.1
3	0.00	2.16	8.2	18.3	32.2	50.1	71.9	97.6
4	0.01	2.23	8.4	18.5	32.5	50.4	72.3	98.0
5	0.01	2.31	8.5	18.7	32.7	50.7	72.7	98.5
6	0.02	2.38	8.7	18.9	33.0	51.1	73.1	99.0
7	0.02	2.45	8.8	19.1	33.3	51.4	73.5	99.4
8	0.03	2.52	8.9	19.3	33.5	51.7	73.9	99.9
9	0.04	2.60	9.1	19.5	33.8	52.1	74.3	100.4
10	0.05	2.67	9.2	19.7	34.1	52.4	74.7	100.8
11	0.06	2.75	9.4	19.9	34.4	52.7	75.1	101.3
12	0.08	2.83	9.5	20.1	34.6	53.1	75.5	101.8
13	0.09	2.91	9.6	20.3	34.9	53.4	75.9	102.3
14	0.11	2.99	9.8	20.5	35.2	53.8	76.3	102.7
15	0.12	3.07	9.9	20.7	35.5	54.1	76.7	103.2
16	0.14	3.15	10.1	20.9	35.7	54.5	77.1	103.7
17	0.16	3.23	10.2	21.2	36.0	54.8	77.5	104.2
18	0.18	3.32	10.4	21.4	36.3	55.1	77.9	104.6
19	0.20	3.40	10.5	21.6	36.6	55.5	78.3	105.1
20	0.22	3.49	10.7	21.8	36.9	55.8	78.8	105.6
21	0.24	3.58	10.8	22.0	37.2	56.2	79.2	106.1
22	0.26	3.67	11.0	22.3	37.4	56.5	79.6	106.6
23	0.28	3.76	11.2	22.5	37.7	56.9	80.0	107.0
24	0.31	3.85	11.3	22.7	38.0	57.3	80.4	107.5
25	0.34	3.94	11.5	22.9	38.3	57.6	80.8	108.0
26	0.37	4.03	11.6	23.1	38.6	58.0	81.3	108.5
27	0.40	4.12	11.8	23.4	38.9	58.3	81.7	109.0
28	0.43	4.22	11.9	23.6	39.2	58.7	82.1	109.5
29	0.46	4.32	12.1	23.8	39.5	59.0	82.5	110.0

Reduction to the Meridian; Values of $k = \frac{2 \sin^2 \frac{1}{2} p}{\sin 1''}$

Sec.	0^m	1^m	2^m	3^m	4^m	5^m	6^m	7^m
	"	"	"	"	"	"	"	"
30	0.49	4.42	12.3	24.0	39.8	59.4	83.0	110.4
31	0.52	4.52	12.4	24.3	40.1	59.8	83.4	110.9
32	0.56	4.62	12.6	24.5	40.3	60.1	83.8	111.4
33	0.59	4.72	12.8	24.7	40.6	60.5	84.2	111.9
34	0.63	4.82	12.9	25.0	40.9	60.8	84.7	112.4
35	0.67	4.92	13.1	25.2	41.2	61.2	85.1	112.9
36	0.71	5.03	13.3	25.4	41.5	61.6	85.5	113.4
37	0.75	5.13	13.4	25.7	41.8	61.9	86.0	113.9
38	0.80	5.24	13.6	25.9	42.1	62.3	86.4	114.4
39	0.83	5.34	13.8	26.2	42.5	62.7	86.8	114.9
40	0.87	5.45	14.0	26.4	42.8	63.0	87.3	115.4
41	0.91	5.56	14.1	26.6	43.1	63.4	87.7	115.9
42	0.96	5.67	14.3	26.9	43.4	63.8	88.1	116.4
43	1.01	5.78	14.5	27.1	43.7	64.2	88.6	116.9
44	1.06	5.90	14.7	27.4	44.0	64.5	89.0	117.4
45	1.10	6.01	14.8	27.6	44.3	64.9	89.5	117.9
46	1.15	6.13	15.0	27.9	44.6	65.3	89.9	118.4
47	1.20	6.24	15.2	28.1	44.9	65.7	90.3	118.9
48	1.26	6.36	15.4	28.3	45.2	66.0	90.8	119.5
49	1.31	6.48	15.6	28.6	45.5	66.4	91.2	120.0
50	1.36	6.60	15.8	28.8	45.9	66.8	91.7	120.5
51	1.42	6.72	15.9	29.1	46.2	67.2	92.1	121.0
52	1.48	6.84	16.1	29.4	46.5	67.6	92.6	121.5
53	1.53	6.96	16.3	29.6	46.8	68.0	93.0	122.0
54	1.59	7.09	16.5	29.9	47.1	68.3	93.5	122.5
55	1.65	7.21	16.7	30.1	47.5	68.7	93.9	123.1
56	1.71	7.34	16.9	30.4	47.8	69.1	94.4	123.6
57	1.77	7.46	17.1	30.6	48.1	69.5	94.8	124.1
58	1.83	7.60	17.3	30.9	48.4	69.9	95.3	124.6
59	1.89	7.72	17.5	31.1	48.8	70.3	95.7	125.1

Reduction to the Meridian; Values of $k = \frac{2 \sin^2 \frac{1}{2} p}{\sin 1''}$

Sec.	8^m	9^m	10^m	11^m	12^m	13^m	14^m
	"	"	"	"	"	"	"
0	125.7	159.0	196.3	237.5	282.7	331.8	384.7
1	126.2	159.6	197.0	238.3	283.5	332.6	385.6
2	126.7	160.2	197.6	239.0	284.2	333.4	386.6
3	127.2	160.8	198.3	239.7	285.0	334.3	387.5
4	127.8	161.4	198.9	240.4	285.8	335.2	388.4
5	128.3	162.0	199.6	241.2	286.6	336.0	389.3
6	128.8	162.6	200.3	241.9	287.4	336.9	390.2
7	129.3	163.2	200.9	242.6	288.2	337.7	391.1
8	129.9	163.8	201.6	243.3	289.0	338.6	392.1
9	130.4	164.4	202.2	244.1	289.8	339.4	393.0
10	131.0	165.0	202.9	244.8	290.6	340.3	393.9
11	131.5	165.6	203.6	245.5	291.4	341.2	394.8
12	132.0	166.2	204.2	246.3	292.2	342.0	395.8
13	132.6	166.8	204.9	247.0	293.0	342.9	396.7
14	133.1	167.4	205.6	247.7	293.8	343.7	397.6
15	133.6	168.0	206.3	248.5	294.6	344.6	398.6
16	134.2	168.6	206.9	249.2	295.4	345.5	399.5
17	134.7	169.2	207.6	249.9	296.2	346.4	400.5
18	135.3	169.8	208.3	250.7	297.0	347.2	401.4
19	135.8	170.4	208.9	251.4	297.8	348.1	402.3
20	136.3	171.0	209.6	252.2	298.6	349.0	403.3
21	136.9	171.6	210.3	253.0	299.4	349.8	404.2
22	137.4	172.2	211.0	253.6	300.2	350.7	405.1
23	138.0	172.9	211.7	254.4	301.0	351.6	406.0
24	138.5	173.5	212.3	255.1	301.8	352.5	407.0
25	139.1	174.1	213.0	255.9	302.6	353.3	408.0
26	139.6	174.7	213.7	256.6	303.5	354.2	408.9
27	140.2	175.3	214.4	257.4	304.3	355.1	409.9
28	140.7	175.9	215.1	258.1	305.1	356.0	410.8
29	141.3	176.6	215.8	258.9	305.9	356.9	411.7

Reduction to the Meridian; Values of $k = \frac{2 \sin^2 \frac{1}{2} p}{\sin 1''}$

Sec.	8^m	9^m	10^m	11^m	12^m	13^m	14^m
	"	"	"	"	"	"	"
30	141.8	177.2	216.4	259.6	306.7	357.7	412.7
31	142.4	177.8	217.1	260.4	307.5	358.6	413.6
32	143.0	178.4	217.8	261.1	308.4	359.5	414.6
33	143.5	179.0	218.5	261.9	309.2	360.4	415.5
34	144.1	179.7	219.2	262.6	310.0	361.3	416.5
35	144.6	180.3	219.9	263.4	310.8	362.2	417.5
36	145.2	180.9	220.6	264.1	311.6	363.1	418.4
37	145.8	181.6	221.3	264.9	312.5	364.0	419.4
38	146.3	182.2	222.0	265.7	313.3	364.8	420.3
39	146.9	182.8	222.7	266.4	314.1	365.7	421.3
40	147.5	183.5	223.4	267.2	315.0	366.6	422.2
41	148.0	184.1	224.1	267.9	315.8	367.5	423.2
42	148.6	184.7	224.8	268.7	316.6	368.4	424.2
43	149.2	185.4	225.5	269.5	317.4	369.3	425.1
44	149.7	186.0	226.2	270.3	318.3	370.2	426.1
45	150.3	186.6	226.9	271.0	319.1	371.1	427.0
46	150.9	187.3	227.6	271.8	319.9	372.0	428.0
47	151.5	187.9	228.3	272.6	320.8	372.9	429.0
48	152.0	188.5	229.0	273.3	321.6	373.8	429.9
49	152.6	189.2	229.7	274.1	322.4	374.7	430.9
50	153.2	189.8	230.4	274.9	323.3	375.6	431.9
51	153.8	190.5	231.1	275.6	324.1	376.5	432.8
52	154.4	191.1	231.8	276.4	325.0	377.4	433.8
53	154.9	191.8	232.5	277.2	325.8	378.3	434.8
54	155.5	192.4	233.2	278.0	326.7	379.3	435.8
55	156.1	193.1	234.0	278.8	327.5	380.2	436.7
56	156.7	193.7	234.7	279.5	328.4	381.1	437.7
57	157.3	194.4	235.4	280.3	329.2	382.0	438.7
58	157.8	195.0	236.1	281.1	330.0	382.9	439.7
59	158.4	195.7	236.8	281.9	330.9	383.8	440.6

Reduction to the Meridian; Values of $k = \dfrac{2 \sin^2 \frac{1}{2} p}{\sin 1''}$

Sec.	15m	16m	17m	18m	19m	20m	21m
	"	"	"	"	"	"	"
0	441.6	502.5	567.2	635.9	708.4	784.9	865.3
1	442.6	503.5	568.3	637.0	709.7	786.2	866.6
2	443.6	504.6	569.4	638.2	710.9	787.5	868.0
3	444.6	505.6	570.5	639.4	712.1	788.8	869.4
4	445.6	506.7	571.6	640.6	713.4	790.1	870.8
5	446.5	507.7	572.8	641.7	714.6	791.4	872.1
6	447.5	508.8	573.9	642.9	715.9	792.7	873.5
7	448.5	509.8	575.0	644.1	717.1	794.0	874.9
8	449.5	510.9	576.1	645.3	718.4	795.4	876.3
9	450.5	511.9	577.2	646.5	719.6	796.7	877.6
10	451.5	513.0	578.4	647.7	720.9	798.0	879.0
11	452.5	514.0	579.5	648.9	722.1	799.3	880.4
12	453.5	515.1	580.6	650.0	723.4	800.7	881.8
13	454.5	516.1	581.7	651.2	724.6	802.0	883.2
14	455.5	517.2	582.9	652.4	725.9	803.3	884.6
15	456.5	518.3	584.0	653.6	727.2	804.6	886.0
16	457.5	519.3	585.1	654.8	728.4	806.0	887.4
17	458.5	520.4	586.2	656.0	729.7	807.3	888.8
18	459.5	521.5	587.4	657.2	730.9	808.6	890.2
19	460.5	522.5	588.5	658.4	732.2	809.9	891.6
20	461.5	523.6	589.6	659.6	733.5	811.3	893.0
21	462.5	524.6	590.8	660.8	734.7	812.6	894.4
22	463.5	525.7	591.9	662.0	736.0	813.9	895.8
23	464.5	526.8	593.0	663.2	737.3	815.2	897.2
24	465.5	527.9	594.2	664.4	738.5	816.6	898.6
25	466.5	528.9	595.3	665.6	739.8	817.9	900.0
26	467.5	530.0	596.5	666.8	741.1	819.2	901.4
27	468.5	531.1	597.6	668.0	742.3	820.5	902.8
28	469.5	532.2	598.7	669.2	743.6	821.9	904.2
29	470.5	533.2	599.9	670.4	744.9	823.2	905.6

Reduction to the Meridian; Values of $k = \dfrac{2 \sin^2 \frac{1}{2} p}{\sin 1''}$

Sec.	15m	16m	17m	18m	19m	20m	21m
	"	"	"	"	"	"	"
30	471.5	534.3	601.0	671.6	746.2	824.6	907.0
31	472.6	535.4	602.2	672.8	747.4	825.9	908.4
32	473.6	536.5	603.3	674.1	748.7	827.3	909.8
33	474.6	537.6	604.5	675.3	750.0	828.6	911.2
34	475.6	538.7	605.6	676.5	751.3	829.9	912.6
35	476.6	539.7	606.8	677.7	752.6	831.2	914.0
36	477.6	540.8	607.9	678.9	753.8	832.6	915.5
37	478.7	541.9	609.1	680.1	755.1	833.9	916.9
38	479.7	543.0	610.2	681.3	756.4	835.3	918.3
39	480.7	544.1	611.4	682.6	757.7	836.6	919.7
40	481.7	545.2	612.5	683.8	759.0	838.0	921.1
41	482.8	546.3	613.7	685.0	760.2	839.3	922.5
42	483.8	547 4	614.8	686.2	761.5	840.7	923.9
43	484.8	548.4	616.0	687.4	762.8	842.0	925.3
44	485.8	549.5	617.2	688.7	764.1	843.4	926.8
45	486.9	550.6	618.3	689.9	765.4	844.7	928.2
46	487.9	551.7	619.5	691.1	766.7	846.1	929.6
47	488.9	552.8	620.6	692.4	768.0	847.5	931 0
48	490.0	553.9	621.8	693.6	769.3	848.9	932.4
49	491.0	555.0	623.0	694.8	770.6	850.2	933.8
50	492.0	556.1	624.1	696.0	771.9	851.6	935.2
51	493.1	557.2	625.3	697.3	773.1	852.9	936.6
52	494.1	558.3	626.5	698.5	774.5	854.3	938.1
53	495.2	559.4	627.6	699.7	775.8	855.7	939.5
54	496.2	560.5	628.8	701.0	777.1	857.1	940.9
55	497.2	561.6	630.0	702.2	778.4	858.4	942.3
56	498 3	562.7	631.2	703.5	779.7	859.8	943.8
57	499.3	563.9	632.3	704.7	781.0	861.1	945.2
58	500.3	565.0	633.5	705.9	782.3	862.5	946.6
59	501.4	566.1	634.7	707.1	783.6	863.9	948.1

Reduction to the Meridian; Values of $k = \frac{2 \sin^2 \frac{1}{2} p}{\sin 1''}$

Seconds.	22m	23m	24m	Seconds.	22m	23m	24m
	"	"	"		"	"	"
0	949.6	1037.8	1129.9	30	993.2	1083.3	1177.5
1	951.0	1039.3	1131.4	31	994.7	1084.8	1179.1
2	952.4	1040.8	1133.0	32	996.2	1086.4	1180.7
3	953.8	1042.3	1134.6	33	997.6	1087.9	1182.3
4	955.3	1043.8	1136.2	34	999.1	1089.5	1183.9
5	956.7	1045.3	1137.8	35	1000.6	1091.0	1185.5
6	958.2	1046.8	1139.3	36	1002.1	1092.6	1187.1
7	959.6	1048.3	1140.9	37	1003.5	1094.1	1188.7
8	961.1	1049.8	1142.5	38	1005.0	1095.7	1190.3
9	962.5	1051.3	1144.0	39	1006.5	1097.2	1191.9
10	963.9	1052.8	1145.6	40	1008.0	1098.8	1193.5
11	965.4	1054.3	1147.2	41	1009.4	1100.3	1195.1
12	966.9	1055.9	1148.8	42	1010.9	1101.9	1196.7
13	968.3	1057.4	1150.4	43	1012.4	1103.4	1198.3
14	969.8	1058.9	1152.0	44	1013.9	1105.0	1199.9
15	971.2	1060.4	1153.6	45	1015.4	1106.5	1201.5
16	972.7	1062.0	1155.2	46	1016.9	1108.1	1203.1
17	974.1	1063.5	1156.8	47	1018.4	1109.6	1204.7
18	975.5	1065.0	1158.3	48	1019.9	1111.2	1206.4
19	977.0	1066.5	1159.9	49	1021.4	1112.7	1208.0
20	978.5	1068.1	1161.5	50	1022.8	1114.3	1209.6
21	979.9	1069.6	1163.1	51	1024.3	1115.8	1211.2
22	981.4	1071.1	1164.7	52	1025.8	1117.4	1212.9
23	982.9	1072.6	1166.3	53	1027.3	1118.9	1214.5
24	984.4	1074.2	1167.9	54	1028.8	1120.5	1216.1
25	985.8	1075.7	1169.5	55	1030.3	1122.0	1217.7
26	987.3	1077.2	1171.1	56	1031.8	1123.6	1219.4
27	988.8	1078.7	1172.7	57	1033.3	1125.1	1221.0
28	990.3	1080.3	1174.3	58	1034.8	1126.7	1222.6
29	991.8	1081.8	1175.9	59	1036.3	1128.3	1224.2

Second Part of the Reduction to the Meridian.

Values of $m = \frac{2 \sin^4 \frac{1}{2} p}{\sin 1''}$

Minutes.	0^s	10^s	20^s	30^s	40^s	50^s
	"	"	"	"	"	"
5	0.01	0.01	0.01	0.01	0.01	0.01
6	0.01	0.01	0.01	0.02	0.02	0.02
7	0.02	0.02	0.03	0.03	0.03	0.04
8	0.04	0.04	0.05	0.05	0.05	0.06
9	0.06	0.07	0.08	0.08	0.08	0.09
10	0.09	0.10	0.11	0.11	0.12	0.13
11	0.14	0.15	0.15	0.16	0.17	0.18
12	0.19	0.20	0.22	0.23	0.24	0.25
13	0.27	0.28	0.30	0.31	0.33	0.34
14	0.36	0.38	0 39	0.41	0.43	0.45
15	0.47	0.49	0.52	0.54	0.56	0.59
16	0.61	0.64	0.67	0.69	0.72	0.75
17	0.78	0.81	0.84	0.88	0.91	0.95
18	0.98	1.02	1.06	1.09	1.13	1.18
19	1.22	1.26	1.30	1.35	1.40	1.44
20	1.49	1.54	1.60	1.65	1.70	1.76
21	1.82	1.87	1.93	1.99	2.06	2.12
22	2.19	2.25	2.32	2.39	2.46	2.54
23	2.61	2.69	2.77	2.85	2.93	3.01
24	3.10	3.18	3.27	3.36	3.45	3.55
25	3 64	3.74	3.84	3.94	4.05	4.15
26	4.26	4.37	4.48	4.60	4.72	4.83
27	4.96	5.08	5.20	5.33	5.46	5.60
28	5.73	5.87	6.01	6.15	6.30	6.44
29	6.59	6.75	6.90	7.06	7.22	7.38
30	7.55	7.72	7.89	8.06	8.24	8.42
31	8.61	8.79	8.98	9.17	9.37	9.57
32	9.77	9.97	10.18	10.39	10.61	10.82
33	11.04	11.27	11.50	11.73	11.96	12.20
34	12.44	12.69	12.94	13.19	13.45	13.71
35	13.97	14.24	14.51	14.78	15.06	15.35

FORM FOR

SURVEY OF .... DETERMINATION OF THE LATITUDE,

and South of

DATE AND STATION.—1843, *October* 13.—*Mouth of the Big Black River,*

NAME OF STAR, γ *Pegasi, South of the Zenith.*

INSTRUMENTS ... { Sextant No. 2197, by *Troughton & Simms,* and *Mean Solar* Chronometer No. 76, by *Charles*

No. for reference.	Times of observation by chronometer.	MERIDIAN DISTANCES, $= p$.		$\frac{2 \sin^2 \frac{1}{2} p}{\sin 1''} = k$	$\frac{\cos l \,.\, \cos D}{\text{Co sine } a}$	Reduction to the meridian, (in arc,) $= x$.
		In mean solar time.	In sidereal time.			
	h. m. s.	m. s.	m. s.	''		' ''
1	10 18 40.4	9 44.2	9 45.8	187.3	Constant multiple, 1.227.	3 49.8
2	19 44.4	8 40.2	8 41.6	148.3		2 51.9
3	20 48	7 36.5	7 37.7	114.2		2 27.3
4	21 46.4	6 38.2	6 39.3	86.9		1 47.6
5	22 44.4	5 40.2	5 41.1	63.4		1 17.8
6	23 54	4 30.5	4 31.2	40.1		0 49.2
7	25 12	3 12.6	3 13.1	20.3		0 24.9
8	26 46	1 38.6	1 38.8	5.2		0 06.3
9	28 16.4	0 08.2	0 08.2	0.0		0 00.0
10	29 42	1 17.4	1 17.6	3.2		0 03.9
11	31 42	3 17.4	3 17.9	21.4		0 26.2
12	32 54.4	4 29.8	4 30.5	40.0		0 49
13	34 18	5 53.4	5 54.3	68.5		1 24
14	36 14.2	7 49.6	7 50.9	123.5		2 31.5
15	38 32.2	10 07.6	10 09.2	202.3		4 08.2
16	40 06	11 41.4	11 43.3	269.9		5 31.1

Observer, *Major J. D. Graham.*
Computer, *do. do.*

RECORD AND COMPUTATION.

from Observed Double Circum-Meridian Altitudes of Stars, North the Zenith.

a tributary to the river Saint John, Maine.

Artificial Horizon of Mercury.

Young.

Observed double circum-meridian altitudes of star.	True circum-meridian altitude of star, as corrected for refraction and errors of instrument, = *a*.	True meridian altitudes deduced, $=(a+x)=$ A.	Latitude deduced from each observation, = L = $(90^\circ + D - A)$.
° ′ ″	° ′ ″	° ′ ″	° ′ ″
114 34 15	57 18 38.5	57 22 28.3	46 56 42.55
36 15	57 19 38.5	57 22 30.4	56 40.45
37 10	57 20 06	57 22 33.3	56 37.55
38 10	57 20 36	57 22 23.6	56 47.25
39 30	57 21 16	57 22 33.8	56 37.05
40 30	57 21 46	57 22 35.2	56 35.65
41 05	57 22 03.5	57 22 28.4	56 42.45
41 50	57 22 26	57 22 32.3	56 34.55
41 50	57 22 26	57 22 26	56 40.85
41 50	57 22 26	57 22 29.9	56 36.95
41 00	57 22 01	57 22 27.2	56 39.65
39 45	57 21 23.5	57 22 12.5	56 58.35
38 40	57 20 51	57 22 15	56 55.85
36 30	57 19 46	57 22 17.5	56 53.55
33 20	57 18 11	57 22 19.2	56 51.85
30 50	57 16 56	57 22 27.1	46 56 39.75

LATITUDE—Deduced from a mean of 16 altitudes of star γ *Pegasi* .... 46° 56′ 43″.4

Deduced from a mean of 10 altitudes of star γ *Cephei*, observed this night with same sextant .. 46 57 10.7

Mean, or latitude adopted .... 46° 56′ 57″

Form for Record and Computation—Continued.

D=apparent declinat'n of star=$14°19'10''.85$ N. log cos	9.98629
l=approximate latitude of place = $46°57'$ . . log cos	9.83418
Sum	19.82048
a=approximate merid. alt. of star=$57°22'10''$. log cos	9.73176
$\frac{\cos l \cos D}{\cos a}$ = constant multiple = 1.227 . . . log	0.08872
Refraction (ther. 28°, bar. 29.14 in.) for mean obs'd alts.	— 0′ 39″
Index-error of sextant	+ 2 40
*Error of eccentricity, &c., of sextant	+ 1 40
Apparent AR. of the star γ Pegasi	$0^h05^m14^s.09$
Sidereal time at mean noon at this station	13 26 20 . 83
Sidereal interval from mean noon of star's culmination	10 38 53 . 16
Retardation of mean on sidereal time	— 1 44 . 96
Mean time of culmination of star γ Pegasi	10 37 08 . 2
Chronometer (C. Y. 76) *slow* of mean time at time of observation	— 08 43 . 6
Time by chronometer of culmination of star γ Pegasi	$10^h28^m24^s.6$

On this night, October 13, 1843, Major Graham obtained for the latitude of this station, from 75 observations on 5 stars south of the zenith, combined with 21 observations on γ Cephei and Polaris, to the north	46° 56′ 56″.3
On the night of October 24, by 43 observations on 4 southern stars, combined with 2 observations on γ Cephei, the latitude deduced was	46 56 57.2
On September 17, 1844, 66 observations on north and south stars gave for the latitude of this station	46 56 60.4

* The error of eccentricity is approximately ascertained by comparing latitudes, well determined by observations on north and south stars, with that which will result from north or south stars individually of various meridional altitudes. It varies with the altitudes observed; that is to say, it is different for different parts of the limb of the instrument.

LVI.—*To Determine the Latitude by an Altitude of a Star Near the Pole, at any hour.*

$$L = A - (\triangle \cos p) + \alpha (\triangle \sin p)^2 \tan A - \beta (\triangle \sin p)^2 (\triangle \cos p)$$

where—

A = the observed altitude, corrected for refraction, &c.;

$\triangle$ = the polar distance of the star in seconds of arc;

$\alpha = \frac{1}{2} \sin 1''$; $\log \alpha = 4.3845449$;

$\beta = \frac{1}{3} \sin^2 1''$; $\log \beta = 8.89403$; and

p = the hour-angle of the star.

$\pm p$ = sidereal time — AR. ✱ = solar time + AR. ☉ — AR. ✱

p is *plus* when the star is west, and *minus* when it is east of the meridian.

The sign of $\cos p$ should also be attended to, for when p is greater than 6^h, or $90°$, the cosine is negative, and the second and fourth terms change the sign *minus* to *plus*.

The fourth term may be generally omitted; its greatest value being only $0''.55$.

This formula is only applicable to stars within a very few degrees of the pole.

For other circumpolar stars—

$$\tan x = \tan \triangle \cos p$$

$$\sin y = \frac{\cos x \sin A}{\cos \triangle}$$

$$L = y \mp x$$

in which the upper sign is used when the star is above the pole; the under when below the pole.

FORM FOR RECORD

SURVEY OF .... DETERMINATION OF THE

DATE AND STATION.—1843, *September* 6—*Woodstock*,

NAME OF STAR.—*Polaris*, (*a Ursæ Minoris*,) *observed on*

INSTRUMENTS ... { Sextant No. 2197, by *Troughton &* / *Mean Solar* Chronometer, No. 2440,

No. for reference.	Times of observation by *mean solar chronometer No.* 2440.	True sidereal times of observation.	MERIDIAN DISTANCES.		$-\triangle \cos p$
			In sid'l time, $=p$.	In arc, $=p$.	
	h. m. s.	*h. m. s.*	*h. m. s.*	° ′ ″	′ ″
1	1 33 02.5	20 05 34.1	4 58 23.2	74 35 48	—24 18.1
2	1 34 28	20 06 59.8	4 56 57.5	74 14 22.5	—24 54.5
3	1 35 42.7	20 08 14.7	4 55 42.6	73 55 39	—25 19.8
4	1 36 38.2	20 09 10.4	4 54 46.9	73 41 43.5	—25 41.4
5	1 39 07.5	20 11 40.1	4 52 17.2	73 04 18	—26 34.7
6	1 41 11.2	20 13 44.1	4 50 13.2	72 33 22.5	—27 27.1
7	1 44 28.2	20 17 01.7	4 46 55.6	71 43 54	—28 40.8

Observer, *Major J. D. Graham.*
Computer, *Do.*

AND COMPUTATION.

LATITUDE, *from Observed Double Altitudes of Polaris.*

New Brunswick, (*Grover's Inn.*)

between four and five hours before its upper meridian passage.

Simms, and Artificial Horizon of Mercury.
by *Parkinson & Frodsham.*

$+\alpha(\triangle \sin p)^2 . \tan A$	$-\beta(\triangle \sin p)^2 . (\triangle \cos p)$	Observed double altitudes of *Polaris* out of the meridian.	True altitudes of star, as corrected for refraction and errors of instrument, = A.	Latitude deduced from each observation, = L.
′ ″	′	° ′ ″	° ′ ″	° ′ ″
+ 1 11.63	— 0.32	93 01.30	46 31 58.6	46 08 51.8
+ 1 11.41	— 0.33	93 02.45	46 32 36	46 08 52.6
+ 1 11.20	— 0.33	93 03.50	46 33 08.6	46 08 59.7
+ 1 11.04	— 0.33	93 04.40	46 33 33.6	46 09 02.9
+ 1 10.63	— 0.34	93 06.15	46 34 21	46 08 56.6
+ 1 10.28	— 0.35	93 08.20	46 35 23.5	46 09 06.3
+ 1 09.68	— 0.37	93 10.50	46 36 38.5	46 09 07

LATITUDE, deduced from a mean of 7 altitudes of star *Polaris*, 46° 08′ 59″.4.

Form for Record and Computation—Continued.

Apparent declination of star = 88° 28′ 30″.5.
Apparent N. P. D. of star = 1° 31′ 29″.5 = 5489″.5 = △

Refraction (ther. 57°; bar. 30.013 inches)	− 55″.4
Index-error of sextant	+ 2′ 50″
Errors of eccentricity, &c., of sextant	+ 1′ 28″
Apparent AR. of the star *Polaris* (*a Ursæ Minoris*)	1^h 03^m $57^s.3$
Sidereal time at mean noon at this station	11 00 27 .1
Sidereal interval from mean noon of star's culmination	10 03 30 .2
Retardation of mean on sidereal time	− 2 18 .2
Mean time of culmination of star *Polaris*	14 01 12
Chronometer No. 2440 fast of mean time at time of observation	4 29 24 .8
Time by chronometer of culmination of star *Polaris*	6^h 30^m $36^s.8$

The reduction of the mean time of observation to sidereal time, in the preceding example, might have been omitted by using table of *AR. in Arc into Mean Time*, pages 198, &c. Thus:

Mean time of observation		1^h 33^m $02^s.5$
Mean time of culmination of *Polaris*		6 30 36 .8
Hour-angle, *p*, in intervals of mean time		4^h 57^m $34^s.3$
Sidereal equivalents, in arc	4^h	= 60° 09′ 51″.39
	57^m	= 14 17 20 .45
	34^s	= 8 31 .40
	$0^s.3$	= 4 .51
p, in arc		= 74° 35′ 47″.75

Computation, First Observation.

1st term.		2d term.		3d term.	
log cos *p* (+) =	9.4242480	sin *p*	= 9.98411		
log △ =	3.7395327	△	= 3.73953		
	3.1637807	△ sin *p*	= 3.72364	. . .	= 3.16378
△ cos *p* =	1458″.1				
1st term =	− 24′ 18″.1	(△ sin *p*)²	= 7.44728	. . .	= 7.44728
		log *a*	= 4.38454	log *β*	= 8.89403
A =	46° 31′ 58″.6	tan A	= 0.02325		
					9.50509
	46 07 40 .5		1.85507	3d term	= − 0″.32
2d term =	+ 1 11 .63		= 71″.63		
		2d term	= + 1′ 11″.63		
	46 08 52 .13				
3d term =	− 0 .32				
Latitude =	46° 08′ 51″.81				

LVI.—*Determination of the Latitude by Transits over the Prime Vertical.*

Suppose a transit-instrument so placed that the transit-axis is on the meridian, or very nearly so, and that the axis is horizontal, and the collimation nothing:

1. Call the time T at which a star whose declination is D passes the middle wire of the instrument on the eastern side of the meridian, the clock-correction to reduce the observed time to the true E, and the right ascension of the star AR.; and let T′ and E′ denote the corresponding quantities for the western transit. Then the two hour angles, in sidereal time, will be, the eastern negative,

$$t = \mathrm{T} + \mathrm{E} - \mathrm{AR.}; \qquad t' = \mathrm{T}' + \mathrm{E}' - \mathrm{AR.}$$

Let the unknown latitude of the place be L, and the azimuth of the line of collimation a. The spherical triangle, formed by great circles connecting the zenith, the pole, and the place of the star, gives the following relations:

$$\cot a = \frac{\cos t \cos \mathrm{D} \sin \mathrm{L} - \sin \mathrm{D} \cos \mathrm{L}}{\cos \mathrm{D} \sin t}$$

$$= \frac{\cos t' \cos \mathrm{D} \sin \mathrm{L} - \sin \mathrm{D} \cos \mathrm{L}}{\cos \mathrm{D} \sin t'}$$

Whence—

$$\tan \mathrm{L} = \tan \mathrm{D} \frac{\cos \frac{1}{2}(t' + t)}{\cos \frac{1}{2}(t' - t)}$$

If the instrument is very nearly on the prime vertical,

$$\cos \tfrac{1}{2}(t' + t) = \cos 0^\circ = 1$$

and—

$$\tan \mathrm{L} = \tan \mathrm{D} \sec \tfrac{1}{2}(t' - t)$$

for the passage over the middle wire of the instrument.

2. Call the time of passage of the star, from a side wire to the middle wire, τ.

Let the distance, in arc, of one of the lateral wires from the middle wire, measured on a great circle, be 15 f; f being the equatorial interval of the wire, in time.

LVI.—*Prime Vertical Transits*—Continued.

Then, to reduce the transit over a side wire to the center wire,

$$\tau = \frac{f}{[\sin(L + D) \cdot \sin(L - D) \pm \frac{15}{2} f]^{\frac{1}{2}}}$$

The upper sign of the term $\pm \frac{15}{2} f$ is to be used for wires crossed by the star earlier than the middle wire in the eastern transit, and later in the western transit, and the lower sign in the opposite cases. An approximate latitude may be used for L.

3. Should the optical axis not coincide with the middle wire, substitute $f \pm c$ for f in the above, according as the error of collimation, c, lies on the same or opposite sides of f.

4. The preceding formula gives the latitude on the supposition that the axis of the instrument is parallel to the horizon. If the instrument is on the prime vertical, but the north end of the axis is, for instance, n seconds too high, the axis is parallel to the horizon of a place whose latitude is n seconds less than where the instrument is placed, and the true latitude is, therefore,

$$L + n$$

5. But should the instrument not be on the prime vertical, the true latitude becomes—

$$L + n \sin a$$

a being the azimuth of the center wire of the telescope, supposed in collimation.

This may be found from the time elapsed between the east and west transits of the same star: thus—

$$\cot u = \tan \tfrac{1}{2} (t' - t) \sin D$$

$$\sin a = \cos D \frac{\sin u}{\cos L}$$

a is taken between 0° and 90° when the north end of the transit-axis is between the north and west, and between 90° and 180° when the same end is between the north and east.

If n is called *plus* when the north end of the axis is too high, and *vice versa*, the signs of the corrections are indicated by those of the quantities resulting from the formula.

When a is nearly 90°, the correction is exceedingly small; so that, when the instrument is placed nearly east and west, we may proceed in all the computations as if it were exactly so.

LVI.—*Prime Vertical Transits*—Continued.

6. The instrument should be set up in the firmest manner. A change of azimuth between the east and west transits of a star will affect the result much less than an equal change of level.

It is better, in order to obtain a close result in the shortest time, to observe several stars on the same evening, and between the first and last observations to determine with the level the inclination of the axis several times, and then to interpolate for transits between the times of observation of the level. It is of course understood that the changes of inclination must be small, which will be the case if the instrument is properly placed.

7. In order to point the telescope rightly, the hour-angles and zenith-distances of the stars to be observed must be computed for the time of transit.

When the telescope is on the prime vertical, calling p the hour-angle, and z the zenith-distance of the star, then—

$$\cos p = \tan D \cot L$$

$$\cos z = \frac{\sin D}{\sin L}$$

An allowance must be made for the time of crossing the first wire, and for change of zenith-distance from the first to the middle wire.

8. To correct, for errors of collimation, irregularity in the pivots, &c., the instrument may be reversed between the transits over each vertical; *i. e.*, the wires on one side of the center wire are observed, the instrument reversed in its Y's, and the transit over the same wires continued, but in an inverse order; so that, in each vertical the same wire is at one time as far north as it is at another south of the optical axis.

Then let—

L = the latitude sought;

D = the apparent declination of the star;

t = the hour-angle, illuminated axis *north;*

= $\frac{1}{2}$ diff. of sidereal time of transit over the same wire, for same position of axis; and

t' = hour-angle, illuminated axis *south.*

$$\tan L = \frac{\tan D}{\cos \frac{1}{2}(t' + t) \, . \, \cos \frac{1}{2}(t' - t)}$$

LVII.—*To determine the Latitude of a Place by observing the Difference of the Meridional Zenith-Distances of Two Stars on Opposite Sides of the Zenith, with the* Zenith and Equal-Altitude Telescope.

Compute an approximate latitude by the formula—

$$L = \tfrac{1}{2}\,[180^\circ - (\Delta + \Delta')] + \tfrac{1}{2}\,(z - z')$$

where Δ and Δ' are the polar distances of the south and north stars, respectively, and $(z - z')$ the quantity measured by the micrometer. Then—

1. The *correction for level* is applied by adding the angle which the vertical axis of the instrument makes with the zenith when the inclination is *southward*, or subtracting it when to the northward. This correction is found by multiplying the value of one division of the level-scale, in arc, by one-half the mean change, in level-divisions, which any one end of the bubble undergoes by reversing the instrument on the meridian; or, if o and e, o' and e', denote the readings of the object and eye-ends of the bubble, for south and north stars, respectively; corrections for level $= \tfrac{1}{4}\,(o' - e') - \tfrac{1}{4}\,(o - e) \times$ the value of one division of the level-scale in arc.

2. The correction for *error of meridional position* of the central vertical wire is found by computing the usual "reduction to the meridian" for each star; then the difference between the reductions for the northern and southern stars is taken, and one-half that difference added or subtracted, according as the reduction for the *northern* star is *greater* or less than that for the southern; or,

$$\text{correction for position,} = \frac{m' - m}{2}$$

m being the reduction for stars south, and m' for stars north of the zenith.

LVII.—*Zenith Telescope*—Continued.

3. When the star is observed off the line of collimation, the instrument remaining in the plane of the meridian,

$$m = \frac{2 \sin^2 \frac{1}{2} p}{\sin 1''} \times \frac{1}{2} \sin 2 \mathrm{D}$$

The correction to the latitude is one-half of this quantity, whether the star be north or south; and if the two stars forming a pair are observed off the line of collimation, two such corrections, separately computed, must be added to the latitude. D *is minus when south.*

Values of m are given in the following table:

D.	10ˢ	15ˢ	20ˢ	25ˢ	30ˢ	35ˢ	40ˢ	45ˢ	50ˢ	55ˢ	60ˢ	D.
°	″	″	″	″	″	″	″	″	″	″	″	°
5	.00	.01	.02	.03	.04	.06	.08	.10	.12	.14	.17	85
10	.01	.02	.04	.06	.08	.11	.15	.19	.23	.28	.34	80
15	.01	.03	.05	.09	.12	.17	.22	.28	.34	.41	.49	75
20	.02	.04	.07	.11	.16	.22	.28	.36	.44	.53	.63	70
25	.02	.05	.08	.13	.19	.26	.34	.42	.52	.63	.75	65
30	.02	.05	.09	.15	.21	.29	.38	.48	.59	.71	.85	60
35	.03	.06	.10	.16	.23	.31	.41	.52	.64	.77	.92	55
40	.03	.06	.11	.17	.24	.33	.43	.54	.67	.81	.97	50
45	.03	.06	.11	.17	.25	.33	.44	.55	.68	.82	.98	45

4. The *correction for refraction* is applied similarly to reduction to meridian (2) but with a contrary sign; or,

$$\text{Correction for refraction} = \frac{r - r'}{2}$$

$r - r'$ being small, no note need be taken of the state of the barometer and thermometer at the time of observation.

LVII.—*Zenith Telescope*—Continued.

The following table gives the correction to the latitude for differential refraction; arguments, one-half difference of zenith-distances on one side and zenith-distance on the top:

$\frac{z - z'}{2}$	Zenith-distance.					
	0°	10°	20°	25°	30°	35°
′ ″	″	″	″	″	″	″
0 0	.00	.00	.00	.00	.00	.00
0 30	.01	.01	.01	.01	.01	.01
1	.02	.02	.02	.02	.02	.02
1 30	.02	.03	.03	.03	.03	.03
2	.03	.03	.04	.04	.04	.05
2 30	.04	.04	.05	.05	.05	.06
3	.05	.05	.06	.06	.07	.08
3 30	.06	.06	.07	.07	.08	.09
4	.07	.07	.08	.08	.09	.10
4 30	.08	.08	.09	.09	.10	.11
5	.08	.09	.10	.10	.11	.13
5 30	.09	.10	.10	.11	.12	.14
6	.10	.10	.11	.12	.13	.15
6 30	.11	.11	.12	.13	.14	.16
7	.12	.12	.13	.14	.15	.18
7 30	.13	.13	.14	.15	.16	.19
8	.13	.14	.15	.16	.18	.21
8 30	.14	.15	.16	.17	.19	.22
9	.15	.16	.17	.18	.20	.23
9 30	.16	.17	.18	.20	.21	.24
10	.17	.18	.19	.21	.23	.26
10 30	.18	.19	.20	.22	.24	.27
11	.18	.19	.21	.23	.25	.28
11 30	.19	.20	.22	.24	.26	.30
12	.20	.21	.23	.25	.27	.31

The sign of the correction is the same as that of the micrometer-difference.

LVII.—*Zenith Telescope*—Continued.

5. To find the value a of one division of the micrometer, note the time by chronometer of the transit of Polaris or other close circumpolar star over the movable wire placed vertically and set successively before the star for each turn or half-turn of the screw. Then let x be the angular distance *from the meridian* at any reading of the screw; p, the hour-angle of the star at the same instant; and D, its declination:

$$\sin x = \cos D \sin p$$

The value of x is computed for each reading, and the differences of these values divided by the differences of the corresponding micrometer-readings give values for the screw.

A better method, as it avoids displacing the micrometer, is to observe a close circumpolar star near its elongation, when rapidly rising or falling, with but slight motion in azimuth. The level should be carefully noted in order to allow for possible changes, and a correction applied for differential refraction.

The sidereal time of elongation and the azimuth of the star can be determined from LXI.

About 40 or more minutes before elongation, transits are noted, the micrometer being set in advance consecutively by whole or half turns of the screw throughout its length. A correction for rate of chronometer should be applied if sensible.

Let—

t = the difference between the time of observation and the time of elongation of the star; and

s'' = the number of seconds of arc from elongation in the direction of the vertical.

$$s'' = 15 \cos D \left[t - \tfrac{1}{6} (15 \sin 1'')^2 t^3\right]$$

where t is expressed in seconds of time.

LVII.—*Zenith Telescope*—Continued.

Values of $\frac{1}{6}$ (15 sin 1″)^2 t^3 for minutes of time from elongation are given in the following table:

t.	Term.	*t.*	Term.	*t.*	Term.	*t.*	Term.
m.	*s.*	*m.*	*s.*	*m.*	*s.*	*m.*	*s.*
5	0.0	15	0.6	25	3.0	35	8.2
6	0.0	16	0.8	26	3.3	36	8.9
7	0.0	17	0.9	27	3.7	37	9.6
8	0.1	18	1.1	28	4.2	38	10.4
9	0.1	19	1.3	29	4.6	39	11.3
10	0.2	20	1.5	30	5.1	40	12.2
11	0.2	21	1.8	31	5.7	41	13.1
12	0.3	22	2.0	32	6.2	42	14.1
13	0.4	23	2.3	33	6.8	43	15.1
14	0.5	24	2.6	34	7.5	44	16.2

It is convenient to apply these values to the observed time of noting, additive to the observed time before, and subtractive after, either elongation.

The correction to be applied to the observed times of noting for change of level is given by the formula—

$$\pm \left\{ \tfrac{1}{2}(N - S) - \tfrac{1}{2}(N_0 - S_0) \right\} \frac{b}{15 \cos D}$$

where N_0, S_0, the north and south readings for a selected state of level, N, S, the readings for any other state, and b the value of one division of level-scale in seconds of arc; the upper sign to be used for western, the lower sign for eastern elongation.

After these two corrections have been applied we have in one column the readings of the micrometer, and in another the corresponding times, such as would have been obtained if the star had moved uniformly in a vertical line, leaving out of consideration for the present the change in refraction and the rate of the chronometer.

Various methods of combination might be adopted for the determination of the turn of the screw. That of subtracting the values resulting from the first operation from those of the middle one; next, those of the second from those of the middle one, *plus* one, and so on, is recommended for its simplicity.

LVII.—*Zenith Telescope*—Continued.

A number of values are thus obtained for the time of a given number of turns or half-turns, from which is deduced the value of one turn: thus—

Mean time for one turn,	$116^s.774$	log = 2.06735
	cos D	log = 8.40750
	15	log = 1.17609
One turn	44″.765	= 1.65094
Correction for refraction....	— .025	
Correction for rate.........	— .003	
Resulting value	44 .737	

The correction for refraction, in seconds of arc, is negative for either eastern or western elongation, and equals the change of refraction for the space equal to one turn; equal to the value of one turn times the difference of refraction for 1′ at star's altitude divided by 60.

6. The value *b* of 1 division of the level-scale will be best found by using, in conjunction with the micrometer, a distant point as a mark, or the central wire of another instrument used as a collimator; for the space above or below the mark, passed over by the horizontal wire of the micrometer, during the bubble's run over the scale, as the telescope's elevation is gradually altered, may afterward be measured by the micrometer-screw. The temperature should be noted, since the result may change with a change of temperature.

Including all corrections, the general expression for latitude is—

$$L = \frac{180-(\triangle+\triangle')}{2} + \frac{(z-z')}{2} \times a + \frac{(o'+e)-(o+e')}{4} \times b + \frac{(m'-m)}{2} + \frac{(r-r')}{2}$$

a and *b* being the arc values of one division of the micrometer and level-scale respectively.

To correct as much as possible an erroneous determination of the value of the micrometer-screw, select stars for observation, such, if practicable, that the greatest zenith-distance of a pair will belong as often to the north star as to the south star; because if the zenith-distance of the north star is the greatest, the observed quantity is subtractive; if least, additive. For, as a general rule, the error of latitude arising from an erroneous value to the micrometer-screw will be the least when in a set of stars—

$$\Sigma z - \Sigma z' = 0$$

FORM FOR RECORD AND COMPUTATION.

Northwestern Boundary Survey.

Latitude, Station Kootenay West,—Observations with Zenith-Telescope by ....

Date, 1860.	Star. No. B. A. C.	Star. N. S.	Micrometer. Reading.	Micrometer. Diff. Z.-D.	Level. N.	Level. S.	Level. Diff. N. – S.	Meridian distance.	Declination.	Sum and half-sum.	Corrections. Microm.	Corrections. Level.	Corrections. Ref.	Corrections. Merid.	Latitude.	Remarks.
			T. D.	T. D.				s.	° ′ ″	° ′ ″	′ ″	″	″	″	° ′ ″	
July 9	6289	N.	10 90.2		19.7	11.3		23	58 43 15.69	98 11 25.37						
	6391	S.	21 69.2	– 10 79.0	13.6	18.0	+ 4.0	14	39 28 09.68	49 05 42.68	–5 47.76	+1.13	–0.10	+0.09	48 59 56.04	
14		N.	10 89.3		27.1	26.1			17.29	28.44						
		S.	21 72.2	83.9	27.8	26.2	+ 2.6	10	11.15	44.22	49.02	+0.73	–0.10	+0.01	55.84	
15		N.	11 94.4		27.8	29.9			17.60	29.04						
		S.	22 75.8	81.4	30.3	28.4	– 0.2	9	11.44	44.52	48.54	–0.06	–0.10	+0.01	55.83	
16		N.	11 03.3		39.7	13.6		8	17.92	29.66						
		S.	21 95.2	91.9	19.3	34.9	+10.5	9	11.74	44.83	51.92	+2.97	–0.10	+0.02	55.80	
											Mean...	....	....	....	48 59 55.88	

LVIII.—*Knowing the Time and the Latitude of the Place, to find the Azimuth of the Sun or a Star.*

$$\tan \tfrac{1}{2} (A + S) = \cot \tfrac{1}{2} p \frac{\cos \tfrac{1}{2} (\triangle - \lambda)}{\cos \tfrac{1}{2} (\triangle + \lambda)}$$

$$\tan \tfrac{1}{2} (A - S) = \cot \tfrac{1}{2} p \frac{\sin \tfrac{1}{2} (\triangle - \lambda)}{\sin \tfrac{1}{2} (\triangle + \lambda)}$$

$$A = \tfrac{1}{2} (A + S) \mp \tfrac{1}{2} (A - S)$$

the *upper* or *negative* sign is used when λ is greater than $\triangle$;

where

A = the azimuth counted from the *north*, which must be subtracted from 180° if counted from the south;

S = the angle at the star, called the angle of variation;

λ = the co-latitude of the place;

$\triangle$ = the north-polar distance of the sun or star; and

p = the hour-angle at the pole.

Without the Use of a Chronometer, by observing the Altitude of the Sun or Star at the Same Instant with the Observation of the Azimuth.

Let Z = the zenith-distance, corrected for refraction, parallax, and semi-diameter; then—

$$\cos^2 \tfrac{1}{2} A = \frac{\sin k \,.\, \sin (k - \triangle)}{\sin Z \sin \lambda}$$

$$2\,k = Z + \triangle + \lambda$$

LIX.—*To find the* AMPLITUDE *of a Celestial Object at its Rising or Setting;* by "Amplitude" is meant the complement of the azimuth, or distance from the east or west points of the horizon.

This is a particular case of the preceding problem. When the object appears to be in the horizon, its zenith-distance, instead of being 90°, is, on account of refraction and parallax, $90° + k$, where—

$$k = \text{horizontal refraction} - \text{horizontal parallax}$$

$$= 36'\ 29'' - \text{horizontal parallax.}$$

For stars, the horizontal parallax = 0 and $k = 90° 36'\ 29''$; for the sun, $k = 36'29'' - 8''.8 = 90°\ 36'\ 20''.2$. The mean refraction and mean horizontal parallax are here used, as these observations are not susceptible of much accuracy.

LX.—*To find the True Meridian by the Method of Equal Altitudes of the Sun.*

The instrument remaining stationary, observe the readings of the horizontal limb when the altitude of the sun's center is the same in the forenoon and afternoon.

Then, the correction to the mean of these two readings for the change in the sun's declination in the interval is—

$$c = \frac{\frac{1}{2}(D - D')}{\cos L \,.\, \sin \frac{1}{2}(t - t')}$$

where—

$D - D'$ = the change in the sun's declination in the interval of the observations;

$(t - t')$ = this interval of time expressed in arc; and

L = the latitude of the place.

LXI.—*To find the Azimuth of* POLARIS, *or other Close Circumpolar Star, at its Greatest Eastern or Western Elongation.*

$$\cos p = \tan \Delta \cot \lambda = \cot D \tan L = \tan L \tan \Delta$$

$$\cos L \sin A = \sin \Delta = \cos D$$

where—

p = the hour-angle of the star;
D = its declination;
Δ = its polar distance;
A = the required azimuth;
L = the latitude of the place; and
λ = the co-latitude.

The first equations give the hour-angle of the star at its greatest elongation; hence the sidereal time of elongation;

The second, the azimuth of the star at its greatest elongation.

The azimuth at *any* hour-angle is found by the methods in LVIII, or by the formula—

$$A \text{ (in seconds)} = \frac{\sin p}{\cos L} \left\{ \Delta + \Delta^2 \sin 1'' \cos p \tan L \right.$$

$$\left. + \tfrac{1}{3} \Delta^3 \sin^2 1'' \left[(1 + 4 \tan^2 L) \cos^2 p - \tan^2 L \right] \right\}$$

If the hour-angle is counted from the lower culmination, change the sign of the second term.

The most approved method is to observe a series of azimuths of the star *about* the elongation, say for not more than 30 minutes before and after, and to reduce them to the elongation. To do this, compute, from the known latitude, the azimuth of the star at its greatest elongation = A, and call the sidereal time from elongation t; the correction to the azimuth will be—

$$c = (112.5)\, t^2 \sin 1'' \tan A$$

$$\log (112.5) \sin 1'' = 6.7367274$$

The quantities found in the table for "Reduction to the meridian"—

$$\left(2 \frac{\sin^2 \frac{1}{2} p}{\sin 1''} \right)$$

correspond very nearly to—

$$(112.5)\, t^2 \sin 1''$$

so, by entering the table with the time from elongation, and

LXI.—*To find the Azimuth of Polaris, &c.*—Continued.

multiplying the tabular quantities by tan A, we obtain the required correction in seconds of arc. This will be found a convenient substitute for the more rigorous method.

The formula may be separately applied to each observation, or the work may be shortened by computing only the azimuth corresponding to the mean hour-angle and applying to it a *correction to mean azimuth.*

Let n be the number of observations on the star; A_1, the azimuth corresponding to the mean hour-angle; and let, also, τ = the difference between the time of any observation and the mean of the times; then, for a circumpolar star,

$$\text{Correction to mean azimuth} = A_1 - \tan A_1 \frac{1}{n} \Sigma \frac{2 \sin^2 \frac{1}{2} \tau}{\sin 1''}$$

Example.

Means of the times of observations by chronometer = $3^h\ 48^m\ 12^s.3$.

Chronometer time of observation.	τ	Tab. quantities, (page 240.)	
h. m. s.	*m. s.*	$''$	
3 42 30.5	5 41.8	63.7	log 27″.4 .... = 1.4380
44 08.0	4 04.3	32.5	log tan A..... = 8.5144
45 52.0	2 30.3	10.7	
47 15.0	57.3	1.8	
48 59.0	46.7	1.2	9.9524
50 34.0	2 21.7	11.0	Correction. = − 0″.90
52 28.0	4 15.7	35.7	
3 53 51.5	5 39.2	62.8	
		$\frac{1}{n}\Sigma = 27.4$	

Azimuth of star at elongation.. 1° 52′ 42″.8	log t^2........... 6.0734174
Chronometer time of elongation. $4^h\ 06^m\ 20^s.5$	log (112.5) sin 1″. 6.7367274
Mean time of obs'n by chron.. $3^h\ 48^m\ 12^s.3$	log tan A 8.5162425
t = 1088″.2 = $18^m\ 8^s.2$	Cor. = 21″.2 =.. 1.3263873

Azimuth corresponding to mean hour-angle = 1° 52′ 21″.6
Correction to mean azimuth .... = 0 .9

Mean azimuth .... = 1° 52′ 20″.7

LXI.—*To find the Azimuth of Polaris, &c.*—Continued.

Azimuths are usually reckoned from the south and in the direction of south to west. When circumpolar stars are observed, it is more convenient to reckon from the north meridian. The determination of primary azimuths supposes the local time to be known; for secondary azimuths observations for time and azimuth may be made together. The sun is only employed in connection with the inferior class of azimuths.

For the purpose of referring azimuths observed at night to the direction of any geodetic signal, a mark is set up, consisting of a perforated box, (about ¾ of a foot cube,) through the front face of which the light of a bull's-eye lantern is shown, appearing of about the size and brilliancy of the star observed upon. The distance of this mark from the place of observation is generally determined by local circumstances, but should not, if possible, be nearer than about a mile, in order that the sidereal focus of the telescope may not require changing. For day observations a vertical black stripe, of the same width as the aperture, is painted upon a white wand placed vertically above it. If the diameter of the aperture is a quarter of an inch, it will subtend at the distance of a mile an angle of a little more than 0″.8. The horizontal angle between the mark and any trigonometrical station is measured in connection with the triangulation by combining it with all other directions radiating from the station observed from.

Observations for azimuth are usually made in sets, commencing with a number on the mark, followed by about an equal number of readings on the star preceded and followed by level-readings. The instrument is then reversed, and the preceding operations are repeated in the reverse order, the number of readings on the star and mark being as before.

In these observations the optical axis of the telescope of the theodolite must be made to describe a truly vertical plane.

If the axis of the telescope is not horizontal, the correction to the azimuth will be—

$$\pm \frac{d}{4}\left\{ (w + w') - (e - e') \right\} \tan L$$

where $d =$ the value of one division of the level-scale in seconds

LXI.—*To find the Azimuth of Polaris, &c.*—Continued.

of arc, w, e, and w', e', the west and east readings of the level before and after reversal.

The circumpolar stars α, δ, λ Ursæ Minoris and 51 Cephei are those almost exclusively used. When δ Ursæ Minoris and 51 Cephei culminate on either side of the pole, Polaris is not far from its elongation; and, on the contrary, when the pole-star culminates, the other two are not far from their elongations on either side of the meridian. λ Ursæ Minoris, from its greater proximity to the pole, and its small size, presents to the larger instruments a finer and steadier object than Polaris.

LXII.—*Correction for* RUN *in reading Microscopes.*

As it is difficult to adjust the microscopes so that five revolutions of the micrometer-screw shall carry the wire exactly over one of the five-minute spaces on the limb of the instrument, (if it be so graduated,) it is preferred to observe the number of revolutions and the part of a revolution made by the screw while the wire passes over the space; then, let—

m = the mean of *first readings*; that is, the readings obtained by turning the screw in the direction of increasing numbers from the zero of the comb; and

m' = the mean of second, or reverse, readings;

then—

$$\text{(mean) run} = r = m - m' + 300$$

and—

$$\text{true (mean) reading} = \frac{300 \cdot m}{r} = \frac{300\,(r + m' - 300)}{r}$$

= the number of minutes and seconds to be added to the degrees and minutes of the limb.

LXIII.—*Longitude by Lunar Culminations.*

1. Make—

l = true longitude sought;

l' = approximate longitude;

m = observed change, in right ascension, of the moon's bright limb between the first meridian and that sought;

m' = computed change in same, by interpolation;

V = rate of motion, in right ascension, of the moon's bright limb, when on the meridian l'; and

I = the constant difference between the values of the independent variable, or arguments, corresponding to the consecutive tabulated values of the right ascension of the moon;

then—

$$l = l' + \mathrm{I} \, . \, \frac{m - m'}{\mathrm{V}} + \quad . \quad . \quad . \quad . \quad (1)$$

2. *Interpolation.*—Take the following scheme, viz:

I	F	Δ_1	Δ_2	Δ_3	Δ_4	Δ_5
t'''	a'''					
		b''				
t''	a''		c''			
		b'		d'		
t'	a'		c'		e'	
		b		d		f
$t_{\prime}$	$a_{\prime}$		$c_{\prime}$		$e_{\prime}$	
		$b_{\prime}$		$d_{\prime}$		
$t_{\prime\prime}$	$a_{\prime\prime}$		$c_{\prime\prime}$			
		$b_{\prime\prime}$				
$t_{\prime\prime\prime}$	$a_{\prime\prime\prime}$					

In which the column I contains the independent variable, or argument, as time, terrestial longitude, degrees, and the like; F, the value of the function of this variable, as found in any set of tables; Δ_1, Δ_2, &c., the first, second, &c., order of differences of these functions.

LXIII.—*Longitude by Lunar Culminations*—Continued.

Make—

$s =$ the value of the function corresponding to any value t_s between t' and $t_,$;

$$\left.\begin{aligned} t &= \frac{t_s - t'}{t_, - t'} \\ a &= \frac{a' + a_,}{2} \\ c &= \frac{c' + c_,}{2} \\ e &= \frac{e' + e_,}{2} \end{aligned}\right\} \quad \ldots \ldots \ldots \quad (2)$$

Then, according to Bessel, Ast. Nach. No. 30—

$$s = a + \frac{t - \frac{1}{2}}{1} \cdot b + \frac{t\,(t-1)}{1 \cdot 2} \cdot c + \frac{t\,(t-1)\,(t-\frac{1}{2})}{1 \cdot 2 \cdot 3} \cdot d$$

$$+ \frac{(t+1) \cdot t \cdot (t-1)\,(t-2)}{1 \cdot 2 \cdot 3 \cdot 4} \cdot e + \&c.$$

Or, making—

$$\left.\begin{aligned} \Delta_1 &= b \\ \Delta_2 &= \frac{c' + c_,}{2} \\ \Delta_3 &= d \\ \Delta_4 &= \frac{e' + e_,}{2} \end{aligned}\right\} \quad \ldots \ldots \ldots \quad (3)$$

$$s = a' + t\,\Delta_1 + \frac{t\,(t-1)}{1 \cdot 2} \cdot \Delta_2 + \frac{t\,(t-1)\,(t-\frac{1}{2})}{1 \cdot 2 \cdot 3} \cdot \Delta_3$$

$$+ \frac{(t+1) \cdot t \cdot (t-1)\,(t-2)}{1 \cdot 2 \cdot 3 \cdot 4} \cdot \Delta_4 + \&c.$$

or, by the ascending powers of t,

$$s = a' + At + Bt^2 + Ct^3 + Dt^4 + \&c. \quad \ldots \quad (4)$$

LXIII.—*Longitude by Lunar Culminations*—Continued.

in which, stopping at the fourth differences,

$$\left.\begin{aligned} A &= \Delta_1 - \tfrac{1}{2}\Delta_2 + \tfrac{1}{12}\Delta_3 + \tfrac{1}{12}\Delta_4 \\ B &= \tfrac{1}{2}\Delta_2 - \tfrac{1}{4}\Delta_3 - \tfrac{1}{24}\Delta_4 \\ C &= \tfrac{1}{6}\Delta_3 - \tfrac{1}{12}\Delta_4 \\ D &= \tfrac{1}{24}\Delta_4 \end{aligned}\right\} \quad . \quad . \quad . \quad . \quad (5)$$

Also,

$$V = \frac{ds}{dt} = A + 2\,B\,t + 3\,C\,t^2 + 4\,D\,t^3 + \;.\;\;. \quad (6)$$

$$m' = s - a' = A\,t + B\,t^2 + C\,t^3 + D\,t^4 + \;.\;\;. \,(7)$$

The value of *m* may be obtained either from observations on the two meridians, or by observations on one and the tabulated results under the head of Moon Culminations in the Nautical Almanac, which may be used as actual observations on the meridian of the ephemeris.

NOTE.—If the lunar tables were perfectly accurate, the true longitude given by the observation would be found at once by comparing the observed right ascension with that of the ephemeris. There are two methods of avoiding or eliminating the errors of the ephemeris. In the first, the observation is compared with a corresponding one on the same day at the first meridian, or at some meridian the longitude of which is well established. In this method the increase of the right ascension in passing from one meridian to the other is directly observed, and the error of the ephemeris on the day of observation is consequently avoided; but observations at the unknown meridian are frequently rendered useless by a failure to obtain the corresponding observations at the first meridian.

In the second method, the ephemeris is first corrected by means of all the observations taken at the fixed observatories during the semi-lunation within which the observation for longitude falls. The corrected ephemeris then takes the place of the corresponding observation, and is even better than the single corresponding observation, since it has been corrected by means of *all* the observations at the fixed observatories during the semi-lunation.—*Chauvenet's Practical Astronomy.*

LXIII.—*Longitude by Lunar Culminations*—Continued.

3. *Example.*—Let—

$$l' = 4^h\ 55^m\ 50^s \text{ west from Greenwich,}$$

and suppose the following transits with a chronometer marking sidereal time. The *error* of the time-keeper is not material, and the transit is very nearly in the meridian, viz:

Feb. 18. ζ Geminorum	$6^h\ 54^m\ 41^s.75$		
δ Geminorum	7 10 38.97		
☽'s first limb ..	— — — —	$7^h\ 38^m\ 06^s.76$	
ζ Cancri.....	8 03 06.11		
	3)22 08 26.83	7 22 48.943	
		0 15 17.817	
Chronometer rate + 3^s daily........		−0.0318	$0^h\ 15^m\ 17^s.785$

The corresponding observations at Greenwich, as given by the Nautical Almanac, are:

Feb. 18. ζ Geminorum	$6^h\ 54^m\ 57^s.41$		
δ Geminorum	7 10 54.36		
☽'s first limb ..	— — — —	$7^h\ 27^m\ 47^s.66$	
ζ Cancri.....	8 03 21.44		
	3)22 09 13.21	7 23 04.403	$0^h\ 04^m\ 43^s.257$
		$m = 634^s.528 =$	0 10 34.528

Next compute this increase from Nautical Almanac. The right ascensions of the moon are given in that work for the upper and lower passages over the meridian of Greenwich. The independent variable is, therefore, terrestrial longitude, of which the unit is one hour, and the intervals between the consecutive tabulated values of its function, 12. The increase to be computed is for the interval of passage from the upper meridian of Greenwich to that $4^h\ 55^m\ 50^s$ west. Now, according to the scheme and equations (2), (3), and (5),

LXIII.—*Longitude by Lunar Culminations*—Continued.

Day.	Culm.	AR. ☽'s first limb, Nautical Almanac.	Δ1	Δ2	Δ3	Δ4	
17	U	$6^h\ 35^m\ 55^s.73$					$t = \frac{h_s - t'}{t_, - t'} = \frac{4^h\ 55^m\ 50^s}{12^h}$
	L	7 01 56.57	$+26^m\ 00^s.54$				
18	U	7 27 47.66	+25 51.39	$9^s.15$			
	L	7 53 28.84	$b = +25$ 41.18	$c' = -$ 10.21	$-1^s.06$	$e' = +0^s.81$	
19	U	8 18 59.56	+25 30.72	$c_, = -$ 10.46	$d = -0.25$	$e_, = +0.84$	$\log t = 9.6137147$
	L	8 44 20.41	+25 20.85	− 9.87	+ 0.59		
Sum of differences			$2^h\ 08^m\ 24^s.68$	$-\ 0^m\ 39^s.69$	$-\ 0^s.72$	$+\ 1^s.65$	$\Delta_1 = 25^m\ 41^s.180$
Top of left column....................			6 35 55.73	26 00.54	− 9.15	− 1.06	$\Delta_2 = -$ 10.335
Check-sum = bottom of left column...			8 44 20.41	25 20.85	− 9.87	+ 0.59	$\Delta_3 = -$ 0.250
							$\Delta_4 = +$ 0.825

$$A = 25^m\ 41^s.180 + 5^s.167 - 0^s.021 + 0^s.068 = +\ 1546^s.394 \qquad \log A = 3.1893180$$

$$B = \quad -\ 5.167 + 0.062 - 0.034 = -\ 5.139 \qquad \log B = \bar{0}.7108786$$

$$C = \quad -\ 0.041 - 0.068 = -\ 0.109 \qquad \log C = \bar{9}.0374265$$

$$D = \quad +\ 0.034 = +\ 0.034 \qquad \log D = 8.5910646$$

LXIII.—*Longitude by Lunar Culminations*—Continued.

Then equation 7—

A log	3.1893180		
t log	9.6137147		
	2.8030327	Nos..	+ $635^{s}.337$
B log	$\bar{0}$.7108786		
t^2 log	9.2274294		
	9.9383080	Nos..	— 0 .867
C log	$\bar{9}$.0374265		
t^3 log	8.8411441		
	$\bar{7}$.8785706	Nos..	— 0 .007
D log	8.5910646		
t^4 log	8.4548588		
	7.0459234	Nos..	+ 0 .001
m' =	....		634 .464

And equation 6—

A	....		$1546^{s}.394$
B log	$\bar{0}$.7108786		
t log	9.6137147		
2 log	0.3010300		
	$\bar{0}$.6256233	Nos..	— 4 .222
C log	$\bar{9}$.0374265		
t^2 log	9.2274294		
3 log	0.4771213		
	$\bar{8}$.7419772	Nos..	— 0 .055
D log	8.5910646		
t^3 log	8.8411441		
4 log	0.6020600		
	8.0342687	Nos..	+ 0 .011
V =	....		1542 .128

LXIII.—*Longitude by Lunar Culminations*—Continued.

And equation (1)—

$I = 12^h$	....	log	4.6354837
$m - m' = 0^s.064$	....	log	8.8061800
$V = 1542^s.128$	....	log ac	6.8118797
$1^s.792$	....		0.2535434

Whence—

$$l = 4^h\ 55^m\ 50^s + 1^s.792 = 4^h\ 55^m\ 51^s.792$$

If m be the *observed* increase of right ascension between any meridian not the first, (but of which the longitude is well known,) and the meridian sought, interpolate the increase $m_{,}$ for the known meridian as well as m' for that sought. Then, for $m - m'$, in equation (1), substitute $m - (m' - m_{,})$, and the result will be the corrected longitude from the first meridian, as before.

It often happens that two observers do not use the same number of wires, or do not observe the same number of stars at the two places. In such cases the observed increase of the right ascension of the moon's limb requires a correction, which Mr. Walker deduces as follows, from Gauss's method:

For the eastern observatory and western station respectively, let—

A' and A = the observed AR. of a star;

$E = A' - A$ for the same star;

E' = a similar value for another star;

l and l' = the number of wires on which each limb was observed;

a and a' = similar values for a star;

$\lambda = \dfrac{l\,l'}{l + l'}$ for the moon's limb;

$u = \dfrac{a\,a'}{a + a'}$ for one star;

u' = a similar value for another star;

Σ = symbol to denote the aggregate of similar quantities; and

ε = the correction required;

LXIII.—*Longitude by Lunar Culminations*—Continued.

then—

$$\varepsilon = \frac{\Sigma\left(E \cdot \frac{\lambda u}{\lambda + u}\right)}{\Sigma \frac{\lambda u}{\lambda + u}}$$

$$L = l + I \cdot \frac{m - m' + \varepsilon}{V}$$

Also, calling W the *weight* of each day's comparison,

$$W = \frac{\sigma \lambda}{(\sigma + \lambda) z^2}.$$

in which z is the same as as $\frac{I}{V}$ and $\sigma = u + u' + u'' + \&c.$

For the weight of the result of all the comparisons, we have—

$$\frac{\sigma \lambda}{(\sigma + \lambda) z^2}$$

Let e denote the probable error of observation, and E the probable error of the final result; then,

$$E = \frac{e}{\sqrt{\Sigma \frac{\sigma \lambda}{(\sigma + \lambda) z^2}}}$$

It also frequently happens that the moon cannot be observed on the middle wire, in which case she is far enough from the meridian to have a sensible parallax in right ascension; and, as it may be very desirable not to lose the observation, this parallax must be computed and applied to the hour-angle from the middle wire, which is supposed to be nearly coincident with the meridian.

Denoting this parallax in right ascension by p, the horizontal parallax by w, the latitude of the place of observation by φ, and the true declination of the moon by δ, we have from the ordinary series for the parallax in right ascension, neglecting the terms after the first, which would in this case be insignificant,

$$p = \theta \sin w \cos \varphi \sec \delta$$

in which θ is the hour-angle, or equatorial interval in sidereal

LXIII.—*Longitude by Lunar Culminations*—Continued.

time from the lateral wire on which the moon is observed to the central wire; so that, at the instant of observation, the actual distance of the moon's limb from the central wire is—

$$\theta - \theta \sin w \cos \varphi \sec \delta$$

and the reduction to meridian or middle wire will be—

$$\pm \frac{\theta}{\cos \delta} \cdot \frac{1 - \sin w \cos \varphi \sec \delta}{1 - 0.00277m}$$

in which m is the motion of the moon in right ascension in one day, expressed in degrees. The upper sign is to be used when the observation is on a wire *before* and the lower *after* the middle wire.

In what precedes the approximate longitude l' is supposed to be known. When this is not the case, it may be found from—

$$l' = 12^{h} \frac{m}{\Delta'}$$

and the interpolation is then to be made for this value of l' to obtain the value of m'.

LXIV.—*Longitude by the Electric Telegraph.*

[From Chauvenet's Practical Astronomy.]

It is evident that the clocks at two stations, A and B, may be compared by means of signals communicated through an electro-telegraphic wire which connects the stations. Suppose at a time T by the clock at A, a signal is made which is perceived at B at the time T′ by the clock at that station. Let ΔT and $\Delta T'$ be the clock-corrections on the times at these stations respectively, (both being solar or both sidereal.) Let x be the time required by the electric current to pass over the wire, then, A being the more easterly station, we have the difference of longitude λ by the formula—

$$\lambda = (T + \Delta T) - (T' + \Delta T') + x = \lambda_1 + x$$

Since x is unknown we must endeavor to eliminate it. For this purpose let a signal be made at B at the clock-time T″, which is perceived at A at the clock-time T‴, then we have—

$$\lambda = (T''' + \Delta T''') - (T'' + \Delta T'') - x = \lambda_2 - x$$

In these formulæ λ_1 and λ_2 denote the approximate values of the difference of longitude, found by signals east-west and west-east, respectively, when the transmission-time x is disregarded, and the true value is—

$$\lambda = \tfrac{1}{2}(\lambda_1 + \lambda_2)$$

Such is the simple and obvious application of the telegraph to the determination of longitudes; but the degree of accuracy of the result depends greatly—more than at first appears—upon the manner in which the signals are communicated and received.

Suppose the observer at A taps upon a signal-key at an exact second by his clock, thereby producing an audible click of the armature of the electro-magnet at B. The observer at B may not only determine the nearest second by his clock when he hears this click, but may also estimate the fraction of a second; and it would seem that we ought in this way to be able to determine a longitude within one-tenth of a second. But before even this

LXIV.—*Longitude by the Electric Telegraph*—Continued.

degree of accuracy can be secured, we have yet to eliminate, or reduce to a minimum, the following sources of error:

1. The personal error of the observer who gives the signal;
2. The personal error of the observer who receives the signal and estimates the fraction of a second by the ear;
3. The small fraction of time required to complete the galvanic circuit after the finger touches the signal-key;
4. The *armature time*, or the time required by the armature at the station where the signal is received, to move through the space in which it plays and to give the audible click;
5. The errors of the supposed clock-corrections, which involve errors of observation and errors in the right ascensions of the stars employed.

For the means of contending successfully with these sources of error we are indebted to our Coast Survey, which, under the superintendence of Professor Bache, not only called into existence the chronographic instruments, but has given us the most efficient method of using them. The "Method of Star-Signals," as it is called, was originally suggested by the distinguished astronomer, Mr. S. C. Walker, but its full development in the form now employed by the Coast Survey is due to Dr. B. A. Gould.

Method of Star-Signals.

The difference of longitude between the two stations is merely the time required by a star to pass from one meridian to the other, and this interval may be measured by means of a single clock placed at either station, but in the main galvanic circuit extending from one station to the other. Two chronographs, one at each station, are also in the circuit, and, when the wires are suitably connected, the clock-seconds are recorded upon both. A good transit-instrument is carefully mounted at each station.

When the star enters the field of the transit-instrument at A, (the eastern station,) the observer, by a preconcerted signal with his signal-key, gives notice to the assistants at both A and B, who at once set the chronographs in motion, and the clock then records its seconds upon both. The instants of the star's transits

LXIV.—*Longitude by the Electric Telegraph*—Continued.

over the several threads of the reticule are also recorded upon both chronographs by the taps of the observer upon his signal-key. When the star has passed all the threads the observer indicates it by another preconcerted signal, the chronographs are stopped, and the record is suitably marked with date, name of the star, and place of observation, to be subsequently identified and read off accurately by a scale. When the star arrives at the meridian of B, the transit is recorded in the same manner upon both chronographs.

Suitable observations having been made by each observer to determine the errors of his transit-instrument and the rate of the clock, let us put—

T_1 = the mean of the clock-times of the eastern transit of the star over all the threads, as read from the chronograph at A;

T_2 = the same, as read from the chronograph at B;

T_1' = the mean of the clock-times of the western transit of the star over all the threads, as read from the chronograph at A;

T_2' = the same, as read from the chronograph at B;

e, e' = the personal equations of the observers at A and B, respectively;

τ, τ' = the corrections of T_1 and T_1' (or of T_2 and T_2') for the state of the transit-instruments at A and B, or the respective "reductions to the meridian;"

δT = the correction for clock-rate in the interval $T_1' - T_1$;

x = the transmission-time of the electric current between A and B; and

λ = the difference of longitude;

then it is easily seen that we have, from the chronographic records at A,

$$\lambda = T_1' + \delta T + \tau' + e' - x - (T_1 + \tau + e)$$

and from the chronographic records at B,

$$\lambda = T_2' + \delta T + \tau' + e' + x - (T_2 + \tau + e)$$

LXIV.—*Longitude by the Electric Telegraph*—Continued.

and the mean of these values is

$$\lambda = [\tfrac{1}{2}(T_1' + T_2') + \tau'] - [\tfrac{1}{2}(T_1 + T_2) + \tau] + \delta T + e' - e$$

which we may briefly express thus:

$$\lambda = \lambda_1 + e' - e$$

in which—

λ_1 = the approximate difference of longitude found by the exchange of star-signals, when the personal equations of the observers are neglected.

This equation would be final if $e' - e$, or the relative personal equation of the observers, were known; however, if the observers now exchange stations and repeat the above process, we shall have, provided the relative personal equation is constant,

$$\lambda = \lambda_2 + e - e'$$

in which λ_2 is the approximate difference of longitude found as before; and hence the final value is—

$$\lambda = \tfrac{1}{2}(\lambda_1 + \lambda_2)$$

I have not here introduced any consideration of the armature-time, because it affects clock-signals and star-signals in the same manner; and therefore the time read from the chronographic fillet or sheet is the same as if the armature acted instantaneously. It is necessary, however, that this time should be constant from the first observation at the first station to the last observation at the second, and therefore it is important that no changes should be made in the adjustment of the apparatus during the interval.

As the observer has only to tap the transit of the star over the threads, the latter may be placed very close together. The reticules prepared by Mr. W. Würdemann for the Coast Survey have generally contained twenty-five threads, in groups or "tallies" of five, the equatorial intervals between the threads of a group being $2^s.5$, and those between the groups, 5^s; with an additional thread on each side at the distance of 10^s, for use in observations by "eye and ear." Except when clouds intervene and render it necessary to take whatever threads may be available, only the

LXIV.—*Longitude by the Electric Telegraph*—Continued.

three middle tallies, or fifteen threads, are used. The use of more has been found to add less to the accuracy of a determination than is lost in consequence of the greater fatigue from concentrating the attention for nearly twice as long.

A large number of stars may thus be observed on the same night; and it will be well to record half of them by the clock at one station and the other half by the clock at the other station, upon the general principle of varying the circumstances under which several determinations are made, whenever practicable, without a sacrifice of the integrity of the method. For this reason, also, the transit-instrument should be reversed during a night's work at least once, an equal number of stars being observed in each position, whereby the results will be freed from any undetermined errors of collimation and inequality of pivots. Before and after the exchange of the star-signals, each observer should take at least two circumpolar stars to determine the instrumental constants, upon which τ and τ' depend. This part of the work must be carried out with the greatest precision, employing only standard stars, as the errors of τ and τ' come directly into the difference of longitude. The right ascensions of the "signal-stars" do not enter into the computation, and the result is, therefore, wholly free from any error in their tabular places; hence, any of the stars of the larger catalogues may be used as signal-stars, and it will always be possible to select a sufficient number which culminate at moderate zenith-distances at both stations, (unless the difference of latitude is unusually great,) so that instrumental errors will have the minimum effect.

A single night's work, however, is not to be regarded as conclusive, although a large number of stars may have been observed and the results appear very accordant; for experience shows that there are always errors which are constant, or nearly so, for the same night, and which do not appear to be represented in the corrections computed and applied. Their existence is proved when the mean results of different nights are compared. Moreover, it is necessary to interchange the observers in order to eliminate their personal equations. The rule of the Coast Survey has been that when fifty stars have been exchanged on not less than three nights, the observers exchange stations, and fifty stars

LXIV.—*Longitude by the Electric Telegraph*—Continued.

are again exchanged on not less than three nights. The observers should also meet and determine their relative personal equation, if possible, before and after each series, as it may prove that this equation is not absolutely constant.

Before entering upon a series of star-signals, each observer will be provided with a list of the stars to be employed. The preparation of this list requires a knowledge of the approximate difference of longitude, in order that the stars may be so selected that transits at the two stations may not occur simultaneously.

Example.—For the purpose of finding the difference of longitude between the Seaton station of the United States Coast Survey and Raleigh, a list of stars was prepared, from which I extract the following for illustration. The latitudes are:

Seaton station, (Washington) .... $\varphi = +38° 53'.4$
Raleigh station, (North Carolina) .... $\varphi = +35\ 47\ .0$
and Raleigh is assumed to be west from Washington $6^m 30^s$.

Star.	Mag.	α			δ		Seaton sidereal time of Raleigh transit.		
		h.	*m.*	*s.*	°	′	*h.*	*m.*	*s.*
No. 5036, B. A. C.	3	15	09	36	+ 33	52	15	16	06
5084	4.3		18	58	37	54		25	28
5131	4½		27	02	31	51		33	32
5192	5		36	35	26	46		43	05
5259	5		45	43	36	07		52	13
5322	4½		55	59	23	12	16	02	29
5388	5	16	04	09	45	19		10	39
5463	3.4		15	21	46	40		21	51

The following table contains the observations made on one of these stars at the above-named stations by the United States Coast-Survey telegraphic party in 1853—April 28—under the direction of Dr. B. A. Gould.

In this table "Lamp W" expresses the position of the rotation-axes of the transit-instruments. The first column contains the symbols by which the fifteen threads of the three middle tallies

LXIV.—*Longitude by the Electric Telegraph*—Continued.

were denoted; the second column, the times of transit of the star over each thread at Seaton, as read from the chronographs at Seaton; the third column, the times of these transits as read from the chronographs at Raleigh; the fourth column, the mean of the second and third columns; the fifth column, the reduction of each thread to the mean of all, computed from the known equatorial intervals of the threads; the sixth column, the time of the star's transit over the mean of the threads, being the algebraic sum of the numbers in the fourth and fifth columns; and the remaining columns, the Raleigh observations similarly recorded and reduced.

Seaton–Raleigh, 1853, *April* 28.—*Star No.* 5259, *B. A. C.*

Thread.	Seaton observations, lamp W.					Raleigh observations, lamp W.				
	T_1	T_2	Mean	Red.	$\frac{T_1+T_2}{2}$	T_1'	T_2'	Mean	Red.	$\frac{T_1'+T_2'}{2}$
	h. m. s.	*s.*	*s.*	*s.*	*s.*	*h. m. s.*	*s.*	*s.*	*s.*	*s.*
C_1	37.97	38.00	37.98	+25.49	3.47		11.00	11.00	+25.45	36.45
C_2	41.37	41.34	41.36	22.21	3.57	14.58	14.50	14.54	22.25	36.79
C_3	44.03	44.21	44.12	19.06	3.18	17.60	17.55	17.58	19.05	36.63
C_4	47.81	47.74	47.78	15.71	3.49	20.88	20.79	20.84	15.85	36.69
C_5	50.76	50.70	50.73	12.71	3.44	23.90	23.87	23.89	12.70	36.59
D_1	56.96	57.10	57.03	6.21	3.24	30.19	30.05	30.12	6.32	36.44
D_2	0.06	0.04	0.05	3.25	3.30	33.34	33.25	33.30	3.18	36.48
D_3	15 46 3.40	3.38	3.39	+ 0.05	3.44	15 52 36.40	36.30	36.35	+ 0.07	36.42
D_4	6.70	6.70	6.70	− 3.03	[3.67]	39.61	39.53	39.57	− 3.16	36.41
D_5	9.58	9.58	9.58	6.28	3.30	43.00	43.00	43.00	6.36	36.64
E_1	16.03	15.93	15.98	12.54	3.44	49.04	48.81	48.92	12.75	[36.17]
E_2	19.26	19.30	19.28	15.83	3.45	52.30	52.33	52.32	15.90	36.42
E_3	22.47	22.45	22.46	18.99	3.47	55.50	55.41	55.46	19.10	36.36
E_4	25.60	25.60	25.60	22.23	3.38	58.73	58.60	58.67	22.20	36.47
E_5	28.60	28.70	28.65	25.33	3.32	2.08	2.08	2.08	25.38	36.70
				Mean .	3.392				Mean .	36.535

The numbers in the last column for each station would be equal if the observations and chronographic apparatus were perfect; and by carrying them out thus individually we can estimate their accuracy. The numbers 3.67 at Seaton and 36.17 at Raleigh are rejected by the application of Peirce's Criterion, (Method of Least Squares,) and the given means are found from the remaining numbers.

LXIV.—*Longitude by the Electric Telegraph*—Continued.

The corrections of the transit-instruments for this star ($\delta = + 36° 6'.9$) were—

for the Seaton instrument.. $\tau = - 0.028$
for the Raleigh instrument $\tau' = - 0.193$

The rate of the clock was insensible in the brief interval $T_1' - T$. Hence, neglecting the personal equations of the observers, the difference of longitude is found as follows:

$$\tfrac{1}{2}(T_1' + T_2') + \tau' = 15^h\ 52^m\ 36^s.342$$
$$\tfrac{1}{2}(T_1 + T_2) + \tau = 15\ \ 46\ \ 3.364$$
$$\lambda_1 = \quad 6\ \ 32.978$$

In this manner seven other stars were observed on the same night, and the results were as follows:

Star.	λ_1	Diff. from mean, $= v$.
	m. s.	*s.*
5036, B. A. C.	6 33.03	+ 0.04
5084, B. A. C.	33.09	+ 0.10
5131, B. A. C.	32.91	— 0.08
5192, B. A. C.	33.00	+ 0.01
5259, B. A. C.	32.98	— 0.01
5322, B. A. C.	33.00	+ 0.01
5388, B. A. C.	33.02	+ 0.03
5463, B. A. C.	32.91	— 0.08
	Mean, $\lambda_1 = 6$ 32.99	

From the residuals, v, we deduce the mean error of a single determination by one star,

$$\varepsilon = \sqrt{\left(\frac{vv}{m - 1}\right)} = \sqrt{\left(\frac{.0256}{7}\right)} = \pm 0^s.06$$

and hence the mean error of the value $6^m\ 32^s.99$ is—

$$\varepsilon_0 = \pm \frac{0^s.06}{\sqrt{8}} = \pm 0^s.02$$

But this error will be somewhat increased by those errors of the instruments which are constant for the night, and not represented in τ and τ', and by the errors of the personal equations yet to be applied. Moreover, a greater number of determinations should be compared in order to arrive at a just evaluation of the mean error.

LXV.—*Formulæ for Probable Error and Precision.*

[Contributed by Lieutenant Mercur, Corps of Engineers.]

1. Let—

$m =$ the number of observations;

n, n', &c. $=$ results found by observation;

$x =$ their arithmetical mean;

v, v', &c. $= (x-n), (x-n')$, &c. $=$ the residual errors of observation;

$\varepsilon =$ the mean error of n, n', &c.;

$\eta =$ the mean of errors, (arithmetical;)

$E_0 =$ the mean error of final result;

$E_1 =$ the mean error of the observation assumed as the standard of excellence;

$r =$ the probable error of a single observation, n, n', &c.;

$R =$ the probable error of the final result, x;

$h =$ the measure of exactness of a single observation, n, n', &c.;

$H =$ the measure of exactness of the final result, x;

p, p', &c. $=$ the weights of different observations or sets of observations;

$\varepsilon', \varepsilon''$, &c. $=$ the mean errors of observations corresponding to p, p', &c.;

$\Sigma =$ symbol representing the sum;

$$\varepsilon = \sqrt{\frac{\Sigma v^2}{(m-1)}}; \qquad \eta = \frac{\Sigma v}{m}; \qquad E_0 = \sqrt{\frac{\Sigma v^2}{m(m-1)}}$$

$$r = 0.6745\,\varepsilon = 0.8453\,\eta = 0.8453\,\frac{\Sigma v}{\sqrt{m(m-1)}}$$

$$R = 0.6745\,E_0 = 0.6745\sqrt{\frac{\Sigma v^2}{m(m-1)}} = \sqrt{\frac{0.45495\,\Sigma v^2}{m(m-1)}} = \frac{r}{\sqrt{m}}$$

$$h = \frac{0.46936}{r}; \qquad H = h\sqrt{m}$$

2. By "the weight of any determination" is meant its relative approximation to the true value.

It may be measured by the number of observations (each of which may be considered as good as the other, and of which one is assumed to represent the unit of excellence) necessary to give a result equally near the true value.

LXV.—*Formulæ for Probable Error, &c.*—Continued.

Then, since the weights of observations are inversely as the squares of their mean or probable errors,

$$p' = \frac{\varepsilon_{,}^{2}}{\varepsilon'^{2}}; \quad \frac{p'}{p''} = \frac{\varepsilon''^{2}}{\varepsilon'^{2}}$$

and when we arbitrarily assign weights to each observation or set of observations,

$$\varepsilon' = \sqrt{\frac{\Sigma p\, v^2}{(m - 1)}}$$

in which m = the number of observations or sets of observations whose weights are p, p', &c.

When observations are combined by weights, the probable error is given by the formula—

$$R = 0.6745 \sqrt{\frac{\Sigma p\, v^2}{(m - 1)\, \Sigma p}}$$

3. PEIRCE's Criterion for the rejection of doubtful observations.

To apply this to any set of observations involving but one unknown quantity: *

Let—

m = the number of observations taken;

n = the number of doubtful observations to be rejected, (to be found by trial;)

ε = the mean error of one observation in the set of m observations;

v, v', &c. = the residual errors of the observations; and

$\varkappa$ = the ratio of the required limit of error for the rejection of n observations to the mean error ε, so that $\varkappa\varepsilon$ is the limiting error.

Find from the table the value of $\varkappa^2$ for $n = 1$, $n = 2$, $n = 3$, &c., in succession, and reject all observations in which $\varkappa\varepsilon > v$; stopping, however, when the value of $\varkappa\varepsilon$ found for any particular value of n does not reject any observations (not already rejected) for a value of n numerically one less.

* For rules for determining mean and probable errors, and for applying the Criterion to cases involving more than one unknown quantity, see Chauvenet's Manual of Spherical and Practical Astronomy, vol. II, pages 469 to 566.

LXV.—*Formulæ for Probable Error, &c.*—Continued.

Example.—To determine the value of one turn of the micrometer of a zenith-telescope, the following "reduced intervals" were obtained, each corresponding to the ten turns of the micrometer:

Between observations.	Reduced intervals.	v	v^2
	m. s.		
1 and 10	20 35.1	11.0	121.00
2 and 12	43.9	2.2	4.84
3 and 13	41.2	4.9	24.01
4 and 14	51.2	5.1	26.01
5 and 15	46.6	0.5	00.25
6 and 16	44.0	2.1	4.41
7 and 17	38.2	7.9	62.41
8 and 18	54.2	8.1	65.61
9 and 19	47.9	1.8	3.24
10 and 20	43.9	2.2	4.84
11 and 21	21 01.4	15.3	234.09
Sum =	228 27.6	$\Sigma v^2 =$	550.71
x =	20 46.1	$\epsilon^2 =$	55.071

$m = 11$
when $n = 1$ $\kappa^2 = 3.707$
$\kappa^2 \epsilon^2 = 204.148$
$\kappa \epsilon = 14.28$
which rejects (11 and 21)

$m = 11$
when $n = 2$ $\kappa^2 = 2.621$
$\kappa^2 \epsilon^2 = 144.341$
$\kappa \epsilon = 12.01$
which rejects none other.

After rejecting the interval between (11 and 21) the probable error is found as follows:

Between observations.	Reduced intervals.	v	v^2
	m. s.		
1 and 11	20 35.1	9.5	90.25
2 and 12	43.9	0.7	.49
3 and 13	41.2	3.4	11.56
4 and 14	51.2	6.6	43.56
5 and 15	46.6	2.0	4.00
6 and 16	44.0	0.6	.36
7 and 17	38.2	6.4	40.96
8 and 18	54.2	9.6	92.16
9 and 19	47.9	3.3	10.89
10 and 20	43.9	0.7	.49
Sum =	207 26.2	$\Sigma v^2 =$	294.72
x =	20 44.62	$\epsilon^2 =$	32.749

$$\varepsilon = \sqrt{\frac{294.72}{10 - 1}} = 5.72$$

Probable error of single set $= 0.6745 \times 5.72$ or

$$r = \pm 3^s.86$$

Probable error of final result $= \dfrac{r}{\sqrt{m}}$, or

$$R = \pm \frac{3^s.86}{\sqrt{10}} = 1^s.220, \text{ or}$$

$$R = \sqrt{\frac{294.72}{10 \times 9}} \times 0.6745 = 1^s.220$$

This is for ten revolutions; for one revolution the probable error is $\pm 0^s.122$.

Peirce's Criterion.

Values of κ^2 for $\mu = 1$.

m	n = 1	2	3	4	5	6	7	8	9
3	1.480								
4	1.912	1.163							
5	2.278	1.439							
6	2.592	1.687	1.208						
7	2.866	1.910	1.409	1.045					
8	3.109	2.112	1.589	1.229					
9	3.327	2.295	1.753	1.388	1.091				
10	3.526	2.464	1.904	1.531	1.242				
11	3.707	2.621	2.045	1.662	1.373	1.122			
12	3.875	2.766	2.176	1.785	1.492	1.249	1.018		
13	4.029	2.902	2.299	1.901	1.604	1.362	1.145		
14	4.173	3.030	2.416	2.009	1.709	1.465	1.255	1.053	
15	4.309	3.151	2.526	2.111	1.807	1.561	1.354	1.163	
16	4.436	3.264	2.630	2.207	1.898	1.651	1.445	1.259	1.080
17	4.555	3.371	2.729	2.300	1.985	1.736	1.529	1.347	1.176
18	4.668	3.475	2.824	2.389	2.069	1.817	1.609	1.428	1.261
19	4.776	3.571	2.914	2.474	2.150	1.895	1.685	1.504	1.341
20	4.878	3.664	3.001	2.556	2.227	1.970	1.757	1.576	1.415
21	4.975	3.755	3.084	2.634	2.301	2.041	1.827	1.644	1.483
22	5.068	3.840	3.164	2.709	2.373	2.109	1.893	1.710	1.549
23	5.157	3.923	3.240	2.782	2.442	2.176	1.957	1.773	1.612
24	5.242	4.002	3.315	2.852	2.509	2.240	2.019	1.833	1.671
25	5.324	4.078	3.387	2.920	2.573	2.302	2.079	1.892	1.729
26	5.403	4.151	3.456	2.986	2.636	2.362	2.137	1.948	1.784
27	5.479	4.222	3.523	3.049	2.697	2.420	2.194	2.003	1.838
28	5.552	4.291	3.588	3.111	2.756	2.477	2.249	2.056	1.891
29	5.622	4.358	3.651	3.171	2.813	2.532	2.302	2.108	1.941
30	5.690	4.422	3.712	3.229	2.869	2.586	2.354	2.158	1.990
31	5.756	4.484	3.772	3.285	2.923	2.638	2.404	2.207	2.038
32	5.820	4.545	3.829	3.340	2.976	2.689	2.454	2.255	2.085
33	5.882	4.604	3.884	3.394	3.028	2.738	2.502	2.302	2.130
34	5.942	4.661	3.939	3.446	3.078	2.787	2.549	2.347	2.175
35	6.001	4.717	3.992	3.497	3.127	2.834	2.594	2.392	2.218
36	6.058	4.771	4.044	3.547	3.174	2.880	2.639	2.436	2.261
37	6.113	4.823	4.095	3.595	3.221	2.926	2.683	2.478	2.302
38	6.167	4.874	4.144	3.643	3.267	2.970	2.726	2.520	2.343
39	6.219	4.925	4.192	3.689	3.312	3.013	2.768	2.561	2.383
40	6.270	4.974	4.239	3.734	3.356	3.055	2.809	2.601	2.422
41	6.320	5.022	4.285	3.779	3.398	3.097	2.849	2.640	2.460
42	6.369	5.069	4.331	3.822	3.440	3.138	2.888	2.678	2.497
43	6.416	5.114	4.375	3.865	3.481	3.178	2.927	2.716	2.534
44	6.463	5.159	4.418	3.906	3.521	3.217	2.965	2.753	2.570
45	6.508	5.202	4.460	3.947	3.561	3.255	3.002	2.789	2.606
46	6.552	5.245	4.501	3.987	3.600	3.293	3.039	2.825	2.641
47	6.596	5.287	4.542	4.026	3.638	3.330	3.075	2.860	2.675
48	6.639	5.328	4.581	4.065	3.675	3.366	3.110	2.894	2.708
49	6.681	5.368	4.620	4.103	3.712	3.401	3.145	2.928	2.741
50	6.720	5.408	4.657	4.140	3.748	3.436	3.179	2.962	2.774
51	6.761	5.447	4.695	4.176	3.784	3.471	3.213	2.994	2.806
52	6.800	5.484	4.732	4.212	3.819	3.505	3.246	3.027	2.838
53	6.838	5.522	4.768	4.247	3.853	3.538	3.279	3.059	2.869
54	6.876	5.559	4.804	4.282	3.887	3.571	3.311	3.090	2.899
55	6.913	5.595	4.839	4.316	3.920	3.603	3.342	3.121	2.929
56	6.950	5.630	4.873	4.349	3.952	3.635	3.373	3.151	2.959
57	6.986	5.665	4.907	4.382	3.984	3.666	3.404	3.181	2.988
58	7.021	5.699	4.941	4.415	4.016	3.697	3.434	3.210	3.017
59	7.056	5.733	4.974	4.447	4.047	3.728	3.463	3.239	3.046
60	7.090	5.766	5.006	4.478	4.078	3.758	3.492	3.268	3.074

Geographical Positions.

	Latitude.	Longitude west from Greenwich.
	° ′ ″	*h. m. s.*
Cambridge, Observatory	42 22 48.1	4 44 31.04
Quebec, Citadel	46 48 17.3	4 44 49.42
New York, Observatory	40 43 48.5	4 55 56.65
Oswego, Court-House	43 27 49.1	5 06 01.05
WASHINGTON, Observatory	38 53 38.8	5 08 12.12
Buffalo, Michigan and Exchange streets	42 52 41.8	5 15 27.58
Detroit, New L. S. Observatory	42 19 58.6	5 32 12.24
Chicago, City Hall	41 53 06.2	5 50 32.08
Saint Louis, Washington University	38 37	6 00 49.02
Saint Paul, Custom-House	44 53	6 12 21.84
Fort Leavenworth, Engineer Observatory	39 21	6 19 39.35
Omaha, Coast-Survey Observatory	41 16	6 23 46.33
Denver, Mint	39 45 01.8	6 59 58.72
Salt Lake, Coast-Survey Observatory	40 46	7 27 35.45
San Francisco, Washington Square	37 47 55.3	8 09 38.23

TABLES AND FORMULÆ.

PART IV.

APPENDIX.

APPENDIX.

LXVI.—*Field Magnetic Observations.*

[By Captain CHAS. W. RAYMOND, Corps of Engineers.]

These observations have for their object the determination of the magnetic declination, dip, and intensity, at any given time and place.

The following instructions have reference to the determination of the magnetic elements on land. It will be supposed that the theodolite magnetometer and dip-circle are the instruments employed; the first to determine the declination and the horizontal component of the intensity, and the second to determine the inclination or dip.

Magnetic Declination.

The magnetometer having been mounted upon its tripod, or upon a sound post firmly imbedded in the ground, the horizontal limb and the rotation-axis of the telescope must be leveled, the vertical wire of the telescope made truly vertical, and its collimation-error reduced. The magnet must then be suspended by as few filaments as possible; four or five are usually required. The magnet is then made horizontal by adjusting the balancing-ring, the position of which should be carefully preserved throughout the experiments. In order to adjust the magnet-scale to the stellar focus of its lens, the telescope must be turned upon the sun or a star, and adjusted to perfectly distinct vision. The telescope must then be turned upon the suspended magnet, and the scale-ring screwed in or out until the scale is seen with perfect distinctness.

The lines of detorsion and collimation must now be brought into the plane of the magnetic meridian by the following method: The magnet being suspended, turn the instrument in azimuth until the scale is seen through the telescope. Remove the magnet and suspend the brass detorsion-cylinder. Bring the axis of the cylinder, by estimation, into the plane of the magnetic meridian by turning the torsion-circle. Remove the cylinder and suspend the declination-magnet. Turn the instrument in azimuth until the vertical wire of the telescope bisects the scale-zero. Remove the declination and suspend the short magnet.

LXVI.—*Field Magnetic Observations, &c.*—Continued.

Turn the torsion-circle until the vertical wire of the telescope coincides with the scale-zero. Exchange the magnets and adjust as before. The time, temperature, and readings of the verniers and torsion-circle should be recorded. This method requires that the magnets and detorsion-cylinder should be of equal weight. With the cylinder alone, the plane of detorsion may be determined to within about one degree, an error which does not seriously affect the accuracy of declinations observed in ordinary field-work.

The instrument is now in adjustment for observations of declination.

The *angular value of one scale-interval* and the *scale-zero* of each magnet employed must be determined at some convenient time. The former is unchangeable. The latter must be occasionally redetermined, as it is liable to change through accident. To determine the angular scale-value, fix the magnet in the position which it occupies when suspended, in such a way that the instrument may be moved in azimuth without disturbing it. Turn the instrument in azimuth until the vertical wire of the telescope coincides with a scale-division. Record the vernier and scale readings. Turn the instrument until the vertical wire coincides with another division, and record as before. Repeat over different parts of the scale until a sufficient number of observations have been obtained. Each pair of observations furnishes a single determination of the required value. The probable error of the mean value may be determined by the method of least squares.

To determine the *scale-zero*, or reading of the magnetic axis, suspend the magnet, turn the instrument in azimuth until the vertical wire of the telescope coincides with a division near the middle of the scale, and record the scale-reading. Invert the magnet, and record the reading corresponding to this position. Move the instrument slightly in azimuth, and repeat this operation until three or four readings with the scale erect, and as many with the scale inverted, have been obtained. The scale-zero may then be determined by the method of alternate means, (see Form A.) These observations should, if practicable, be made about the epoch of the day when the magnet is stationary.

LXVI.—*Field Magnetic Observations, &c.*—Continued.

The *co-efficient of torsion* may be determined as follows: The declination-magnet being suspended and the instrument in adjustment for observations of declination, record the readings of the scale and torsion-circle. Turn the torsion-circle through an angle of 90°, and record this difference of arc and the corresponding scale-reading. Turn the torsion-circle 180° in the reverse direction, and record as before. Finally, turn the torsion-circle back to its original position, and repeat the readings and record.

Computation.

$$\frac{H}{F} = \frac{u}{90^\circ - u}$$

$1 + \frac{H}{F}$ = co-efficient of torsion; and

u = difference in scale-readings (reduced to arc) corresponding to a change of direction of the magnet caused by twisting the suspension-thread through an angle of 90°.

The mean value of u deduced from the observations is employed. For convenience the co-efficient of torsion is usually applied to the angular value [a] of the scale-interval, (see Form B.)

At some convenient time, while the instrument is in position, the vernier-readings corresponding to the astronomical meridian must be determined, either by turning the telescope upon a point of which the azimuth is known, or directly, by any suitable method.

The preliminary adjustments and determinations having been made, the instrument is left in position for observations of the diurnal variations in declination. At some time early in the morning, the north end of the magnet attains its most easterly position, which is called the *morning eastern elongation.* The record of scale-readings must be commenced early enough to include this elongation. When this point is fairly passed, or the north end of the magnet has fully commenced its westerly motion, the readings may be discontinued until about noon, when they must be resumed and continued until the western elongation has been observed, and the easterly motion has fairly set in.

LXVI.—*Field Magnetic Observations, &c.*—Continued.

The telescope should be reversed at each observation, and a mean of the readings in the two positions should be taken. The temperature should be noted. The readings should be made half or quarter hourly during the periods of observation. The observations of the first day will determine the approximate times of elongation, or *turning-hours*, in accordance with which the periods of observation are to be subsequently regulated.

Computation.

$$\delta = \delta' \pm a \left(1 + \frac{H}{F}\right)(e - s + fr)$$

δ = value of declination for the day;

δ' = declination at instant of final instrumental adjustment, which is $\left\{\begin{matrix} + \\ - \end{matrix}\right\}$ when the magnetic meridian is $\left\{\begin{matrix} \text{west} \\ \text{east} \end{matrix}\right\}$ of the true north meridian;

a = angular scale-value;

$1 + \frac{H}{F}$ = co-efficient of torsion;

e = mean of scale-readings at the elongations, which is $\left\{\begin{matrix} + \\ - \end{matrix}\right\}$ when $\left\{\begin{matrix} \text{greater} \\ \text{less} \end{matrix}\right\}$ than s;

s = scale-reading of magnetic axis, (zero of magnet-scale,) which is $\left\{\begin{matrix} + \\ - \end{matrix}\right\}$ when $\left\{\begin{matrix} \text{less} \\ \text{greater} \end{matrix}\right\}$ than e;

r = difference between scale-readings at elongations, or *daily range*.

f = factor for reduction to the mean of *hourly* observations. It may be taken, with its sign, from the following table:

f					
January......	— 0.089	May........	— 0.013	September ...	— 0.044
February	— 0.040	June	+ 0.010	October......	— 0.096
March	— 0.019	July	+ 0.005	November ...	— 0.096
April	— 0.068	August	— 0.023	December....	— 0.154

LXVI.—*Field Magnetic Observations, &c.*—Continued.

The correction to δ' is positive or negative according as it indicates a motion from or toward the true meridian. (See Form C.)

Magnetic Intensity.

For the determination of the *horizontal intensity* two distinct series of experiments are required—*experiments of deflection* and *experiments of oscillation.*

The *experiments of deflection* are made as follows: The deflection-bar is made fast in its position, and the copper damper placed within the box. The instrument is then adjusted as for observations of declination, the short magnet being suspended.

The experiments should be made, when practicable, at three distances. The first position of the deflector should be at about three times its length from the suspended magnet; the third, at a distance about one-third greater; and the second midway between the other two. These distances are measured from center to center. At the beginning, the verniers are read and time noted, in order to follow changes of declination. The temperature is recorded at each observation. The long magnet is placed upon its carriage on the deflection-bar, on either side of the suspended magnet, and at the nearest distance. The instrument is then turned in azimuth until the vertical wire of the telescope coincides with the scale-zero. The time is then noted and the verniers read. The carriage is then moved to the next distance, and finally to the greatest distance, and the observation is repeated at each position.

The deflector is then reversed on its carriage, and the observations are repeated at the three distances, beginning with the greatest and ending with the least. At the nearest distance the magnet is again reversed, and the complete set of observations is repeated, in order to obtain a double set of results. The deflector is now placed on the opposite side of the suspended magnet at the nearest distance. A double set of experiments similar to that already described is then made.

At suitable intervals special observations should be made to measure changes of declination. For this purpose the deflector is removed, and the instrument is turned in azimuth until the vertical wire of the telescope coincides with the scale-zero. The

LXVI.—*Field Magnetic Observations, &c.*—Continued.

verniers are then read and the temperature and time noted. From these observations we may, by simple interpolation, determine with sufficient accuracy corrections for the reduction of the observed angles of deflection to the same declination.

Computation.

$$\frac{m}{X} = \tfrac{1}{2}r^3 \sin u \left(1 - \frac{P}{r^2}\right)$$

m = magnetic moment of the deflector;
X = horizontal intensity;
r = distance between the centers of the magnets; and
P = constant depending on the distribution of magnetism in the magnets.

The value of $\frac{m}{X}$ must be computed for each distance separately and a mean adopted. (See Form D.)

The correction depending upon the constant P may be neglected when there is a considerable difference between the lengths of the magnets. Its value is greatest when the magnets are equal in length, and zero when the lengths are in the proportion of 1 to 1.224.

To determine the value of P, deflections are made alternately at two different distances, which should be in the proportion of 1 to 1.32. About twenty-five corresponding sets should be obtained.

Computation.

$$P = \frac{A - A_1}{\frac{A}{r^2} - \frac{A_1}{r_1^2}}$$

$$\left.\begin{matrix} A \\ A_1 \end{matrix}\right\} = \text{value of } \frac{m}{X} \text{ for} \left\{\begin{matrix} \text{shorter} \\ \text{longer} \end{matrix}\right\} \text{distance} \left\{\begin{matrix} r \\ r_1 \end{matrix}\right.$$

To reduce the angles of deflection determined from different sets to the same temperature:

$$\sin u = \frac{\sin u_0}{1 - (t_0 - t)q}$$

u_0 = observed angle at temperature t_0;
u = angle reduced to standard temperature t; and
q = temperature-constant, determined as explained hereafter.

LXVI.—*Field Magnetic Observations, &c.*—Continued.

The *experiments of oscillation* are made as follows: The instrument having been adjusted as for observations of declination, with the long magnet suspended, and the co-efficient of torsion having been determined, the magnet is made to oscillate horizontally by attracting or repelling one of its poles. The impulse should be sufficient to make it oscillate beyond the limits of the scale for at least ten minutes, as steadiness of motion is thus acquired. All vertical oscillations must be carefully checked. When the amplitude is sufficiently reduced, the scale is read at the limits of an oscillation, and a division near the mean of these readings is selected as the zero or point at the passage of which the times are to be noted. The intervals of oscillation may now be noted in the following way: The approximate interval corresponding to six oscillations is noted for convenience. The instant of passage is then noted at every sixth oscillation up to the sixtieth; then at the hundredth, two hundreth, and three hundreth; then at every sixth oscillation up to the three hundred and sixtieth; and then at the four hundredth, if it be desirable to prolong the observations to this extent. The number of oscillations timed should depend on the length of the interval. The entire time of observation should not exceed a quarter of an hour. The approximate time of ten oscillations is computed for convenience at the sixtieth. For the semi-oscillations timed the magnet will always move in the same direction. The amplitude of oscillation at the beginning and end of the experiments should be noted. At suitable intervals the mean reading of the scale should be observed, since the zero is liable to changes due to variations of declination.

The temperature should be observed at intervals, the bulb of the thermometer being placed within the box.

Computation.

$$mX = \frac{\pi^2 K}{T^2}.$$

m = magnetic moment of the magnet;

X = horizontal intensity;

π = 3.14159;

K = moment of inertia of magnet and stirrup;

LXVI.—*Field Magnetic Observations, &c.*—Continued.

T = corrected time of oscillation;

$$T^2 = T'^2\left(1 + \frac{H}{F}\right)\left(1 - (t' - t)\,q\right)$$

T′ = observed time of oscillation;

$\left.\begin{matrix} t' \\ t \end{matrix}\right\}$ = temperature of magnet when $\left\{\begin{matrix}\text{oscillating;} \\ \text{deflecting;}\end{matrix}\right.$

q = temperature-constant, or change in magnetic moment for a change in temperature of 1° Fahrenheit. (See Form E.)

To determine the *moment of inertia of the magnet and stirrup*, [K], the moment of inertia of the inertia-ring must first be computed. For this purpose accurate measurements of its outer and inner radii (in decimals of a foot) and the weight of the ring (in grains) are required. These data are usually furnished by the maker.

The instrument being in adjustment with the long magnet suspended, the inertia-ring is balanced upon the magnet by means of the balancing-blocks. The time of a single oscillation is then determined as before described. The load is then removed and the time of oscillation is again determined. At least twelve sets of these experiments should be made. A separate determination of the co-efficient of torsion must be made for the loaded magnet. The temperature must be recorded throughout the experiments.

Computation.

$$K = K'' \frac{T^2}{T'^2 - T^2}$$

$$K'' = \tfrac{1}{2}\,(r^2 + r'^2)\,w$$

K = moment of inertia of the suspended mass;

K″ = moment of inertia of the ring;

$\left.\begin{matrix} T' \\ T \end{matrix}\right\}$ = corrected time of single oscillation of $\left\{\begin{matrix}\text{loaded} \\ \text{unloaded}\end{matrix}\right\}$ magnet;

$\left.\begin{matrix} r \\ r' \end{matrix}\right\} = \left\{\begin{matrix}\text{outer} \\ \text{inner}\end{matrix}\right\}$ radius of ring, in feet; and

w = weight of ring, in grains.

LXVI.—*Field Magnetic Observations, &c.*—Continued.

The values of π^2 K for different temperatures should be tabulated. The co-efficient of expansion for brass (0.00001) may be employed for their computation.

The *change in magnetic moment for a change in temperature of* 1° *Fahrenheit*, [q], is best determined by the method of deflections. The magnet for which q is to be determined is the deflector. At least three consecutive sets of deflections should be made, the first and third being at about the same temperature, and the intermediate set at a very different one. A mean of the results from the first and third sets must be compared with the result from the intermediate set. The required difference of temperature may be produced by a jacket of ice and hot water, or advantage may be taken of extreme natural temperatures.

Computation.

$$q = \frac{a\,n \cot u}{t - t_0}$$

q = temperature-constant;

$\left.\begin{matrix} t \\ t_0 \end{matrix}\right\} = \left\{\begin{matrix} \text{higher} \\ \text{lower} \end{matrix}\right\}$ temperature;

n = difference of scale-readings corresponding to $t - t_0$;

a = angular value of one scale-interval; and

u = angle of deflection corresponding to t_0.

Computation of the Horizontal Intensity and Magnetic Moment.

$$X = \sqrt{\frac{\alpha}{\beta}}$$

$$m = \sqrt{\alpha\,\beta}$$

X = absolute horizontal intensity;

m = magnetic moment of deflecting and oscillating magnet;

α = mX, determined by experiments of oscillation; and

$\beta = \frac{m}{X}$, determined by experiments of deflection.

Strictly, the values of α and β should be corrected for the effect of induction. These corrections are very small, and require

LXVI.—*Field Magnetic Observations, &c.*—Continued.

special apparatus for their determination. They are therefore neglected in field-work. (See Form E.)

To reduce m to a Standard Temperature.

$$m_0 = m\,(1 + (t - t_0)\,q)$$

m_0 = value of m at standard temperature t_0;
t = temperature at time of experiments; and
q = temperature-constant.

Computation of the Total Intensity.

$$\varphi = \text{X} \sec \theta$$

φ = total intensity;
X = horizontal intensity; and
θ = inclination or *dip*.

To convert measures of intensity expressed in English units into their equivalents expressed in the metric system, multiply by 0.46108, (log = 9.66378.)

To convert measures of intensity expressed in metric units into their equivalents expressed in English units, multiply by 2.1688, (log = 0.33622.)

Magnetic Inclination.

The dip-circle having been mounted on its tripod or post, the horizontal limb must be leveled. The needle is charged by means of the magnetizing-bars, and then suspended as follows: Raise the Y's, and placing the needle in them, lower it gently upon the agate supports. Turn the vertical circle slowly in azimuth around the entire circle, and see whether the needle plays freely, and whether its face lies in the plane of the face of the vertical circle in all azimuths. If necessary, the agate supports must be re-adjusted. The face of the vertical circle is then brought into the plane of the magnetic meridian by the following method: Turn the vertical circle in azimuth until the needle is vertical. Record the reading of the azimuth-circle. Reverse the needle on its supports; make it again vertical by a slight movement in azimuth, and record as before. Turn the vertical circle 180° in azimuth, and repeat the double observation. A mean of the four readings is the reading of the magnetic prime-vertical, from which the settings of the magnetic meridian are obtained by

LXVI.—*Field Magnetic Observations, &c.*—Continued.

adding and subtracting 90°. Set the vertical circle at one of these readings. (See Form F.)

The observations for the determination of the inclination are made as follows: Record the reading of the azimuth-circle. Record the polarity of the needle, (marked end *north* or *south*,) the position of the vertical-circle, (face *east* or *west*,) a mean of the readings of the vertical circle at the ends of the needle, and the temperature Fahrenheit. Raise the needle, reverse it on its supports, and repeat the observations. Bring the needle back to its original position, and observe as before. Reverse it again, and repeat the observations. Turn the vertical circle 180° in azimuth, and repeat the observations. Remove the needle, and reverse its polarity. Suspend it again, and repeat the observations with the circle and needle in both positions, as before.

Computation.

$$\theta = \frac{\alpha + \beta}{2} + c$$

θ = magnetic inclination;

$\left.\begin{matrix}\alpha\\ \beta\end{matrix}\right\}$ = mean of observed values of θ $\left\{\begin{matrix}\text{before}\\ \text{after}\end{matrix}\right\}$ reversal of polarity;

c = constant correction for errors of axle and limb.

The *constant correction* (c) is determined as follows: A complete set of experiments must be made in the plane of the magnetic meridian. The vertical circle is then turned in either direction about 45° in azimuth, and a similar series of experiments is made. The vertical circle is then turned 90° in azimuth in the opposite direction, and the experiments are again repeated.

Computation.

$$c = \theta - \theta_1;\ \cot^2 \theta = \cot^2 \theta' + \cot^2 \theta''$$

$\left.\begin{matrix}\theta'\\ \theta''\end{matrix}\right\}$ = observed inclinations in planes at right angles to each other; and

θ_1 = observed inclination in the plane of the magnetic meridian.

FORM A.

Determination of the Zero of the Magnet-Scale.

Station, Fort Yukon, Alaska.—Date, August 12, (p. m.,) 1869.—Observer, C. W. R.—Recorder, C. W. R.

Scale-zero of declination-magnet.

Position of scale.	Reading of scale.	Alternate means.	Zeros.	Mean zero.
Erect	45.00			
Inverted	57.00	45.50	51.25	
Erect	46.00	55.50	50.75	50.18
Inverted	54.00	44.75	49.37	
Erect	43.50	55.25	49.37	
Inverted	56.50			

FORM B.

Determination of the Co-efficient of Torsion.

Station, Fort Yukon, Alaska.—Date, August 13, 1869.—Observer, C. W. R.—Recorder, C. W. R.

Declination-magnet suspended.

Circle-readings.	Scale-readings.	Diff. of arc.	Diff. of scale.	Mean for 90°.
		°		
3.95	51.25			
30.95	30.50	90	20.75	
12.95	79.50	180	49.00	23.56
3.95	55.00	90	24.50	

Computation.

$$90^\circ = 324000''.0$$

$$u = 23^{d}.56 = 699\ .7 \qquad \log\ 2.84491$$

$$90^\circ - u = 323300\ .3 \qquad \log\ 5.50961$$

$$\frac{H}{F} = 0.00216 \qquad \log\ 7.33530$$

$$1 + \frac{H}{F} = 1.00216$$

$$a^* \left(1 + \frac{H}{F}\right) = 29''.76 \qquad \frac{H}{F} = \frac{u}{90^\circ - u}$$

$^*a = 29''.70$

Form C.

Declination—Record of Observations.

Station, Fort Porter, Buffalo, N. Y., 222 feet north of flag-staff.—Date, June 14, 1872.—Observer, A. N. L.

Azimuth-reading at adjustment, ver. B	158°	09′	00″
ver. A	338	09	30
Reading of mark on flag-staff..ver. B	166	21	00
ver. A	346	21	00
Azimuth of flag-staff, west of north	175	35	53

Magnet C_2 suspended; scale-zero, 21.7.

Time.	Scale.	Remarks.
h. m.		
6 00 a. m.	21.7	
6 20	22.3	
6 45	22.6	
7 00	22.7	Maximum.
7 30	22.4	
8 00	22.0	
11 30	15.6	
12 00 m.	14.75	
12 30 p. m.	14.35	Minimum.
12 45	14.65	
1 00	15.05	
1 15	15.7	

Computation.

Correction to δ'.			
$e = -18.5$ $s = -21.7$	$e - s = 3.2$		Astr. merid .. 161° 56′ 53″ Az. at adjustment...... 158 09 15
		$e - s + fr = 3.28$	$\delta' = +\ 3\ 47\ 38$
$r = 8.35$ $f = +\ .01$	$fr = +0.08$	$a\left(1 + \frac{H}{F}\right) = 166''$	Corr. to $\delta' = +\ 9\ 04.5$ $\delta = +\ 3\ 56\ 42.5$

FORM D.

Horizontal Intensity—Experiments of Deflection.

Station, Fort Yukon, Alaska.—Date, August 14, 1869.—Observer, C. W. R.—Recorder, C. W. R.

Magnets at right angles to each other; long magnet deflecting; short magnet suspended.

Magnet at 1′.09 east and west.

Position of deflector.	No. of observations.	Marked end.	Temp., F.	Verniers E. and W.	Means of verniers.	Time.	Means corr. for change of declination.	Arc of deflection; semi-arc.
			°	° ′	° ′	h. m.	° ′	° ′
E.	3	E.	81.0	71 18 251 18	161 18	2 10	161 17.5	
	4	W.	82.0	55 33 235 33	145 33	2 18	145 33.7	15 43.8 7 51.9
	9	E.	81.0	71 21 251 21	161 21	3 02	161 19.7	15 46.0 7 53.0
	10	W.	82.0	55 35 235 35	145 35	3 08	145 36.4	15 43.3 7 51.6
							Mean....	7 52.2
W.	15	E.	85.5	71 34 251 34	161 34	4 13	161 34.1	
	16	W.	87.0	55 30 235 30	145 30	4 22	145 29.8	16 04.3 8 02.1
	21	E.	84.0	71 36 251 36	161 36	5 22	161 36.5	16 06.7 8 03.3
	22	W.	85.0	55 42 235 42	145 42	5 35	145 41.4	15 55.1 7 57.5
	Mean	for E.	and	W.....	7 56.6		Mean....	8 01.0

Remarks.

1^h 34^m p. m.—Turned instrument on 50.18 declination-scale. Verniers, E., 63° 51′; W., 243° 51′. Temperature, 78°.5 F., (attached.)

3^h 45^m p. m.—Deflector away to show changes of declination. Temperature, 87° F. Verniers, E., 63° 49′; W., 243° 49′.

6^h 00^m p.m.—End of experiments. Temperature, 82° F.; scale, 58.50. Verniers, E., 63° 54′; W., 243° 54.′

Computation.

		log.
$u = 7°56'.6$	$\sin u$	9.14049
$r = 1^f.09$	r^3	0.11228
$P = 0$	$1 - \frac{P}{r^2}$	
	$\frac{1}{2}$	9.69897
	$\frac{m}{X}$	8.95174

FORM E.

Horizontal Intensity—Experiments of Oscillation.

Station, Fort Yukon, Alaska.—Date, August, 16, 1869.—Chronometer, Bliss & Creighton, 1609, M. T.—Daily rate, unknown.—Observer, C. W. R.—Recorder, J. J. M.

Long magnet suspended without load.

No. of observations.	Times by chronometer.	Temp., F.	Diff. of times.	Time of 10 oscillations.	Remarks.
	h. m. s.	°	*s.*	*s.*	
0	1 16 05.0	82.5			
6	16 40.8				Approximate time of 6 oscillations at the beginning 35^s.
12	17 16.5				
18	17 52.3				
24	18 28.3				Scale reading noted just before 200th oscillation to detect changes of declination. Reading, 65°.00; local time, $12^h\ 15^m$ p. m.
30	19 04.0		179.0	59.67	
36	19 39.7		178.9	59.63	
42	20 15.5		179.0	59.67	
48	20 51.3		179.0	59.67	
54	21 27.0		178.7	59.57	
60	22 02.8		178.8	59.60	
100	26 01.5	81.0	238.7	59.67	
200	35 57.5	81.5	596.0	59.60	

Time of 10 oscillations, $59^s.64$; time of 1 oscillation, $5^s.964$.

Computation.

q	0.00015	T'	Logarithms.
$t'-t$	0°.06		
		T'^2	0.77554
$(t'-t)q$	0.000009	$1+\frac{H}{F}$	1.55108
$1-(t'-t)q$	0.99991	$1-(t'-t)q$	0.00103
	Logarithms.	T^2	9.99996
$\frac{m}{X}$*	8.95174	$\pi^2 K$	1.55207
mX	9.83242	mX	1.38449
m^2	8.78416	m	9.83242
			9.39208
		X	
			0.44034

* Experiments of deflection.

Form F.

Inclination—Determination of the Dip.

Station, Willet's Point, N. Y.—Date, August 5, 1872.—Observer, C. W. R.—Recorder, C. W. R.—Dip-circle by Würdemann; Lloyd's needle.

Meridian observations.—Settings for magnetic meridian.

Face of circle.	Face of needle.	Readings of hor. limb.
		° ′
S.	S.	157 27
S.	N.	158 42
N.	S.	337 01
N.	N.	338 12

Magnetic prime-vertical, 157° 50′; settings, 247° 50′, 67° 50′.

Marked end.	Face of circle.	Face of needle.	Means of N. and S. ends.	Means.	Means.
N.	E.	E.	73° 04′	73° 02′ 00″	72° 36′ 30″
		E.	73 00		
		W.	72 12	72 11 00	
		W.	72 10		
	W.	E.	72 30	72 29 00	72 39 00
		E.	72 28		
		W.	72 49	72 49 00	
		W.	72 49		
Mean ...	$[\alpha]$				72 37 45
S.	E.	E.	72° 43′	72° 44′ 00″	72° 55′ 00″
		E.	72 45		
		W.	73 00	73 06 00	
		W.	73 12		
		E.	72 49	72 48 00	73 01 00
		E.	72 47		
		W.	73 14	73 14 00	
		W.	73 14		
Mean ...	$[\beta]$				72 58 00
			Resulting inclination	$\left[\frac{\alpha+\beta}{2}\right]$	72 47 52

Computation.

$$\theta = \frac{\alpha+\beta}{2} + c;\ \frac{\alpha+\beta}{2} = 72^\circ\ 47'\ 52'';\ c = +\ 18'';\ \theta = 72^\circ\ 48'.$$

www.ingramcontent.com/pod-product-compliance
Lightning Source LLC
LaVergne TN
LVHW020225110826
845151LV00003B/832